VOLUME NINETY FIVE

# Advances in GENETICS

## Genetics, Genomics and Fish Phenomics

# ADVANCES IN GENETICS, VOLUME 95

Serial Editors

**Theodore Friedmann**
Department of Pediatrics, University of California at San Diego, School of Medicine, CA, USA

**Jay C. Dunlap**
Department of Genetics, The Geisel School of Medicine at Dartmouth, Hanover, NH, USA

**Stephen F. Goodwin**
Department of Physiology, Anatomy and Genetics, University of Oxford, Oxford, UK

VOLUME NINETY FIVE

# Advances in GENETICS

## Genetics, Genomics and Fish Phenomics

Edited by

**NICHOLAS S. FOULKES**

*Institute of Toxicology and Genetics, Karlsruhe Institute of Technology, Eggenstein-Leopoldshafen, Germany; Centre for Organismal Studies, Ruprecht-Karls-Universität Heidelberg, Heidelberg, Germany*

ELSEVIER

AMSTERDAM • BOSTON • HEIDELBERG • LONDON
NEW YORK • OXFORD • PARIS • SAN DIEGO
SAN FRANCISCO • SINGAPORE • SYDNEY • TOKYO
Academic Press is an imprint of Elsevier

Academic Press is an imprint of Elsevier
50 Hampshire Street, 5th Floor, Cambridge, MA 02139, United States
525 B Street, Suite 1800, San Diego, CA 92101-4495, United States
125 London Wall, London EC2Y 5AS, United Kingdom
The Boulevard, Langford Lane, Kidlington, Oxford OX5 1GB, United Kingdom

First edition 2016



## Notices

Knowledge and best practice in this field are constantly changing. As new research and experience broaden our understanding, changes in research methods, professional practices, or medical treatment may become necessary.

Practitioners and researchers must always rely on their own experience and knowledge in evaluating and using any information, methods, compounds, or experiments described herein. In using such information or methods they should be mindful of their own safety and the safety of others, including parties for whom they have a professional responsibility.

To the fullest extent of the law, neither the Publisher nor the authors, contributors, or editors, assume any liability for any injury and/or damage to persons or property as a matter of products liability, negligence or otherwise, or from any use or operation of any methods, products, instructions, or ideas contained in the material herein.

ISBN: 978-0-12-804800-9
ISSN: 0065-2660



*Publisher:* Zoe Kruze
*Acquisition Editor:* Zoe Kruze
*Editorial Project Manager:* Hannah Colford
*Production Project Manager:* Vignesh Tamil
*Designer:* Greg Harris

Typeset by TNQ Books and Journals

# CONTENTS

# CONTRIBUTORS

**K. Asakawa**
National Institute of Genetics and SOKENDAI (The Graduate University for Advanced Studies), Mishima, Japan

**E.L. Benard**
Institute of Biology, Leiden University, Leiden, The Netherlands

**C. Bertolucci**
Università di Ferrara, Ferrara, Italy

**D. Casane**
Laboratory EGCE, CNRS and University of Paris-Sud, Gif-sur-Yvette, France; Paris Diderot University, Sorbonne Paris Cité, France

**K.C. Cheng**
The Pennsylvania State University College of Medicine, Hershey, PA, United States

**Y. Ding**
The Pennsylvania State University College of Medicine, Hershey, PA, United States

**N.S. Foulkes**
Institute of Toxicology and Genetics, Karlsruhe Institute of Technology, Eggenstein-Leopoldshafen, Germany; Centre for Organismal Studies, Ruprecht-Karls-Universität Heidelberg, Heidelberg, Germany

**M. Hibi**
Nagoya University, Nagoya, Japan

**M. Itoh**
Chiba University, Chiba, Japan

**S.R. Katz**
The Pennsylvania State University College of Medicine, Hershey, PA, United States

**K. Kawakami**
National Institute of Genetics and SOKENDAI (The Graduate University for Advanced Studies), Mishima, Japan

**B. Klotz**
University of Würzburg, Würzburg, Germany

**A.Y. Lin**
The Pennsylvania State University College of Medicine, Hershey, PA, United States

**A.H. Meijer**
Institute of Biology, Leiden University, Leiden, The Netherlands

**F. Müller**
University of Birmingham, Birmingham, United Kingdom

**A. Muto**
National Institute of Genetics and SOKENDAI (The Graduate University for Advanced Studies), Mishima, Japan

**P.I. Racz**
Institute of Biology, Leiden University, Leiden, The Netherlands

**S. Rastegar**
Institute of Toxicology and Genetics, Karlsruhe Institute of Technology, Eggenstein-Leopoldshafen, Germany

**J. Regneri**
University of Würzburg, Würzburg, Germany

**S. Rétaux**
Paris-Saclay Institute of Neuroscience, CNRS and University Paris-Sud, Gif-sur-Yvette, France

**J. Rougeot**
Institute of Biology, Leiden University, Leiden, The Netherlands

**M. Schartl**
University of Würzburg, Würzburg, Germany; University Clinic Würzburg, Würzburg, Germany; Texas A&M University, College Station, TX, United States

**H.P. Spaink**
Institute of Biology, Leiden University, Leiden, The Netherlands

**U. Strähle**
Institute of Toxicology and Genetics, Karlsruhe Institute of Technology, Eggenstein-Leopoldshafen, Germany; Ruprecht-Karls-Universität Heidelberg, Heidelberg, Germany

**D. Vallone**
Institute of Toxicology and Genetics, Karlsruhe Institute of Technology, Eggenstein-Leopoldshafen, Germany

**H. Wada**
National Institute of Genetics and SOKENDAI (The Graduate University for Advanced Studies), Mishima, Japan

**D. Whitmore**
UCL, London, United Kingdom

**J. Wragg**
University of Birmingham, Birmingham, United Kingdom

**X. Xin**
The Pennsylvania State University College of Medicine, Hershey, PA, United States

CHAPTER ONE

# Studying the Evolution of the Vertebrate Circadian Clock: The Power of Fish as Comparative Models

**N.S. Foulkes*,**,[1], D. Whitmore[§], D. Vallone* and C. Bertolucci[¶]**
*Institute of Toxicology and Genetics, Karlsruhe Institute of Technology, Eggenstein-Leopoldshafen, Germany
**Centre for Organismal Studies, Ruprecht-Karls-Universität Heidelberg, Heidelberg, Germany
[§]UCL, London, United Kingdom
[¶]Università di Ferrara, Ferrara, Italy
[1]Corresponding author: E-mail: nicholas.foulkes@kit.edu

## Contents



## Abstract

The utility of any model species cannot be judged solely in terms of the tools and approaches it provides for genetic analysis. A fundamental consideration is also how its biology has been shaped by the environment and the ecological niche which it occupies. By comparing different species occupying very different habitats we can learn how molecular and cellular mechanisms change during evolution in order to optimally adapt to their environment. Such knowledge is as important as understanding how

*Advances in Genetics*, Volume 95
ISSN 0065-2660
http://dx.doi.org/10.1016/bs.adgen.2016.05.002

these mechanisms work. This is illustrated by the use of fish models for studying the function and evolution of the circadian clock. In this review we outline our current understanding of how fish clocks sense and respond to light and explain how this differs fundamentally from the situation with mammalian clocks. In addition, we present results from comparative studies involving two species of blind cavefish, *Astyanax mexicanus* and *Phreatichthys andruzzii*. This work reveals the consequences of evolution in perpetual darkness for the circadian clock and its regulation by light as well as for other mechanisms such as DNA repair, sleep, and metabolism which directly or indirectly are affected by regular exposure to sunlight. Major differences in the cave habitats inhabited by these two cavefish species have a clear impact on shaping the molecular and cellular adaptations to life in complete darkness.

# 1. INTRODUCTION

## 1.1 What Are Circadian Clocks and Why Should We Study Them?

The environment of our planet is far from constant. It is dominated by cyclic changes, namely the day–night cycle, the seasons, tides, and lunar cycles that are all driven by the geophysical properties of the Earth. Depending on where an organism lives, these cycles can pose a major threat to survival. For example, the surface temperature in a desert can differ by as much as 30–40°C between the day and night. Therefore, during evolution a frequent survival strategy that plants and animals have adopted has been to develop mechanisms that allow anticipation of these regular challenges. Central to this strategy are endogenous timing mechanisms, whereby organisms can optimally coordinate physiological or behavioral adaptations required for survival in advance of the actual environmental changes. One of the most commonly encountered biological timing systems is the circadian clock which allows anticipation of the day–night cycle. The circadian rhythms generated by this clock dominate most aspects of the biology of plants, animals, fungi, and even unicellular organisms (Pittendrigh, 1993). Like any good clock, the circadian clock mechanism is able to continue ticking even under constant conditions such as constant darkness and constant temperature. However, in this artificial state it generates rhythms with a period of about 24 h (hence, circadian: *circa*, about; *diem*, a day). As a result, the clock generated rhythm progressively shifts relative to the 24 h cycle of the outside world and is said to "free run." Therefore, normal clock function also involves regular resetting by environmental signals such as light and temperature indicative of the actual time of day (so-called "zeitgebers" or time givers), to ensure its synchronization with the natural day–night cycle (Roenneberg & Foster, 1997; Roenneberg, Daan, & Merrow, 2003). In turn, the clock communicates with most aspects of physiology and behavior via a complex combination of cell autonomous as well as systemic signals such as rhythmic hormone production (Schibler et al., 2015).

Since the mid-1990s, our understanding of clocks, how they function and how they are regulated has advanced considerably. This has been in large part due to genetic analysis of the clock mechanism initially by unbiased forward genetic screens in *Arabidopsis*, *Neurospora*, Cyanobacteria, *Drosophila*, and the mouse that lead to the identification of the first clock genes and ultimately allowed us to build a unified, coherent picture of clock function (Hardin, 2011; Loros & Dunlap, 2001; Lowrey & Takahashi, 2011; Mackey, Golden, & Ditty, 2011; Mas, 2005).

At the molecular and cellular level, what is a clock? In most organisms, the clock mechanism consists of a core of interconnected transcription–translation negative feedback loops that is made up of so-called core clock genes and proteins and requires about 24 h to complete one cycle (Reppert & Weaver, 2001). In the case of vertebrates, two key core clock proteins are CLOCK and BMAL. These are both bHLH PAS domain transcription factors which form heterodimers and thereby bind to E-box enhancer elements and activate transcription. E-box enhancers are located in the promoters of a broad range of genes including the negative elements of the core clock mechanism: the *period* (*per*) and *cryptochrome* (*cry*) genes. Upon activation of *per* and *cry* expression and their translation, the PER and CRY proteins translocate to the nucleus where they interact with the CLOCK/BMAL heterodimers, interfering with their transcriptional activation function and thereby down-regulating E-box-driven activation. As a result, *per* and *cry* gene expression is reduced and levels of these negative regulators fall. At a critical reduced level of PER and CRY proteins, CLOCK and BMAL are once again able to direct activation and the cycle can be repeated. This whole process takes about 24 h to complete one cycle and by stringent genetic analysis, this regulatory network has been pinpointed as the primary genetic source of circadian rhythmicity. However, it is now clear that the clock is not so simple. The *bmal* gene itself is rhythmically expressed under the control of a second "stabilizing" feedback loop involving the two orphan nuclear receptors, REV-ERB$\alpha$ and ROR$\alpha$ (Preitner et al., 2002). Furthermore, reinforcing this core cycling gene-regulatory network, additional layers of regulation provided by posttranscriptional and posttranslational modifications (Lim & Allada, 2013; Vanselow & Kramer, 2007), epigenetic regulation, and metabolic networks (Asher & Sassone-Corsi, 2015) serve to reinforce the core transcription–translation feedback loop and make it more robust. Clock genes and a functional circadian clock are characteristically expressed in most tissues and cell types. However in vertebrates, certain dedicated clock structures, for example, the suprachiasmatic nucleus (SCN) of the hypothalamus in mammals, and the pineal gland in birds, reptiles, amphibia, and fish, appear to serve as "master" clocks (Menaker, Moreira, & Tosini, 1997) that coordinate the activity of the other "peripheral" tissue clocks via a complex set of cycling systemic signals (Mohawk, Green, & Takahashi, 2012; Schibler & Sassone-Corsi, 2002). Therefore, the circadian timing system can be considered like an orchestra with the master clock serving as the conductor.

One of the most impressive features of the circadian clock is that its basic properties are highly conserved across immense distances of evolution. Furthermore, the circadian timing system represents one of the few biological systems where its basic functionality (circadian rhythmicity) is manifest at all levels of biological organization: from the basic molecular level, through cell biology, neurobiology, physiology, behavior, and even to the level of ecology. From the biomedical perspective, it is also now apparent that regular disruption of this clock as a result of modern life styles (eg, shift work) is a major risk factor for many important pathologies such as metabolic disorders, cancer, and cardiovascular illness (Brown, 2016; Martino & Young, 2015; Savvidis & Koutsilieris, 2012). Furthermore, the refinement of drug-delivery protocols to optimally match the natural circadian rhythmicity of metabolism and cellular physiology (so-called "chronotherapy") holds great potential for improving the efficacy of treatment for diseases such as cancer (Dallmann, Okyar, & Levi, 2016). Therefore, the circadian clock remains a topic of intense interest with many basic questions related to the regulation and function of this timing mechanism still unanswered.

## 1.2 Why Use Fish Genetic Models to Study Clocks?

Given the wealth of information that we have now obtained from a relatively small but diverse set of genetic models, how can genetic analysis in fish species help us to better understand the circadian clock? Of course as is evident from this review volume, the zebrafish and medaka increasingly offer a powerful repertoire of genetic tools for functional analysis, which combined with their suitability for high-throughput screening and in vivo imaging, do make them suitable models for detailed functional analysis of core clock mechanisms. Nevertheless, given the pace and sheer scale of research into basic clock mechanisms in models such as mice and *Drosophila*, it is not an unreasonable question to ask how much impact the use of fish genetic models at this stage can have on significantly improving our understanding of circadian clocks. However, as we will attempt to illustrate in this review, fish *do* have an important role to play. It is not only the genetics and screening tools that they offer, but also their biology which makes them highly attractive and relevant models. Specifically there are a number of unique and remarkable features which actually make them much better suited than mice as models to address many fundamental questions in vertebrate circadian clock biology. In particular,

1. Fish are particularly well suited to studying photoreception. As well as eyes, fish possess a photosensitive pineal gland, deep brain photoreceptors, and dermal melanophores that are light responsive (Doyle & Menaker, 2007). Furthermore, the peripheral clocks of fish are directly entrained by light (Sassone-Corsi, Foulkes, & Whitmore, 2000). Thus tissues can be explanted into a culture dish and then simply by exposing them to light, their internal clocks are reset. This is certainly not the case in mammals, where the only light-sensitive cells responsible for resetting the clock are restricted to the retina, specifically the intrinsically photosensitive retinal ganglion cells (ipRGCs), cells which express the photoreceptor, melanopsin (Lucas, Lall, Allen, & Brown, 2012). These photoreceptors together with input from the rod and cone visual photoreceptor cells, signal to the circadian timing system via indirect neuronal projections to the hypothalamic SCN. Then, in turn, timing information is relayed to the remaining peripheral clocks via a complex set of systemic signals (Schibler & Sassone-Corsi, 2002). Thus, fish and in particular fish cell lines have become fascinating models for studying the mechanisms whereby more generally clocks respond to light.
2. From a more global perspective, fish are ideal for exploring how the clock adapts over the course of evolution in response to changes in environmental conditions. Teleosts are the largest, most diverse group of vertebrates, with between 25,000 and 30,000 known species successfully adapted to a full range of aquatic habitats. In many cases, these habitats differ considerably in terms of the ambient lighting conditions. For example, depending on the geographical location, quality and depth of water, the day length, and wavelengths and intensity of light that fish experience can vary considerably (Loew & McFarland, 1990). Furthermore, fish have also successfully colonized extreme, perpetually dark habitats such as subterranean caves and abyssal marine environments

where they may also experience extremes of temperature and food deprivation. By comparing the circadian timing system of such nontraditional model species inhabiting diverse habitats with "normal" surface living species such as zebrafish and medaka, which are also accessible to detailed functional genetic analysis, we can potentially learn much about how circadian clocks evolve in response to their environmental lighting conditions. The ever-expanding application of "omics" tools also facilitates detailed functional analysis of even nontraditional models. Genome and transcriptome sequences as well as large-scale metabolome and proteome datasets can now be routinely obtained for even the most genetically inaccessible models. Furthermore, combining tools such as fish-derived cell cultures, the CRISPR/CAS system for genome editing, and comparative studies with the zebrafish and medaka, there is also the opportunity to perform detailed functional studies with these fascinating nontraditional fish species.

In the following part of this review we attempt to illustrate the general utility of fish models for studying how the circadian clock senses light and their eminent suitability for exploring the evolution of the circadian clock and other elements of physiology and behavior that directly or indirectly rely upon regular exposure to sunlight.

# 2. HOW DO FISH CLOCKS DIRECTLY SENSE LIGHT?

## 2.1 Light-Inducible Gene Expression

The demonstration that the peripheral clocks in adult zebrafish organs such as the heart and kidney could be directly entrained in vitro by simply exposing explanted tissue to light–dark (LD) cycles came as a big surprise (Sassone-Corsi et al., 2000). Indeed, at that time, studies in rodents were pointing to a complete reliance on retinal photoreceptor cells for the entrainment of the circadian clock by light (Lucas et al., 2012). Interestingly, the directly light-entrainable peripheral clocks were also documented in primary cell cultures derived from zebrafish embryos or adult fin tissue (Fig. 1) (Sassone-Corsi et al., 2000; Vallone, Gondi, Whitmore, & Foulkes, 2004).

Again, this situation differed considerably from what had emerged in studies of mammalian cell lines which were clearly not light responsive. Mammalian cultured cell clocks can only be synchronized by acute temperature changes, transient pharmacological treatments such as serum shocks, or by applying specific activators of a range of signaling pathways (Balsalobre,

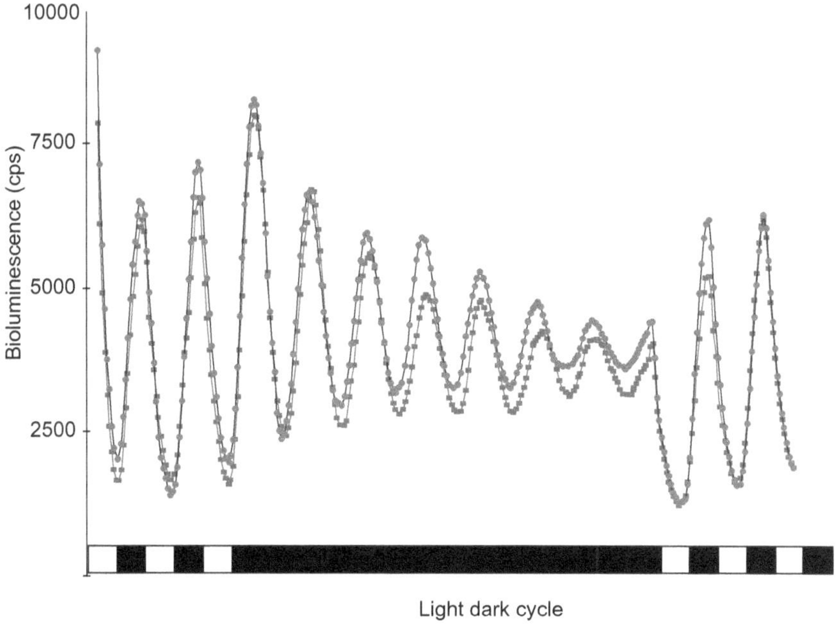


**Figure 1** Zebrafish cell lines models for studying clock and light-regulated gene expression. In vivo bioluminescence assay of Pac2 zebrafish cells transfected with the clock-regulated, luciferase reporter construct, *per1b:luc*. The bioluminescence emitted by two independently transfected sets of cells were monitored in real time after addition of luciferin to the culture medium and under various lighting conditions using an automated scintillation counter. White horizontal bars correspond to light and black bars, to dark periods. (See color plate)

Damiola, & Schibler, 1998; Balsalobre, Marcacci, & Schibler, 2000; Brown, Zumbrunn, Fleury-Olela, Preitner, & Schibler, 2002). The results obtained with fish cell lines led to the prediction that fish possess widely expressed photopigments that are coupled with the circadian clock and that the peripheral photoreception function has been lost during mammalian evolution. One attractive theory to explain this shift toward an exclusively retina-based circadian photoreception system during mammalian evolution invokes the existence of a "nocturnal bottleneck" (Gerkema, Davies, Foster, Menaker, & Hut, 2013). More specifically, the early ancestors of mammals evolved during the age of the dinosaurs. To avoid predation from diurnal carnivorous dinosaurs, these early mammals are predicted to have adopted a nocturnal existence, hiding in subterranean burrows during the day time. In this situation, any benefits from having a distributed photoreception system were lost and photoreception was focused in the retina. In this way only the animal's head would need to experience sunlight at the opening of the burrow for efficient clock entrainment and this would avoid the need to emerge completely from the burrow and risk attack from predators (Gerkema et al., 2013).

Actually, the notion that fish cells might be light sensitive was not a new discovery. It had been previously shown that all vertebrates with the curious exception of placental mammals, contain a DNA repair pathway termed photoreactivation that enables them to harness blue light photons to catalyze the repair of UV-damaged DNA (Kim & Sancar, 1993; Sancar, 2003). This pathway is based on a group of enzymes called photolyases; ancient, highly conserved proteins which are close relatives of the cryptochrome proteins and incorporate flavin (FAD) and methenyltetrahydrofolate (MTHF) as chromophores (Sancar, 2003). The photolyase proteins are not only light responsive themselves, but also the mRNA expression of photolyase genes was shown to be induced upon exposure to blue light in *Xenopus* as well as in goldfish and in zebrafish cells (Fukushima, Naito, & Ryoji, 2009; Mitani, Uchida, & Shima, 1996; Tamai, Vardhanabhuti, Foulkes, & Whitmore, 2004). These results already pointed to the presence of widely expressed photoreceptors in nonmammalian vertebrates. However, an obvious question that remains is whether the DNA repair machinery and the circadian clock share common photoreceptors.

More clues as to how light is perceived by peripheral clocks in fish came from the discovery that light exposure induced the expression of a battery of genes in zebrafish. DNA microarray analysis of the transcriptome in cultured heart tissue, the zebrafish Pac2 cell line, 9-h-old zebrafish embryos as well as in 5-day-old zebrafish larvae revealed several classes of gene that were consistently up-regulated in the various systems upon exposure to light (Gavriouchkina et al., 2010; Weger et al., 2011). These included genes involved in DNA repair (eg, *6-4 DNA photolyase*, *DASH photolyase*, *ddb2*, and *xpc*), stress responses (eg, the heat shock chaperones *hsp90a2*, *hspd1* and their regulating transcription factor, *hsf2*) as well as core circadian clock elements and notably the two clock genes: *Period2* (*per2*) and *Cryptochrome1a* (*Cry1a*). These important findings pointed to a more general impact of light on fish cell biological processes rather than just the circadian clock.

The light-induced *cry1a* and *per2* clock genes encode elements of the negative limb of the core clock transcription–translation loop with CRY1a serving as a potent inhibitor of CLOCK-BMAL activation via preventing heterodimer formation (Tamai, Young, & Whitmore, 2007) and PER2 playing a more complex role as either transcriptional coactivator or corepressor depending on its transcriptional regulatory targets (Wang, Zhong, Zhong, Zhang, & Wang, 2015). The induction of *per2* and *cry1a* mRNA expression by light is reminiscent of the situation in the mammalian clock, where expression of the two genes *per1* and *per2* is induced in the SCN upon phase shifting light pulses (Albrecht, Sun, Eichele, & Lee, 1997). The transient increase in the levels of these negative elements of the core clock feedback loop is predicted to cause resetting of the phase of the clock. A bioinformatics comparison of the promoter regions of the light-regulated genes in zebrafish revealed an enrichment of the D-box enhancer sequence (Gavriouchkina et al., 2010; Weger et al., 2011). Furthermore, unbiased promoter analysis of the *per2* and *cry1a* promoters using a transient transfection assay in the zebrafish Pac2 cell line has confirmed the importance of the integrity of the D-box enhancer for light-regulated gene expression (Mracek et al., 2012; Vatine et al., 2009). Interestingly, in the case of the *per2* promoter, the D-box is sited in close proximity to an E-box element and both D- and E-boxes are essential for light-inducible gene expression (Vatine et al., 2009). Finally, the D-box enhancer sequence in the context of an artificial heterologous promoter directs robust light-inducible gene expression both in transfected zebrafish cell lines as well as in transgenic fish embryos confirming the importance of this enhancer for light-regulated gene expression (Mracek et al., 2012; Vatine et al., 2009; Weger et al., 2011).

In mammals, the D-box is the DNA-binding site for a group of three basic leucine zipper (bZip) transcription factors which also share a so-called PAR domain (proline and acidic amino acid rich). These factors are the thyrotrope embryonic factor (TEF), the hepatic leukemia factor (HLF), and the D-box binding protein (DBP). They are all transcriptional activators that can function in various homo- and heterodimer combinations (Gachon, 2007). In addition, a fourth bZip factor called E4BP4 can also bind to D-boxes and thereby acts as a transcriptional repressor (Cowell, 2002). These transcription factors have been extensively studied in the mouse and knockout experiments point to them serving in the regulation and coordination of various metabolic detoxification pathways (Green & Takahashi, 2006). In addition, the expression of these PAR factors is clock regulated, and so they serve as clock output factors in mammals, relaying clock-generated timing information to its regulatory targets. In fish, D-box regulation clearly shows some major differences. Namely, as a result of genome duplication events in the ancestors of zebrafish, there are many more D-box—binding factors (TEF1, TEF2, HLF1, HLF2, DBP1, DBP2, and six E4BP4 homologs) (Ben-Moshe et al., 2010).

Considering homo- and heterodimer formation as well as the evidence for tissue specificity in the expression of these factors, it is clear that the regulation of the D-box exhibits a higher level of complexity in fish. Furthermore, while in mammals the D-box represents a component of clock output pathways, in fish they serve in the clock input pathway linking light with the core clock mechanism. This remarkable finding illustrates how during evolution, the function of whole regulatory networks can be altered as part of an adaptive strategy. The precise regulatory mechanisms operating at the level of D-box–binding transcription factors remains unclear. The MAPK pathway, more specifically ERK/MEK, has been reported to play a negative regulatory role upon exposure of cells to blue light (Mracek et al., 2013). In addition, a systematic pharmacological study aiming to dissect the signaling pathways underlying the response of a zebrafish cell line (ZEM-2S) to light has implicated a blue light–dependent mechanism involving the phosphoinositide pathway as well as cross-talk with nitric oxide (NO) and MAPK signaling (Ramos, Moraes, Poletini, Lima, & Castrucci, 2014). Despite all these findings, more work is now required to clarify which signal transduction pathways serve to couple photoreceptors with the regulation of gene expression.

## 2.2 Peripheral Photoreceptors

This last challenge leads us logically on to the next major unanswered question: What are the widely expressed photoreceptors that characterize fish cells and tissues? Given the absence of directly light-entrainable peripheral clocks in mammals, it is reasonable to assume that either these photoreceptors have been lost during mammalian evolution or alternatively they have been recruited to serve other functions. The strongest candidates to date for the fish peripheral photoreceptors are opsins. These seven-transmembrane, G-protein-coupled receptors (GPCRs) which are conjugated with a vitamin A–based retinaldehyde chromophore, serve as the principle photoreceptors in rod and cone-mediated visual photoreception in the retina. Furthermore, melanopsin has been demonstrated to act as a key photoreceptor in the nonvisual, circadian clock response to light of ipRGC cells in the mammalian retina (Hughes, Jagannath, Hankins, Foster, & Peirson, 2015; Schmidt et al., 2011). Melanopsin is also highly expressed in dermal melanophores of *Xenopus*, where it was originally cloned (Provencio, Jiang, De Grip, Hayes, & Rollag, 1998). A report in zebrafish has identified a staggering 32 nonvisual opsin genes. These opsins show a broad and overlapping tissue pattern of expression in adults, with the brain itself possessing more than 20 nonvisual opsins (Davies et al., 2015). A photoreceptive function has been confirmed in most of them, and therefore they appear to represent a source of good candidates for peripheral photoreceptor function. Of course, given this large number, the obvious question is why do fish need so many different photoreceptors? Is there overlap and redundancy of function or is it possible that these opsins can serve other nonphotoreceptive functions. In this regard opsins have been implicated in the thermotaxis of mammalian sperm (Pérez-Cerezales et al., 2015). Consistently, *Drosophila* opsins have also been demonstrated to play a role in temperature discrimination by larvae (Shen et al., 2011) pointing to a general property of opsins to serve as both photo- and thermoreceptors. Clearly, in the future, one obvious strategy to assign function to these multiple fish opsins will be a systematic reverse genetic approach where each opsin gene needs to be mutated and the resulting phenotypes examined. While the growing use of CRISPR/CAS technology should greatly facilitate this task, the need to test combinations as well as single gene mutations will still make this a major undertaking.

A final observation that seems to add even more complexity to our model for the light-sensing pathways of fish cells is that even without specialized photoreceptor molecules, cells can respond to light. Specifically, blue light exposure can raise the levels of reactive oxygen species (ROS) as the result of its direct effect on flavin-containing oxidases (Aggarwal, Quintanilha, Cammack, & Packer, 1978; Lockwood et al., 2005). This effect is observed in a wide variety of cells, even mammalian tissue culture cells that do not possess any dedicated photoreceptor function. Light-induced ROS production has been shown to induce clock gene expression in zebrafish cells via the activation of the MAPK pathway, and its effect on the immediate early transcription factors c-Fos and c-Jun and thereby regulation of genes by AP1 enhancer function (Hirayama, Cho, & Sassone-Corsi, 2007). Light-induced expression of the enzyme catalase has also been proposed to control the accumulating levels of ROS and thereby to limit the duration of the signaling response (Hirayama et al., 2007). It has been speculated that this response is sufficient to explain all the effects of light on the clock. However, there are a number of issues that still need to be addressed in order to improve our understanding of the involvement of ROS in peripheral photoreception in fish. First, the involvement of MAPK and AP1 enhancers in the regulation of gene expression by light is questioned by systematic promoter analysis which instead has pointed to the D-box serving as the primary light responsive enhancer and has failed to detect any effect of light on transcriptional regulation via AP1 enhancers (Mracek et al., 2013, 2012). Furthermore, the role of MAPK as an activator of clock gene expression has also been questioned based on the results of a genetic approach using dominant negative and dominant active forms of ERK and MEK (Mracek et al., 2013). Instead it has been suggested that ERK/MEK plays a down-regulatory role in light-inducible gene expression. However, there is a clear requirement for ROS signaling in light-induced clock gene expression (Hirayama et al., 2007), suggesting at least some degree of cross-talk between ROS and other signaling pathways linked with photoreception.

Clearly many questions still remain concerning how fish peripheral clocks are able to sense and respond to light. One strategy that has been exploited to identify the key elements in peripheral clock photoreception has been comparative studies involving blind cavefish. The rationale for these studies is that during evolution for millions of years under perpetual darkness, as well as their striking loss of visual function, blind cavefish may also loose photic entrainment function of their circadian clocks. In this scenario, loss of function mutations may potentially pinpoint key genetic elements linking clocks with light. However, evolution under the extreme environmental conditions frequently encountered in subterranean caves, also entails many other physiological and behavioral adaptations. Therefore, studying these fascinating fish also has the immense potential to shed light on other key genetic and physiological mechanisms whereby animals directly or indirectly respond to the effects of sunlight exposure. This next section reviews some of the progress in this area.

## 3. HOW DO FISH (AND CLOCKS) ADAPT DURING EVOLUTION UNDER PERPETUAL DARKNESS?

The extensive presence of photopigments and the wide range of light-impacted biological processes raises an interesting question—if light is so critical to fish biology then what happens to animals that have evolved in complete darkness, either in deep-sea environments or within cave complexes? How have these animals "compensated" or evolved aspects of their physiology to cope with this lack of light-regulated biology?

### 3.1 Different Cave Environments

Subterranean environments are typically characterized by continuous darkness, constant temperature, and a scarcity of food and predators. Both terrestrial and aquatic animals which have evolved in these extreme conditions show common features known as "troglomorphisms," including enhanced longevity, low metabolic rate, depigmentation, and most strikingly, complete eye loss (anophtalmia) (Alunni et al., 2007; Stemmer, Schuhmacher, Foulkes, Bertolucci, & Wittbrodt, 2015; Yoshizawa, Yamamoto, O'Quin, & Jeffery, 2012). The aquatic invertebrates and vertebrates which are obliged to live in such subterranean environments, typically inhabit freshwater aquifers such as cave waters, wells, and phreatic layers. Importantly, the cave environment is not simply a "loss of light" environment. The cave habitat has many other dramatically changed conditions, which can act as major natural selective pressures. Not least of which is a lack of food, loss of predation pressure, dramatic changes in water quality, and severe anoxia. Furthermore, each cave will differ in the balance and nature of specific environmental changes that it contains. It is a mistake to consider all caves as identical dark environments. Again one must be cautious in evaluating whether or not any daily cues from outside of the cave impinge on the underground environment, in the form of daily rain, dramatic temperature shifts, or animal migrations—as with bats or many insects. However, considered at a simpler, more general level, in all cave environments the loss of light exposure means that the main entraining signal, zeitgeber to the clock is lost.

Worldwide, subterranean waters are the home to around 200 species of so-called hypogean fish, more commonly known as cavefish (Trajano, Bichuette, & Kapoor, 2010). One of the more accessible and "useful" species for such analysis is *Astyanax mexicanus*, the blind Mexican cavefish. These animals are typically found in the Northeastern region of Mexico, and exist as a series of isolated cave populations, which numbers around 29 identified strains. About a million years ago, with a global decline in sea levels, rivers than ran under the limestone mountains of Eastern Mexico "dried up," creating a series of isolated underground pools. This trapped separate populations of *Astyanax* in distinct underground pools. Individuals then began the process of evolving to this new dark environment, but in relative isolation (Bradic, Beerli, García de León, Esquivel-Bobadilla, & Borowsky, 2012; Coghill, Hulsey, Chaves-Campos, de León, & Johnson, 2014). Undoubtedly, in periods of flood, new individuals would enter these caves from local rivers, providing a fresh input to these isolated gene pools. However, in many caves this gene flux would not occur, or rarely, a fact that could contribute to an interesting diversity among specific cave strains of *Astyanax*.

Perhaps the most significant "advantage" of *Astyanax* as a model system is that the "original" starting population still exists within the neighboring rivers. Surface and cave strains can still interbreed, and so speciation has not progressed so sufficiently as to produce distinct new cave species. However, the existence of the surface, river populations allows in a research sense for a superb comparison and analysis of adaptations, and specific molecular and anatomical changes that have occurred following evolution in a constantly dark environment. This is clearly an over-simplification, and several issues do need to be stressed. For one, clearly the surface fish population has continued to evolve itself over the last million years, and so all differences seen might not simply be a reflection of evolution in a cave environment.

These advantages are not offered by the Somalian cavefish *Phreatichthys andruzzii*. This species colonized phreatic layers beneath the Somalian desert about 5 million years ago, and initially the cave populations may have maintained some level of gene flow with surface-dwelling forms (Calderoni et al., 2016). This situation may have lasted for about 2 million years ago, until the Somalia desertification led to *Phreatichthys* complete hypogean isolation (Hernández Fernández & Vrba, 2006). Therefore, the cave environment of *Phreatichthys* differs fundamentally from that of *Astyanax* and in particular the greater extent and degree of isolation in its subterranean habitat is predicted to have led to the observed, much stronger troglomorphic phenotype of *Phreatichthys* (Fig. 2).

## 3.2 Blind Clocks in *Phreatichthys andruzzii*

Studies of circadian clocks in cavefish species have shed new light on the evolution of the circadian clock. Since the main function of this biological clock is to entrain behavioral and physiological processes to local time, and the most common zeitgeber is the natural LD cycle, researchers predicted that the disappearance of a functional circadian clock in cavefish would be the logical consequence of exposure to constant darkness over the course of millions of years. This prediction was largely confirmed by molecular and behavioral analysis of the circadian clock of *Phreatichthys*. Exposure of these cavefish to an artificial LD cycle resulted in an arrhythmic pattern of both locomotor activity and clock gene expression in central and peripheral tissues and in a primary cell line derived from adult fins (Cavallari et al., 2011). However, cell culture experiments confirmed that these animals still do possess peripheral circadian clocks (see later) (Cavallari et al., 2011). Therefore, these fish still have clocks, but these clocks are completely "blind."

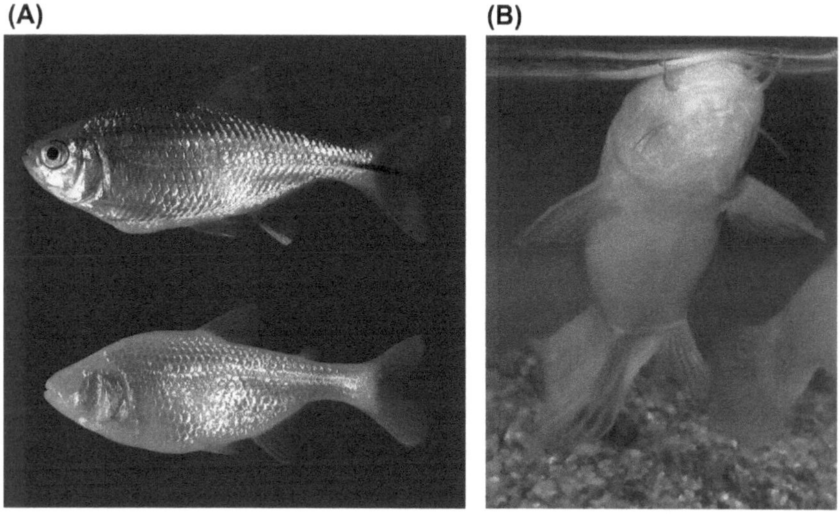


**Figure 2** Blind cavefish used in circadian clock studies. (A) Adult surface (upper) and blind cave (lower) forms of the Mexican tetra, *Astyanax mexicanus* (*Photograph by T. Katherine Tamai.*). (B) Adult *Phreatichthys andruzzii* (*Photograph by Andrea Margutti.*). (See color plate)

The molecular basis of the blindness of the *Phreatichthys* circadian clocks was tracked down by the cloning and characterization of nonvisual opsin genes (Cavallari et al., 2011; Tarttelin et al., 2012). Two nonvisual opsins, the melanopsin homolog (Opn4m2) and the teleost multiple tissue (TMT) opsin, are widely expressed in fish tissues and in *Phreatichthys*, both genes carried mutations that lead to truncation of the proteins in the fifth membrane spanning domain and thereby loss of the crucial seventh transmembrane spanning segment to which retinaldehyde is covalently linked. Therefore, the truncated protein would be predicted not to bind to the chromophore and thereby not to serve as a functional photopigment (Cavallari et al., 2011). Using an in vitro reporter assay it was shown that the corresponding wild-type Opn4m2 and TMT opsin cDNA sequences derived from the zebrafish are able to restore light-induced expression of the promoter of the *per2* clock gene (Lahiri et al., 2005; Vatine et al., 2009; Ziv, Levkovitz, Toyama, Falcón, & Gothilf, 2005). Taken together, these results are consistent with these opsin gene mutations at least in part contributing to clock blindness. Interestingly, however, the cavefish data also point to the existence of additional mutations affecting other photopigments (Cavallari et al., 2011). While TMT and melanopsin both react to blue and green light wavelengths, zebrafish cells respond to a much broader range of wavelengths extending to red wavelengths (Mracek et al., 2013). Thus, this predicts the existence of additional photoreceptors that have also accumulated mutations in the cavefish. However, this does not reflect a global loss of opsin function. For example, the nonvisual opsin exorhodopsin has a normal sequence and appears to have retained its function in mediating negative phototaxis, whereby adult cavefish tend to avoid illuminated environments (Tarttelin et al., 2012). Therefore, by pinpointing loss-of-function mutations in the large repertoire of teleost opsin genes, it should be possible to significantly reduce the list of candidates for opsin photoreceptors contributing to peripheral clock photoreception. This approach will thereby serve to better focus the reverse genetic analysis of opsin function in zebrafish and medaka models.

The consequences for the circadian clock during evolution in the cave environment appear to be more complex in the case of *Astyanax*.

## 3.3 Differences Between Surface and Cave Clocks in *Astyanax*

It is possible to study clock function in *Astyanax* in a lab setting, where the variables of light exposure, feeding, and temperature can be tightly regulated. However, collecting biological samples, with which to perform molecular clock analysis, is both somewhat tricky and limited in scope, not least as these animals are harder to breed under lab conditions than zebrafish, for example. Consequently, to date, most molecular clock studies has been limited to taking caudal fin samples at different circadian time points and lighting conditions, and extracting RNA for subsequent quantitative PCR analysis of gene expression. The fact that the circadian system of fish is highly decentralized, and that the clock mechanism is apparently very conserved, across species and tissues, means that this analysis, though limited, does provide some interesting insights.

Data collected so far shows that *Astyanax* cave strains maintained under LD cycles all still possess sustained oscillations in RNA levels for the *period1* (*per1*) gene, a core circadian clock component. These oscillations in gene expression persist in constant dark, following entrainment on a LD cycle (Beale et al., 2013). This fact alone reveals several key features of *Astyanax* circadian clock biology, including the fact that the circadian oscillator must still be entrained to a LD cycle, meaning that some aspects of the input or entrainment pathway remain functional. Secondly, a core clock mechanism still exists, which is capable of generating an oscillation with a period close to 24 h. In other words, clock input and core clock function are retained in cave strains of *Astyanax* when observed under lab conditions. There are, however, certain specific and interesting differences between cave clocks and surface rhythms. The two cave strains examined to date, Pachon and Chica, show a very similar waveform in their clock molecular oscillations, but the timing of this rhythm (the phase angle relative to the LD cycle) is very different to that in surface fish. The *per1* transcriptional rhythm is clearly delayed by about 6 h in cave strains, peaking in the early day versus the late night peak seen for surface fish populations. It is not clear that there is any biological advantage to this phase change in cave strains, as clearly under natural conditions in the wild these animals do not experience a LD cycle with which to entrain. However, this observation does provide interesting insight as to the changes/mutations that must have occurred in the clock mechanism following evolution in the dark. A change in phase angle can reflect either a difference in clock free-running period, and/or an alteration in the response to the entraining light cycle (a change in the underlying phase-response curve, PRC). Or in other words, either there are alterations in the core clock mechanism or in the way the cave circadian clock responds to light. The answer to this question is far from definitive at this point, and it is quite likely that alterations/mutations have occurred in both conceptual parts of the clock system. Only a few core clock genes have been cloned so far in *Astyanax* and so it is impossible to describe the precise changes that have occurred in the core clock mechanism. It is also extremely likely that mutations have also occurred within the DNA regulatory elements/promoters of these key genes. But with modern whole genome sequencing, this issue should be relatively easily resolved in the near future. One strategy, in essence, is to view the cavefish populations' as one would view any typical forward genetics study, and apply the appropriate techniques to determining the critical molecular changes in the clock mechanism.

## 3.4 Has the *Astyanax* Light-Response Altered Within Cave Strains?

As mentioned earlier, we do not yet know how the core clock mechanism has altered in precise detail. However, we do have evidence that the light input pathway to the cavefish clock has dramatically changed in cave strains when compared to surface populations. Both the *per2* and *cry1a* genes are normally acutely induced by phase shifting light pulses and appear to have a direct impact on core clock function (Tamai et al., 2007; Vatine et al., 2009). In *Astyanax* surface fish both of these genes are robustly light induced, as in zebrafish (Beale et al., 2013). Curiously, however, in cave strains this light induction is significantly reduced. Although presumably this reduced induction is sufficient to entrain the clock in a laboratory setting (if perhaps more slowly), and of course excluding a role for other unknown key factors as cavefish clocks can be set in the lab. The nature of this acute gene expression change is perhaps more interesting. It is not the fact that the light induction itself is reduced, but that the basal expression level of these two genes is dramatically raised in the dark. Light cannot produce much more transcript because the starting levels are already so high. This raised basal level has a number of major implications. Not least related to the fact that these two proteins are powerful transcriptional regulators, if not directly acting at the DNA level. For example, it is well known in zebrafish that overexpression of both *cry1a* and/or *per2* has a strong repressive action on the function of CLOCK/BMAL in the core clock mechanism and strongly reduces the amplitude of the clock, lengths period, or in fact possibly "stops" the rhythm completely (Beale et al., 2013; Tamai et al., 2007). We hypothesize that this raised level of *cry1a* and *per2* is one of the major reasons why the cave strain circadian molecular rhythm appears damped or repressed, and is likely to underpin the phase-angle change described earlier. Though, to date, there has not been a complete analysis of photopigment biology in *Astyanax*, the few opsins that have been sequenced and compared show none of the mutations or truncations seen in the Somalian cavefish, *Phreatichthys* (Cavallari et al., 2011). However, more work is needed to determine if alteration in gene sequence or expression pattern of some of the newly described photopigments might play a role in this altered *Asytanax* light response.

Is this increase in basal expression for *per2/cry1a* limited to just these clock light input factors, or is it a more general phenomenon? Although this idea has not been tested exhaustively at this time, it does seem to be true for the restricted extra cases examined. In fact, it is especially true for one of the other major light-activated teleost gene expression responses, the light-driven increase in DNA repair enzyme expression, especially that for the photolyases, *6-4* and *CPD photolyase*. DNA repair proteins, especially the photolyases, are extremely robustly light induced in zebrafish, which leads to enhanced DNA repair during the day when animals are exposed to increased UV damage (Tamai et al., 2004). The surface strain of *Astyanax* shows precisely the same degree of acute light induction of *CPD photolyase*, with a robust increase during the day (Beale et al., 2013). In the case of the cave strains, precisely the same response is seen as for *cry1a* and *per2*, the relative light induction of *CPD photolyase* is reduced, but only because the basal levels are significantly raised. Although one should not over generalize, it does appear as if light responsive genes are generally "turned on" in cave populations, which has allowed us to speculate that *Astyanax* cavefish look more as if they are living in constant light conditions at a molecular level than the actual constant darkness of the cave.

The reasons for this interesting alteration in basic biology are hard to determine, but it is interesting to speculate that perhaps there is a critical advantage for animals in the dark to have basally increased DNA repair enzyme expression, even if this is at the cost of a less functional or repressed circadian oscillator. We have shown that these light-induced photolyases can function in the dark to enhance the actual process of DNA repair, perhaps by recruiting other DNA repair machinery to the site of DNA damage. Hence, this increase in photolyase expression in the dark is not leading to the production of a functionless protein, but can have a beneficial effect. In an evolutionary context, when the original animals were trapped within the cave, or perhaps when "new" animals are washed in by flooding, these individuals will lose most of their ability to repair mutations through DNA damage, as they no longer experience the light necessary to activate these pathways. Any mutation that spontaneously occurs in this dark-trapped population that leads to an endogenous increase in DNA repair enzyme expression is likely to give those animals possessing that mutation a clear selective advantage over nonmutant individuals. As there is no light to detect, and perhaps little advantage in possessing a precise circadian timing mechanism, the disadvantages of turning on light-driven pathways is very minimal versus the clear benefits of possessing constantly active DNA repair processes. In fact, of course, having a dysfunctional circadian clock could itself be an advantage, as we discuss later.

One of the interesting aspects of working with both surface and cave strains of *Astyanax* is that they are not yet distinct species, and so it is possible to create F1 hybrid animals. Mating surface and cave strains produces F1 animals with restored, small eyes and an increase in pigmentation. Interestingly, in terms of the circadian and DNA repair light response, these F1 animals have almost identical light responses to those measured in surface fish population (Beale et al., 2013). Hence, the mutations that have occurred can be "rescued" within one generation, and presumably, therefore, represent recessive mutations in the light input pathway. The clock rhythm phase difference is also restored to surface fish timing, but the waveform of the molecular clock oscillation is a perfect "mix" of both the surface and cave strains. The added complexity here is not unexpected as the core clock is made up of a large number of genes/proteins, and so a simple rescue might be less likely. Of course, this approach means that a quantitative trait locus (QTL) approach for looking at clock and light input pathway changes is a feasible strategy in the F2 generation.

## 3.5 *Astyanax* in the Field

Another major advantage of working with *A. mexicanus* as a model system is the potential to work with these animals under natural conditions in the wild, in Mexico. Surface fish can be caught in many locations in the rivers surrounding the caves of North Eastern Mexico, and several caves can be accessed relatively easily (Fig. 3). A few cave populations have been examined for their circadian rhythmicity, but most of the focus has been on Chica cave just south of the town of Cuidad Valles, in part because of the ease with which time points can be performed at this location, but also because of the large bat population this cave possesses, which exhibits a very robust rhythm in daily activity (Beale et al., 2013). The data, collected over three years in

(A) (B)

**Figure 3** Studying *Astyanax* in their natural habitat. (A) Dr. Victor Reynoso studying surface forms in the Micos River, NE Mexico. (B) Entrance to the Tinaja Cave. (See color plate). *(Photographs by David Whitmore)*

the field is quite conclusive. There is no evidence of any molecular *per1* rhythm in the samples collected under natural conditions. In fact, the expression levels of *per1* are extremely low, and this gene appears to be even more strongly repressed under cave conditions. This is despite the fact that the bat population might well have provided a robust rhythmic entrainment cue due to their daily activity and rhythmic defecation, the major energetic input into the cave and source of food for the fish. A major caveat to this observation, of course and in fact to these studies as a whole, is that they are a population analysis of molecular rhythms within these animals. We have no individual animal data at this time. Consequently, although under natural cave conditions there appears to be absolutely no daily oscillation, each animal could still have a clock, but this clock remains unsynchronized when examined across the population. Although feasible, we view this conclusion as unlikely when individual animal activity is considered (see later).

As shown in the lab studies, light-induced gene expression is also basally raised under cave conditions. In fact, the change in *CPD photolyase* expression is even more exaggerated, with a considerable level of overexpression compared to the surface fish population. This might not only represent changes due to mutations in the light input pathway as described earlier, but also a response to the harshness of the actual cave environment, compared to the artificial conditions in the lab. The water in these caves is extremely hypoxic, among many other changes, which could drive an increase in the activity of DNA repair mechanisms.

The results for *Astyanax* in the laboratory as generally confirmed under natural wild conditions, and if anything the responses seen under these conditions are even stronger, with the animals showing less suggestion of any clock rhythmicity and dramatically enhanced "light-regulated" gene transcriptional levels.

Comparing all the data obtained for *Astyanax* and *Phreatichthys* reveals marked differences in the circadian clock of these two species. Indeed, while *Phreatichthys* showed no rhythms at the molecular or behavioral levels both in LD and DD as well as an infradian cell culture clock, *Astyanax* cavefish still possesses a circadian clock that can be entrained by light. However, in the *Astyanax* clock, differences in the regulation of light-regulated clock genes do point to some abnormalities in clock input pathways. It is therefore tempting to speculate that the weaker hypogean clock phenotype of *Astyanax* reflects a shorter period of isolation and/or a less stringent isolation within the Mexican compared with the Somalian cave habitat.

## 3.6 Cavefish Sleep, Activity, and Metabolism

Food supply in a cave environment is typically extremely limited due in large part to the absence of green plants which are completely reliant on photosynthesis. Food is either washed passively into the cave with the flow of water, or actively carried in with the daily migration of animals such as bats into and out of cave systems. Therefore a number of basic strategies seem to have evolved in cavefish to survive these challenging conditions. One adaptation appears to rely on the use of food and feeding activity to entrain the circadian clock in order to anticipate regular, daily food availability with increased food anticipatory activity and thereby to achieve more effective exploitation of the limited food supply. Other strategies center on reducing the levels of metabolism and energy expenditure, which include changes to sleeping patterns.

In *Phreatichthys* the absence of a light-entrained circadian oscillator is clearly compensated for by the presence of a food-entrainable circadian oscillator (Cavallari et al., 2011). Adult *Phreatichthys* show a strong entrainment of rhythmic locomotor activity to regular daily feeding. As a consequence, they exhibit characteristic food anticipatory activity, namely an increase in the level of locomotor activity a few hours prior to the anticipated mealtime. The entrainment of behavioral circadian clock output is matched by rhythmic *Clk1a* and *Per1b* gene expression in the brain, heart, fin, and liver. These results point to *Phreatichthys* having a functional circadian clock that is entrainable by feeding but not by LD cycles.

The effect of a metabolic flux on the circadian clock has also been tested in vitro in a primary cell line derived from fin clips of *Phreatichthys* that was transfected with a clock-regulated zebrafish *per1b:luc* reporter construct. Transient treatment of this reporter cell line with dexamethasone, an agonist of the glucocorticoid receptor, induced a bioluminescence rhythm that persisted for almost three cycles with an infradian period of 43 h. These results reveal the existence of an abnormal peripheral circadian clock in *Phreatichthys* that displays an infradian period under constant conditions that contrasts with the circadian period shown by the food-entrained central circadian clock at the behavioral level (Cavallari et al., 2011). Interestingly, a behavioral study (Duboué & Borowsky, 2012) of cave adapted balitorid loaches from Southeast Asia (*Schistura jaruthanini*, *Schistura oedipus*, *Schistura Spiesi*, and *Nemacheilus troglocataractus* from Thailand, and *Schistura kaysonei* from Laos) revealed that *S. oedipus* showed an infradian rhythm of locomotor with a free running period of 38–40 h. This shift of periodicity, from circadian to infradian has not been found at the behavioral level in *Phreatichthys* in which the locomotor activity in constant conditions is either arrhythmic or under a regular, daily feeding regime, circadian (Cavallari et al., 2011). Indeed, the properties of the food-entrained behavioral rhythms in *Phreatichthys* implicate the presence of a central food-entrainable clock which is able to drive and thereby override, rather than simply synchronize, peripheral clock rhythmicity. Given the paucity of information on the location and function of the food-entrained circadian clock, this data from cavefish potentially provides important new insight.

A pilot investigation performed in the laboratory using juvenile *Astyanax* (Zafar & Morgan, 1992) revealed entrainment of swimming activity to daily food administration. In addition, a field study tested feeding entrainment of the circadian clock in the cavefish population of the Chica Cave which is also used as a roost by a large bat colony that flies into and out of the cave two times per day, and provides a major food source for cavefish, guano. Even though this cavefish population is exposed to this daily, regular source of food, its molecular clock is not entrained to the food availability and there is no significant oscillation in the levels of *Per1* and *Per2* expression (Beale et al., 2013). The dichotomy in these data would seem to question the efficacy of feeding as a Zeitgeber in *Astyanax*. However, considering the general importance of feeding in the evolution of the circadian clock, the nature of feeding entrainment in *Astyanax* clearly deserves more investigation.

Simply "sitting and watching" *Astyanax* behavior in a tank can tell you quite a lot about the activity and general behavioral changes that have occurred between surface and cave strains. Surface fish shoal, moving around the tank as a group and exhibit very robust diurnal activity with little movement during the night. While it is immediately apparent that cave strains move around individually, with little social interaction, at a lower apparent velocity, and exhibit no clear-cut day–night differences in activity. Activity monitoring in young and old adults confirms this general observation in a more quantitative manner, that surface fish have a strong circadian rhythm in activity, whereas this is weak at best in cave populations (Beale et al., 2013; Duboué & Borowsky, 2012; Duboué, Keene, & Borowsky, 2011). This fact is true even under LD conditions, where the endogenous circadian clock can still be entrained by the environmental cycle in both populations. Clearly, in cave strains, although they are capable of generating a molecular circadian oscillation, this is weakly coupled to the circadian output of swimming. The same conclusion has generally been drawn regarding sleep behavior, though a "trickier" phenotype to assay than global circadian rhythms in activity. Most cave strains, especially in the larval/fry form, appear to show a sleep-loss phenotype compared to surface fish equivalents. This result is not quite so absolute and clear in the adult forms, however. It has been shown, for example, that adults from Tinaja cave and Cueva Los Sabinos do not show such a reduction in sleep behavior (Keene, Yoshizawa, & McGaugh, 2015). This could represent a complex difference between adult and larval forms, as well as possible cave-specific changes in the sleep phenotype. Nevertheless, *Astyanax* is likely to serve as a valuable model system for the future study of sleep, not least because of these specific surface to cave differences, which can in some senses be viewed as the equivalent as identifying sleep mutants through a forward genetic mutant screen in a standard laboratory animal.

The reasons why cave populations might lack a sleep phenotype, or at least have adapted a continuously active lifestyle, are likely to be numerous. Lack of easy food availability in certain caves can be extreme, especially those without a significant bat population. Maybe it is simply necessary to keep moving and foraging for food in an environment where there is almost nothing to eat. Secondly, of course, cave populations do not suffer from predation, which is a major selective pressure on surface fish populations, where predation from other fish, birds, and amphibians is considerable selective pressure. Another reason relates to metabolic issues, a factor which is obviously connected to the lack of abundant nutrition. The presence of eyes has generally been thought to be highly energy demanding, and so one argument behind the loss of a functional retina in cave strains relates to the "energy-saving" advantages of an eyeless phenotype. However, there have been relatively few studies that have tested the actual metabolic differences between cave and surface strains of *Astyanax*. Perhaps the most thorough of these studies is the work of Moran , Softley, and Warrant (2014). To perhaps oversimplify their findings, surface fish show about 20% increase in oxygen demand during the day compared to night, even when swimming activity is controlled for by the nature of the metabolic measurements employed. In the case of cavefish, this daily rhythm in oxygen demand is eliminated under both LD and constant dark conditions. Even in the presence of an entrained molecular core clock, there is no subsequent rhythm in a clock-regulated process, in this case metabolism. This again supports the idea of an uncoupling between the core clock and its normally regulated output processes. By eliminating the daily rhythm in oxygen demand, cavefish, in particular the Pachon strain, use 27% less energy in a day than the equivalent surface fish. In addition, both *Phreatichthys* and *Barbopsis devecchii*, another hypogean species endemic to Somalia, have oxygen consumption and a metabolic rate 2–3 times lower than geographically related epigean species (Ercolini, Berti, Chelazzi, & Messana, 1987). Hence, there is in fact a clear selective advantage for a cave animal to be arrhythmic, rather than to possess daily controlled activity and metabolic processes.

## 4. OUTLOOK

In conclusion, *Astyanax* and *Phreatichthys* have proven to be powerful model systems in which to study adaptations to the cave environment. In the text earlier, we have focused on aspects relating to circadian clocks and related biological processes, but of course this combination of models clearly lends itself well to the study of many other aspects of basic cell and neurobiology, sensory physiology, anatomy, and metabolic functions. What is more than apparent, however, is that one can expect a dramatic increase in progress in studying these species over the next few years, accelerated by the advent of convenient and cost-effective genomic and transcriptomic sequencing approaches. In addition, the potential to study metabolomics and epigenetics in a detailed and thorough way will yield exciting new findings in the comparison between surface and cave populations of *A. mexicanus* as well as other diverse cavefish species.

## ACKNOWLEDGMENTS

C.B. acknowledges research grants from the University of Ferrara (FAR2014). The research of D.V. and N.S.F. was funded through the Helmholtz funding program BIFTM. We also acknowledge support by the Deutsche Forschungsgemeinschaft and the open access publishing fund of the Karlsruhe Institute of Technology. D.W. thanks Dr. Andrew Beale and Kathy Tamai for their work and effort, and the BBSRC for grant support.

## REFERENCES

Aggarwal, B. B., Quintanilha, A. T., Cammack, R., & Packer, L. (1978). Damage to mitochondrial electron transport and energy coupling by visible light. *Biochimica et Biophysica Acta, 502*(2), 367–382.

Albrecht, U., Sun, Z. S., Eichele, G., & Lee, C. C. (1997). A differential response of two putative mammalian circadian regulators, mper1 and mper2, to light. *Cell, 91*(7), 1055–1064.

Alunni, A., Menuet, A., Candal, E., Pénigault, J.-B., Jeffery, W. R., & Rétaux, S. (2007). Developmental mechanisms for retinal degeneration in the blind cavefish *Astyanax mexicanus*. *Journal of Comparative Neurology, 505*(2), 221–233. http://doi.org/10.1002/cne.21488.

Asher, G., & Sassone-Corsi, P. (2015). Time for food: the intimate interplay between nutrition, metabolism, and the circadian clock. *Cell, 161*(1), 84–92. http://doi.org/10.1016/j.cell.2015.03.015.

Balsalobre, A., Damiola, F., & Schibler, U. (1998). A serum shock induces circadian gene expression in mammalian tissue culture cells. *Cell, 93*(6), 929–937.

Balsalobre, A., Marcacci, L., & Schibler, U. (2000). Multiple signaling pathways elicit circadian gene expression in cultured Rat-1 fibroblasts. *Current Biology: CB, 10*(20), 1291–1294.

Beale, A., Guibal, C., Tamai, T. K., Klotz, L., Cowen, S., Peyric, E., … Whitmore, D. (2013). Circadian rhythms in Mexican blind cavefish *Astyanax mexicanus* in the lab and in the field. *Nature Communications, 4*, 1–10. http://doi.org/10.1038/ncomms3769.

Ben-Moshe, Z., Vatine, G., Alon, S., Tovin, A., Mracek, P., Foulkes, N. S., & Gothilf, Y. (2010). Multiple PAR and E4BP4 bZIP transcription factors in zebrafish: diverse temporal and spatial expression patterns. *Chronobiology International, 27*(8), 1509–1531. http://doi.org/10.3109/07420528.2010.510229.

Bradic, M., Beerli, P., García de León, F. J., Esquivel-Bobadilla, S., & Borowsky, R. L. (2012). Gene flow and population structure in the Mexican blind cavefish complex (*Astyanax mexicanus*). *BMC Evolutionary Biology, 12*(1), 9. http://doi.org/10.1186/1471-2148-12-9.

Brown, S. A. (2016). Circadian metabolism: from mechanisms to metabolomics and medicine. *Trends in Endocrinology & Metabolism*, 1–12. http://doi.org/10.1016/j.tem.2016.03.015.

Brown, S. A., Zumbrunn, G., Fleury-Olela, F., Preitner, N., & Schibler, U. (2002). Rhythms of mammalian body temperature can sustain peripheral circadian clocks. *Current Biology: CB, 12*(18), 1574–1583.

Calderoni, L., Rota-Stabelli, O., Frigato, E., Panziera, A., Kirchner, S., Foulkes, N. S., … Fuselli, S. (2016). Relaxed selective constraints drove functional modifications in peripheral photoreception of the cavefish *Phreatichthys andruzzii* and provide insight into the time of cave colonization. *Heredity*.

Cavallari, N., Frigato, E., Vallone, D., Fröhlich, N., Lopez-Olmeda, J. F., Foà, A., … Foulkes, N. S. (2011). A blind circadian clock in cavefish reveals that opsins mediate peripheral clock photoreception. *PLoS Biology, 9*(9), e1001142. http://doi.org/10.1371/journal.pbio.1001142.

Coghill, L. M., Hulsey, C. D., Chaves-Campos, J., de León, F. J. G., & Johnson, S. G. (2014). Next generation phylogeography of cave and surface *Astyanax mexicanus*. *Molecular Phylogenetics and Evolution, 79*(C), 368–374. http://doi.org/10.1016/j.ympev.2014.06.029.

Cowell, I. G. (2002). E4BP4/NFIL3, a PAR-related bZIP factor with many roles. *BioEssays: News and Reviews in Molecular, Cellular and Developmental Biology, 24*(11), 1023–1029. http://doi.org/10.1002/bies.10176.

Dallmann, R., Okyar, A., & Levi, F. (2016). Dosing-time makes the poison: circadian regulation and pharmacotherapy. *Trends in Molecular Medicine, 22*(5), 430–445. http://doi.org/10.1016/j.molmed.2016.03.004.

Davies, W. I. L., Tamai, T. K., Zheng, L., Fu, J. K., Rihel, J., Foster, R. G., … Hankins, M. W. (2015). An extended family of novel vertebrate photopigments is widely expressed and displays a diversity of function. *Genome Research, 25*(11), 1666–1679. http://doi.org/10.1101/gr.189886.115.

Doyle, S., & Menaker, M. (2007). Circadian photoreception in vertebrates. *Cold Spring Harbor Symposia on Quantitative Biology, 72*(1), 499–508. http://doi.org/10.1101/sqb.2007.72.003.

Duboué, E. R., & Borowsky, R. L. (2012). Altered rest-activity patterns evolve via circadian independent mechanisms in cave adapted balitorid loaches. *PLoS One, 7*(2). e30868–5 http://doi.org/10.1371/journal.pone.0030868.

Duboué, E. R., Keene, A. C., & Borowsky, R. L. (2011). Evolutionary convergence on sleep loss in cavefish populations. *Current Biology: CB, 21*(8), 671–676. http://doi.org/10.1016/j.cub.2011.03.020.

Ercolini, A., Berti, R., Chelazzi, L., & Messana, G. (1987). Oxygen consumption in hypogean and epigean cyprinids (pisces). *Monitore Zoologico Italiano, 22*(1), 23–30. http://doi.org/10.1080/03749444.1987.10736719.

Fukushima, N., Naito, Y., & Ryoji, M. (2009). Induction of (6-4) photolyase gene transcription by blue light in *Xenopus* A6 cells. *Biochemical and Biophysical Research Communications, 383*(2), 231–234. http://doi.org/10.1016/j.bbrc.2009.03.158.

Gachon, F. (2007). Physiological function of PARbZip circadian clock-controlled transcription factors. *Annals of Medicine, 39*(8), 562–571. http://doi.org/10.1080/07853890701491034.

Gavriouchkina, D., Fischer, S., Ivacevic, T., Stolte, J., Benes, V., & Dekens, M. P. S. (2010). Thyrotroph embryonic factor regulates light-induced transcription of repair genes in zebrafish embryonic cells. *PLoS One, 5*(9), e12542. http://doi.org/10.1371/journal.pone.0012542.

Gerkema, M. P., Davies, W. I. L., Foster, R. G., Menaker, M., & Hut, R. A. (2013). The nocturnal bottleneck and the evolution of activity patterns in mammals. *Proceedings of the Royal Society: Biological Sciences, 280*(1765), 20130508. http://doi.org/10.1098/rspb.2013.0508.

Green, C. B., & Takahashi, J. S. (2006). Xenobiotic metabolism in the fourth dimension: PARtners in time. *Cell Metabolism, 4*(1), 3–4. http://doi.org/10.1016/j.cmet.2006.06.002.

Hardin, P. E. (2011). Molecular genetic analysis of circadian timekeeping in *Drosophila*. *Advances in Genetics, 74*, 141–173. http://doi.org/10.1016/B978-0-12-387690-4.00005-2.

Hernández Fernández, M., & Vrba, E. S. (2006). Plio-pleistocene climatic change in the Turkana basin (East Africa): evidence from large mammal faunas. *Journal of Human Evolution, 50*(6), 595–626. http://doi.org/10.1016/j.jhevol.2005.11.004.

Hirayama, J., Cho, S., & Sassone-Corsi, P. (2007). Circadian control by the reduction/oxidation pathway: catalase represses light-dependent clock gene expression in the zebrafish. *Proceedings of the National Academy of Sciences of the United States of America, 104*(40), 15747–15752. http://doi.org/10.1073/pnas.0705614104.

Hughes, S., Jagannath, A., Hankins, M. W., Foster, R. G., & Peirson, S. N. (2015). Photic regulation of clock systems. *Methods in Enzymology, 552*, 125–143. http://doi.org/10.1016/bs.mie.2014.10.018.

Keene, A. C., Yoshizawa, M., & McGaugh, S. (2015). *Biology and evolution of the Mexican cavefish*. Elsevier.

Kim, S. T., & Sancar, A. (1993). Photochemistry, photophysics, and mechanism of pyrimidine dimer repair by DNA photolyase. *Photochemistry and Photobiology, 57*(5), 895–904.

Lahiri, K., Vallone, D., Gondi, S. B., Santoriello, C., Dickmeis, T., & Foulkes, N. S. (2005). Temperature regulates transcription in the zebrafish circadian clock. *PLoS Biology, 3*(11), e351. http://doi.org/10.1371/journal.pbio.0030351.

Lim, C., & Allada, R. (2013). Emerging roles for post-transcriptional regulation in circadian clocks. *Nature Publishing Group, 16*(11), 1544–1550. http://doi.org/10.1038/nn.3543.

Lockwood, D. B., Wataha, J. C., Lewis, J. B., Tseng, W. Y., Messer, R. L. W., & Hsu, S. D. (2005). Blue light generates reactive oxygen species (ROS) differentially in tumor vs. normal epithelial cells. *Dental Materials, 21*(7), 683–688. http://doi.org/10.1016/j.dental.2004.07.022.

Loew, E. R., & McFarland, W. N. (1990). The underwater visual environment. In R. H. Douglas, & M. B. P. Djamgoz (Eds.), *The visual system of fish* (pp. 1–43) (London).

Loros, J. J., & Dunlap, J. C. (2001). Genetic and molecular analysis of circadian rhythms in neurospora. *Annual Review of Physiology, 63*(1), 757–794. http://doi.org/10.1146/annurev.physiol.63.1.757.

Lowrey, P. L., & Takahashi, J. S. (2011). Genetics of circadian rhythms in mammalian model organisms. *Advances in Genetics, 74*, 175–230. http://doi.org/10.1016/B978-0-12-387690-4.00006-4.

Lucas, R. J., Lall, G. S., Allen, A. E., & Brown, T. M. (2012). How rod, cone, and melanopsin photoreceptors come together to enlighten the mammalian circadian clock. *Progress in Brain Research, 199*, 1–18. http://doi.org/10.1016/B978-0-444-59427-3.00001-0.

Mackey, S. R., Golden, S. S., & Ditty, J. L. (2011). The itty-bitty time machine genetics of the cyanobacterial circadian clock. *Advances in Genetics, 74*, 13–53. http://doi.org/10.1016/B978-0-12-387690-4.00002-7.

Martino, T. A., & Young, M. E. (2015). Influence of the cardiomyocyte circadian clock on cardiac physiology and pathophysiology. *Journal of Biological Rhythms, 30*(3), 183–205. http://doi.org/10.1177/0748730415575246.

Mas, P. (2005). Circadian clock signaling in *Arabidopsis thaliana*: from gene expression to physiology and development. *International Journal of Developmental Biology, 49*(5–6), 491–500. http://doi.org/10.1387/ijdb.041968pm.

Menaker, M., Moreira, L. F., & Tosini, G. (1997). Evolution of circadian organization in vertebrates. *Brazilian Journal of Medical and Biological Research, 30*(3), 305–313.

Mitani, H., Uchida, N., & Shima, A. (1996). Induction of cyclobutane pyrimidine dimer photolyase in cultured fish cells by UVA and blue light. *Photochemistry and Photobiology, 64*(6), 943–948.

Mohawk, J. A., Green, C. B., & Takahashi, J. S. (2012). Central and peripheral circadian clocks in mammals. *Annual Review of Neuroscience, 35*(1), 445–462. http://doi.org/10.1146/annurev-neuro-060909-153128.

Moran, D., Softley, R., & Warrant, E. J. (2014). Eyeless Mexican cavefish save energy by eliminating the circadian rhythm in metabolism. *PLoS One, 9*(9), e107877. http://dx.plos.org/10.1371/journal.pone.0107877.

Mracek, P., Pagano, C., Fröhlich, N., Idda, M. L., Cuesta, I. H., Lopez-Olmeda, J. F., … Foulkes, N. S. (2013). Erk signaling regulates light-induced gene expression via D-box enhancers in a differential, wavelength-dependent manner. *PLoS One, 8*(6), e67858. http://doi.org/10.1371/journal.pone.0067858.

Mracek, P., Santoriello, C., Idda, M. L., Pagano, C., Ben-Moshe, Z., Gothilf, Y., … Foulkes, N. S. (2012). Regulation of per and cry genes reveals a central role for the D-box enhancer in light-dependent gene expression. *PLoS One,* 7(12), e51278. http://doi.org/10.1371/journal.pone.0051278.

Pérez-Cerezales, S., Boryshpolets, S., Afanzar, O., Brandis, A., Nevo, R., Kiss, V., & Eisenbach, M. (2015). Involvement of opsins in mammalian sperm thermotaxis. *Scientific Reports, 5*, 1–18. http://doi.org/10.1038/srep16146.

Pittendrigh, C. S. (1993). Temporal organization: reflections of a Darwinian clock-watcher. *Annual Review of Physiology, 55*(1), 16–54. http://doi.org/10.1146/annurev.ph.55.030193.000313.

Preitner, N., Damiola, F., Lopez-Molina, L., Zakany, J., Duboule, D., Albrecht, U., & Schibler, U. (2002). The orphan nuclear receptor REV-ERBalpha controls circadian transcription within the positive limb of the mammalian circadian oscillator. *Cell, 110*(2), 251–260.

Provencio, I., Jiang, G., De Grip, W. J., Hayes, W. P., & Rollag, M. D. (1998). Melanopsin: an opsin in melanophores, brain, and eye. *Proceedings of the National Academy of Sciences of the United States of America, 95*(1), 340–345.

Ramos, B. C. R., Moraes, M. N. C. M., Poletini, M. O., Lima, L. H. R. G., & Castrucci, A. M. L. (2014). From blue light to clock genes in zebrafish ZEM-2S cells. *PLoS One, 9*(9). e106252–12 http://doi.org/10.1371/journal.pone.0106252.

Reppert, S. M., & Weaver, D. R. (2001). Molecular analysis of mammalian circadian rhythms. *Annual Review of Physiology, 63*(1), 647–676. http://doi.org/10.1146/annurev.physiol.63.1.647.

Roenneberg, T., Daan, S., & Merrow, M. (2003). The art of entrainment. *Journal of Biological Rhythms, 18*(3), 183–194.

Roenneberg, T., & Foster, R. G. (1997). Twilight times: light and the circadian system. *Photochemistry and Photobiology, 66*(5), 549–561.

Sancar, A. (2003). Structure and function of DNA photolyase and cryptochrome blue-light photoreceptors. *Chemical Reviews, 103*(6), 2203–2237. http://doi.org/10.1021/cr0204348.

Sassone-Corsi, P., Foulkes, N. S., & Whitmore, D. (2000). Light acts directly on organs and cells in culture to set the vertebrate circadian clock. *Nature, 404*(6773), 87–91. http://doi.org/10.1038/35003589.

Savvidis, C., & Koutsilieris, M. (2012). Circadian rhythm disruption in cancer biology. *Molecular Medicine (Cambridge, MA), 18*, 1249–1260. http://doi.org/10.2119/molmed.2012.00077.

Schibler, U., Gotic, I., Saini, C., Gos, P., Curie, T., Emmenegger, Y., … Franken, P. (2015). Clock-talk: interactions between central and peripheral circadian oscillators in mammals. *Cold Spring Harbor Symposia on Quantitative Biology*, 027490. http://doi.org/10.1101/sqb.2015.80.027490.

Schibler, U., & Sassone-Corsi, P. (2002). A web of circadian pacemakers. *Cell, 111*(7), 919–922.

Schmidt, T. M., Do, M. T. H., Dacey, D., Lucas, R., Hattar, S., & Matynia, A. (2011). Melanopsin-positive intrinsically photosensitive retinal ganglion cells: from form to function. *Journal of Neuroscience: The Official Journal of the Society for Neuroscience, 31*(45), 16094–16101. http://doi.org/10.1523/JNEUROSCI.4132-11.2011.

Shen, W. L., Kwon, Y., Adegbola, A. A., Luo, J., Chess, A., & Montell, C. (2011). Function of rhodopsin in temperature discrimination in *Drosophila. Science, 331*(6022), 1333–1336. http://doi.org/10.1126/science.1198904.

Stemmer, M., Schuhmacher, L.-N., Foulkes, N. S., Bertolucci, C., & Wittbrodt, J. (2015). Cavefish eye loss in response to an early block in retinal differentiation progression. *Development (Cambridge, England), 142*(4), 743–752. http://doi.org/10.1242/dev.114629.

Tamai, T. K., Vardhanabhuti, V., Foulkes, N. S., & Whitmore, D. (2004). Early embryonic light detection improves survival. *Current Biology: CB, 14*(3), R104–R105.

Tamai, T. K., Young, L. C., & Whitmore, D. (2007). Light signaling to the zebrafish circadian clock by Cryptochrome 1a. *Proceedings of the National Academy of Sciences of the United States of America, 104*(37), 14712–14717. http://doi.org/10.1073/pnas.0704588104.

Tarttelin, E. E., Frigato, E., Bellingham, J., Di Rosa, V., Berti, R., Foulkes, N. S., … Bertolucci, C. (2012). Encephalic photoreception and phototactic response in the troglobiont Somalian blind cavefish *Phreatichthys andruzzii. Journal of Experimental Biology, 215*(Pt 16), 2898–2903. http://doi.org/10.1242/jeb.071084.

Trajano, E., Bichuette, M. E., & Kapoor, B. G. (2010). In E. Trajano, M. E. Bichuette, & B. G. Kapoor (Eds.), *Biology of subterranean fishes* (1st ed.). Boca Raton: CRC Press.

Vallone, D., Gondi, S. B., Whitmore, D., & Foulkes, N. S. (2004). E-box function in a period gene repressed by light. *Proceedings of the National Academy of Sciences of the United States of America, 101*(12), 4106–4111. http://doi.org/10.1073/pnas.0305436101.

Vanselow, K., & Kramer, A. (2007). Role of phosphorylation in the mammalian circadian clock. *Cold Spring Harbor Symposia on Quantitative Biology, 72*(1), 167–176. http://doi.org/10.1101/sqb.2007.72.036.

Vatine, G., Vallone, D., Appelbaum, L., Mracek, P., Ben-Moshe, Z., Lahiri, K., … Foulkes, N. S. (2009). Light directs zebrafish period2 expression via conserved D and E boxes. *PLoS Biology*, 7(10), e1000223. http://doi.org/10.1371/journal.pbio.1000223.

Wang, M., Zhong, Z., Zhong, Y., Zhang, W., & Wang, H. (2015). The zebrafish period2 protein positively regulates the circadian clock through mediation of retinoic acid receptor (RAR)-related orphan receptor α (Rorα). *Journal of Biological Chemistry, 290*(7), 4367–4382. http://doi.org/10.1074/jbc.M114.605022.

Weger, B. D., Sahinbas, M., Otto, G. W., Mracek, P., Armant, O., Dolle, D., … Dickmeis, T. (2011). The light responsive transcriptome of the zebrafish: function and regulation. *PLoS One, 6*(2), e17080. http://doi.org/10.1371/journal.pone.0017080.

Yoshizawa, M., Yamamoto, Y., O'Quin, K. E., & Jeffery, W. R. (2012). Evolution of an adaptive behavior and its sensory receptors promotes eye regression in blind cavefish. *BMC Biology, 10*(1), 108. http://doi.org/10.1186/1741-7007-10-108.

Zafar, N. P., & Morgan, P. (1992). Feeding entrains an endogenous rhythm of swimming activity in the blind Mexican cave fish. *Journal of Interdisciplinary Cycle Research, 23*, 165–166.

Ziv, L., Levkovitz, S., Toyama, R., Falcón, J., & Gothilf, Y. (2005). Functional development of the zebrafish pineal gland: light-induced expression of period2 is required for onset of the circadian clock. *Journal of Neuroendocrinology, 17*(5), 314–320. http://doi.org/10.1111/j.1365-2826.2005.01315.x.

CHAPTER TWO

# Genomic and Transcriptomic Approaches to Study Cancer in Small Aquarium Fish Models

J. Regneri*[,a], B. Klotz*[,a] and M. Schartl*[,§,¶,1]
*University of Würzburg, Würzburg, Germany
§University Clinic Würzburg, Würzburg, Germany
¶Texas A&M University, College Station, TX, United States
[1]Corresponding author: E-mail: phch1@biozentrum.uni-wuerzburg.de

## Contents



## Abstract

Zebrafish and medaka that develop tumors have become valuable tools for experimental cancer research. With the advent of microarrays and new sequencing technologies it has become feasible to perform whole genome, exome, and transcriptome analyses in these fish models. Analyses that compare the two fish models with each other and with data from human tumors have revealed a plethora of important insights. An unexpected high degree of comparability of molecular features of fish and human tumors has been detected. Furthermore, analyses of the fish model data have uncovered molecules that have not received appropriate attention in studies on their human tumor counterparts and thus have provided valuable candidates for novel biomarkers and therapeutic targets.

[a] These authors contributed equally to this work.

*Advances in Genetics*, Volume 95
ISSN 0065-2660
http://dx.doi.org/10.1016/bs.adgen.2016.04.001

# 1. INTRODUCTION

Cancer is a devastating disease that affects all human societies worldwide and is the second most frequent cause of human death (http://www.cdc.gov/nchs/data/nvsr/nvsr64/nvsr64_02.pdf). Differently from other major chronic human diseases, such as cardiovascular diseases or diabetes, no general concepts for prevention and treatment are available, due to many etiologies and a huge variety of pathologies. This complexity is also reflected in often unpredictable responses to treatment regimes and unforeseeable clinical course of the disease. Thus, even more than other major chronic diseases, cancer is a target for personalized medicine and individualized therapies. Such strategies require two major preconditions. First, a deep understanding of the molecular basis of the different types of cancer is required to develop a multitude of drugs that will interfere with the action of the molecular drivers of tumorigenesis, for example, mutant or overexpressed oncoproteins. Second, precise and comprehensive diagnostic tools are key to providing the information about the pathological features and molecular characteristics of the individual cancer and to match them with known mutations or expression signatures for classification and selection of the appropriate treatment. Such in-depth and individualized diagnostics have to be economic to be reasonable, and with the amazing developments of new sequencing technologies, such prospects are becoming more and more realistic. Personal cancer exomes or even full genomes and high-resolution transcriptomes from RNA-seq are on their way to be used for diagnostics in clinical trials (Schilsky, 2014; Simon & Roychowdhury, 2013). Experimental studies which include deep-sequencing analysis for detecting the mutational landscape of various cancer types and for establishing expression signatures also generate huge amounts of data that challenge the capacity of bioinformatics to keep track. From such analyses, the comparison of cancer cells with what is regarded as their normal counterparts has revealed a multitude of cancer-associated gene mutations and innumerable gene expression differences. However, in most cases it is almost impossible to decide which of the many changes are related to the primary events of tumorigenesis, which of those may be more involved in cancer progression, and which might just be a reflection of the chaotic molecular biology of autonomously and fast dividing cells (which for many cancers also exhibit a high degree of genomic instability).

This situation is much better in animal cancer models, where the primary event of tumor initiation is known. Furthermore, it is argued that molecular changes that are shared by cancers of phylogenetically distant species will be more critical and relevant for the pathological condition than others, because they are evolutionary conserved. For decades animal models have been used to better understand the function and biochemistry of cancer-initiating genes. In particular, small aquarium fish, the zebrafish and medaka, have been used to establish transgenic cancer models. Most are based on the transgenic expression of a mutant oncogene, disruption of a known tumor suppressor gene or altered expression of a gene known to be affected in human cancer. On the basis of the full-genome sequence of both fish species (Howe et al., 2013; Kasahara et al., 2007), microarray and deep-sequencing methods have been used in genomic and transcriptomic approaches to analyze cancer development. We review here some exemplary studies and their contribution to building a better understanding of human cancer.

## 2. CANCER MODELS IN FISH

Several cancer models have been established in small aquarium fish species that have provided useful tools for experimental research with genomic and transcriptomic approaches (Table 1).

The first stable transgenic cancer models were established in zebrafish. These lines develop T-cell acute lymphoblastic leukemia (T-ALL), because they carry the mouse *c-Myc* gene under transcriptional control of the zebrafish *rag2* (recombination-activating gene 2) promoter that directs gene expression specifically to the lymphoid cell lineage (Langenau et al., 2003).

In the process of generating transgenic lines that would develop hematopoietic cancers, one line that carries the human $KRAS^{G12D}$ oncogene transcribed from the *rag2* promoter developed tumors of the skeletal muscle compartment, classified as rhabdomyosarcoma (Langenau et al., 2007). An aberrant activation of the *rag2* promoter in satellite and myoblast cell populations leads to malignant muscle tumors that already develop at 10 days postfertilization. In humans, rhabdomyosarcoma is the most common childhood soft-tissue sarcoma with often fatal outcome (Arndt & Crist, 1999).

**Table 1** Genomic and transcriptomic approaches in fish models of cancer

| Cancer | Species | Genotype | Omics type of study | References |
|---|---|---|---|---|
| Melanoma | Zebrafish | *mitfa:BRAF*$^{V600E}$; *p53* $^{-/-}$ | BAC array CGH<br>Exome<br>Microarray transcriptome | Freeman et al. (2009), Yen et al. (2013)[a], and White et al. (2011) |
| | | *mitfa:NRAS*$^{Q61K}$; *p53*$^{-/-}$ | Exome<br>Microarray transcriptome | Yen et al. (2013)[a] and Dovey et al. (2009) |
| | Medaka | *mitfa:xmrk* | RNA-seq transcriptome | Schartl et al. (2012, 2015) |
| Rhabdomyo-sarcoma | Zebrafish | *rag2:KRAS*$^{G12D}$ | BAC array CGH<br>Oligonucleotide array CGH<br>Microarray transcriptome | Freeman et al. (2009), Chen et al. (2005), and Langenau et al. (2007) |
| T ALL | Zebrafish | *rag2:cMyc* | BAC array CGH<br>Oligonucleotide array CGH | Freeman et al. (2009) and Rudner et al. (2011) |
| | | *hlk* mutant<br>*srk* mutant<br>*otg* mutant | Oligonucleotide array CGH | Rudner et al. (2011) |
| MPNST | Zebrafish | *p53*$^{-/-}$ or several mutations in ribosomal protein genes | Oligonucleotide array CGH and whole-genome sequencing | Zhang et al. (2010, 2013) |
| Hepatocellular carcinoma | Zebrafish | Carcinogen induced | Microarray transcriptome | Lam et al. (2006) |
| | | *fabp10: kras*$^{G12V}$ | Microarray transcriptome<br>RNA-seq transcriptome | Nguyen et al. (2011) and Zheng et al. (2014) |
| | | *fabp10:Myc* | RNA-seq transcriptome | Li et al. (2013) and Zheng et al. (2014) |
| | | *fabp10:xmrk* | RNA-seq transcriptome | Li et al. (2014) and Zheng et al. (2014) |
| | | *fabp10:Myc/xmrk* | RNA-seq transcriptome | Li et al. (2015) |

[a]Exomes were also produced from fish, which in addition to the *BRAF* and *NRAS* driver oncogenes carried various mutations or additional transgenes.

One of the oldest animal models for cancer research is the *Xiphophorus* melanoma model, which had already been discovered in the late 1920s (Häussler, 1928; Kosswig, 1928). In this model, interspecies hybrids of platyfish (*Xiphophorus maculatus*) and swordtail (*Xiphophorus hellerii*) spontaneously develop melanomas when backcrossed to swordtails. The reason for the development of malignant melanoma in these fish is the pigment cell-specific overexpression of Xmrk, which is a mutationally activated version of the epidermal growth factor (EGF) receptor (reviewed in Meierjohann & Schartl, 2006). This oncogene was used in medaka to establish a transgenic model for melanoma development (Schartl et al., 2010) (Fig. 1). In this model the *xmrk* cDNA from the platyfish *X. maculatus* is expressed under the transcriptional control of the pigment cell-specific *mitfa* (microphthalmia-associated transcription factor a) promoter of medaka, which directs oncogene expression in melanocytes and xanthophores. Depending upon the genetic background of the transgenic medakas, the pathological features and the tumor type vary. The same transgene leads to the development of less aggressive exophytic xanthoerythrophoroma (XE; a tumor derived from red and yellow pigment cells), superficially spreading, nodular or highly aggressive invasive melanoma at different sites of the body or to melanoma of the eye (uvea melanoma).

Several transgenic melanoma models also exist in zebrafish, which express different oncogenes under the control of the zebrafish *mitfa* promoter. In 2005, Patton et al. generated the first $BRAF^{V600E}$ model for melanoma in zebrafish. The mutated $BRAF^{V600E}$ has been found with a high frequency in almost all human melanomas, but it was in zebrafish that its melanoma-inducing function was shown for the first time in an animal model (Patton, Mathers, & Schartl, 2011; Patton et al., 2005). In 2009, another transgenic zebrafish melanoma model involving

**Figure 1** Transgenic medaka model for melanoma development. Adult *mitfa:xmrk* transgenic medaka showing malignant melanoma development in the tail and anal fin invading the adjacent body musculature. Scale bar represents 10 mm. (See color plate)

*mitfa*-controlled expression of the $NRAS^{Q61K}$ transgene has been described (Dovey, White, & Zon, 2009). The exclusive expression of both human transgenes ($BRAF^{V600E}$ and $NRAS^{Q61K}$) in zebrafish melanocytes results in a hyperpigmentation phenotype, but for developing malignant melanomas additional loss-of-function mutations in p53 are required (Berghmans et al., 2005; Dovey et al., 2009; Patton et al., 2005). In human melanoma, p53 mutations are not very common, although p53 is frequently mutated in many other human cancers (Box & Terzian, 2008).

Pigment cells, except for retinal pigment cells, are of neural crest origin. Schwann cells are also derived from the neural crest and with the pigment cells share a common precursor. Intriguingly, at the molecular and histological level, several links have been detected between certain types of tumors that develop from both cell lineages, namely melanoma and malignant peripheral nerve sheath tumors (MPNSTs) (Van Raamsdonk & Deo, 2013). In particular zebrafish appears to be highly susceptible to MPNST. MPNSTs arise in zebrafish as result of missense mutations in the p53 gene (Berghmans et al., 2005) or mutations in several ribosomal protein genes (Amsterdam et al., 2004).

The zebrafish has also been used to generate several models for liver tumor development. Initially, it was shown that liver tumors of hepatocellular or cholangiocellular origin occur in zebrafish after treatment with carcinogens, for example with DMBA and DBP (Lam & Gong, 2006; Lam et al., 2006). Since 2011, several transgenic zebrafish liver tumor models have been established, which express different oncogenes, namely zebrafish $kras^{G12V}$ (Nguyen et al., 2011, 2012), mouse *Myc* (Li et al., 2013) or the *Xiphophorus xmrk* (Li et al., 2012, 2014), under control of the liver-specific zebrafish *fabp10* (fatty acid-binding protein 10) promoter. In the first transgenic model, a constitutive expression system was used to drive $kras^{G12V}$-induced liver tumor development (Nguyen et al., 2011). This led to the development of liver tumors with variable degrees of severity ranging from hyperplastic lesions (HL) to hepatocellular carcinoma (HCC). However, due to the constitutive expression system, the model has certain drawbacks, which limits its use, for example, for drug screening. A high level of constitutive $kras^{G12V}$ expression was shown to result in premature mortality, whereas low $kras^{G12V}$ expression levels were connected with low tumor incidence (about 30%) and a relatively late tumor onset. To overcome these limitations, a mifepristone-inducible $kras^{G12V}$-transgenic zebrafish line was generated in a subsequent study (Nguyen et al., 2012), which allowed dose-dependent and temporally controlled *kras*$^{G12V}$ expression. Upon mifepristone induction, fish developed liver tumors with 100% penetrance and tumors rapidly progressed from hyperplasia to HCC. In further studies, inducible *xmrk*- (Li et al., 2012) and *Myc*-transgenic (Li et al., 2013) zebrafish liver tumor models were established using the Tet-On system. As also seen in the *kras*$^{G12V}$-transgenic zebrafish, *xmrk*-transgenic fish developed HCC with 100% penetrance upon doxycycline induction, indicating that the exclusive expression of both oncogenes (*kras*$^{G12V}$ and *xmrk*) is sufficient to induce HCC in zebrafish. *Myc*-transgenic zebrafish, in contrast, developed mostly hepatocellular adenoma (HCA), which rarely progressed to carcinoma. Interestingly, withdrawal of mifepristone or doxycycline led to tumor regression in all three inducible zebrafish models, indicating an "oncogene-addition" of these liver tumors. By crossing the *xmrk* and *Myc* single transgenic lines, a doxycycline-inducible *Myc/xmrk* double transgenic line was established in a subsequent study, which was characterized by a faster and more severe hepatocarcinogenesis (Li, Zheng, Li, Li, & Gong, 2015).

Besides the above-mentioned cancer models, a number of other transgenic lines have been produced that also model human tumors and provide excellent tools for experimental research; however, these have not been explored by genomic or transcriptomic approaches so far.

## 3. GENOMIC APPROACHES

Cancer is clearly a genetic-based disease with only very few exceptions for which infectious agents are causative. Tumors are characterized by mutations, which occur somatically, again with only rare exceptions where cancers arise on the basis of hereditary mutations. Finding the mutations that are mainly responsible for the initiation and the development of a tumor (the "driver" genes) and those that have a more mild effect, but still guide the course of the disease and its clinical appearance (the "tumor modifier" genes) is a major goal of current cancer research. With the advent of new deep-sequencing technologies, a plethora of human cancer genomes from biopsies or cell lines have been produced and compared with the corresponding genomes from healthy tissues of the same patient or with reference genomes (eg, The Human Cancer Genome Atlas, https://tcga-data.nci.nih.gov/tcga/ ). Usually a large number of mutations are found in the tumor DNA and only a handful of these affect well-documented oncogenes and tumor suppressor genes. To single out those mutations that most likely are relevant for the development of the tumor and the progression of the disease from the background noise of "passenger" mutations, the criterion of "recurrent" mutations is used, meaning that only those genes are considered which are mutated in a reasonable number of samples. Background mutations are particularly frequent in those cancer types that are marked by a high heterogeneity due to genomic instability (Lengauer, Kinzler, & Vogelstein, 1998), such as in melanoma (Hawryluk & Tsao, 2014; Kaufmann et al., 2014) or colorectal cancer (Ciombor, Haraldsdottir, & Goldberg, 2014). In order to set the threshold frequency, above which a recurrently mutated gene is regarded as being relevant, remains an issue for consideration by the respective community of researchers and no general rules exist. The recurrence criterion is also used for fish cancer models when genomic approaches are used to confirm human cancer mutations or detect novel drivers. So far, genome-wide studies have been restricted to zebrafish because the necessary tools were only available for this more widely used model fish.

Variations in gene copy number are frequently associated with cancer (Fanciulli, Petretto, & Aitman, 2010). In order to detect such changes with a whole-genome approach, comparative genome hybridization (CGH) was developed. It was initially based on an array of large insert clones that cover the entire genome or only a selected, ordered subset. For zebrafish, the first array to be produced (Freeman et al., 2009) consisted of 286 BAC clones spotted on a chip. The clones included in the array were selected on the basis of their homology with known human oncogenes and tumor suppressor genes. Then, DNA from normal tissues and tumors of several cancer models was used for hybridization. The tumor and the control DNA carried a different fluorescent label and were hybridized competitively to the BAC-DNAs on the chip. Copy number alterations (CNAs) between the two probes are then visible by differences in the fluorescence intensity at the BAC clone that contains this sequence.

In *KRAS*$^{G12D}$-driven rhabdomyosarcoma between six and 14, changes (both gains and losses) were recorded. Interestingly, recurrent gain in a larger region covered by two BACs, was noted for a centromeric region of chromosome 17 (five of seven), and for a subtelomeric region of chromosome 22 (two of seven) (Freeman et al., 2009). In a follow-up study, a large number of recurrent (>three of ten) low copy amplifications was found in those rhabdomyosarcomas (Chen et al., 2013). Homologous regions of the human genome were also changed in a set of *RAS*-mutated human rhabdomyosarcomas. Genes found in these amplified regions include *cyclin D1* (*ccnd1*), *homeobox protein C6* (*hoxc6*), and *plexin A1* (*plxna1*), all of which were shown to play a role in rhabdomyosarcoma cell proliferation (Chen et al., 2013), and vascular endothelial growth factor (*vegf*), an important regulator of tumor angiogenesis. In *Myc*-induced T-ALL up to 17 alterations were seen, with common gains and losses distributed over various chromosomes (Freeman et al., 2009). Another study on T-cell ALL using the same transgenic line as well as three mutant lines that show a heritable T-ALL predisposition (*hlk*, *srk*, *otg*) employed high-density oligonucleotide arrays instead of the previously used BAC array (covering only about 40,000 kb of the 1.5 Gb zebrafish genome) and thus had a much higher resolution. More than 1000 amplifications and deletions were found (Rudner et al., 2011). The genes affected by these alterations were compared with CGH data from human T-ALLs. This revealed more than 60 genes that are commonly altered in zebrafish and human leukemia, including TGFβ, which had previously been implicated in T-cell ALL (Remke et al., 2009). Curiously, some oncogenes and tumor suppressor genes already implicated in human T-ALL, such as *CDKN2A, CDKN2B, PTEN, LEF1, WT1*, and *MYB/AHI1*, were not affected in zebrafish. Their absence was explained by the low frequency of copy number changes affecting these genes in humans, which then might be below the sensitivity threshold of the method used in this study. Finally, using the BAC array in the *mitfa:BRAF*$^{V600E}$ melanoma model, between one and 28 genomic changes were recorded, in line with the well-known low genomic stability of melanoma. Half (three of six) of the samples had gains in genomic regions encoded in five BAC clones (Freeman et al., 2009). Although the resolution of the array (about 4-5 Mb) was not high enough to precisely pinpoint the genes that were altered in copy number, inferences were made. This was because those BACs that gave signals for copy number variations were from genomic regions harboring interesting candidate genes for further studies. These included the proto-oncogene *myc*, which plays a widespread role in tumor formation (Meyer & Penn, 2008), *kit*, a frequently mutated gene in human melanoma (Cancer Genome Atlas, 2015), *mitfa*, whose expression is altered in the majority of human melanoma (Wellbrock & Arozarena, 2015), and *braf*, the proto-oncogene to the driver oncogene. The study by Freeman et al. also noted several genomic regions that were commonly subjected to alteration in rhabdomyosarcoma, T-cell ALL and melanoma, indicating that those regions may contain genes that function in a more general way in tumorigenesis.

All of the CGH array studies discussed so far were hampered by the fact that even for the oligonucleotide gene arrays, only a fraction (of varying extent depending on the genome representation in the array) of the entire genome could be assayed. New large-scale sequencing technologies now provide the opportunity to sequence many individual genomes for an affordable price and to conduct genome-wide analyses if a well-annotated reference genome of high contiguity is available. Indeed, two consecutive studies on MPNSTs have made use of massive parallel sequencing in addition to high-density oligonucleotide gene arrays for CGH.

In the first attempt with 36 tumor/normal tissue pairs, it was found that these tumors have recurrently overrepresented genomic regions containing genes that were also amplified in human MPNSTs, including the receptor tyrosine kinase gene *met*, *cyclinD2* (*ccnd2*), *slc45a3*, and cyclin-dependent kinase 6 (*cdk6*) (Zhang et al., 2010). Moreover, this study revealed a novel factor for MPNST development, namely fibroblast growth factor 6 (*fgf6a*), and provided evidence that Fgf6a may be a driver for establishment of this cancer in fish and possibly also in humans. For three samples, in addition to array CGH, Illumina sequencing was performed and provided the same result (but at much higher resolution) as the array and Southern blots with probes from the amplified regions. In a follow-up study, 147 tumors were sequenced at low coverage (Zhang et al., 2013). A large number of large and focal CNAs were recorded. The recurrent large-scale changes accounted for almost 60% of the genome, with nine chromosomes (Dre9, 10, 11, 13, 19, 20, 22, 23, 25) being overrepresented and six chromosomes (Dre2, 5, 8, 15, 17, 24) that were underrepresented. About 1% of all chromosomes (47 chromosomes from all 147 tumors) showed chromothripsis, a catastrophic shattering of the chromosome followed by faulty rejoining of fragments, which results in multiple genomic rearrangements. This phenomenon has been described as a hallmark of many cancer genomes (Stephens et al., 2011). Focal CNAs contained orthologs of known oncogenes, such as *jun*, *pdgfra*, *kita*, *mycn*, *ccnd2a*, *met*, *mdm2*, *cdk6*, *hrasa*, and *kras*. When compared to human MPSNT (also known as "malignant schwannoma," "neurofibrosarcoma," or "neurosarcoma"), only one-quarter of the identified zebrafish genes had orthologs in human that were similarly gained or lost. The validity of this approach to identify tumor driver genes was tested by crossing fish with mutant alleles of candidate genes to those which have the MPNST-initiating mutations (in ribosomal proteins or in p53) and to look for synergisms with respect to tumor formation. The *neurofibromin 2* (*nf2*) gene, known as a driver for schwannoma in humans (Zhang et al., 2014) and whose zebrafish ortholog is located within the recurrently underrepresented region of zebrafish chromosome 5, indeed showed such synergy. However, two other mutant genes that had been categorized as passenger mutations due to their discordant CNA spectrum in fish and humans, had no significant effect on the development of MPNST. Interestingly, some miRNAs, which were grouped by their membership of seed families, were also shown to be commonly gained (eight seed families) or lost (nine families) in zebrafish and human MPNST.

An exome-sequencing approach has been used to examine copy number changes and point mutations in a set of 53 melanomas derived from several zebrafish lines that are transgenic for the human oncogenes $BRAF^{V600E}$ or $NRAS^{Q61K}$. These oncogenes are found to be mutated in human melanomas in approximately 60% of all cases, where they drive tumor development through constitutive signaling of the MAPkinase pathway (Davies et al., 2002). However, most nevi also carry such mutations, despite remaining benign for decades or never transforming to melanoma (Pollock et al., 2003). Obviously, additional changes are necessary for the full-blown cancerous disease to occur. Analogously, the $BRAF^{V600E}$ or $NRAS^{Q61K}$-transgenic zebrafish develop only nevus-like lesions and have to be crossed into a p53 deficient background in order to obtain malignant melanoma formation (Dovey et al., 2009; Patton et al., 2005). Despite p53 not being considered as a major factor in melanoma initiation in humans, the zebrafish model has great promise to identify factors that cooperate with $BRAF^{V600E}$ or $NRAS^{Q61K}$ in human melanocyte transformation. Thus, melanoma with various genetic constitutions were subjected to in-solution exon capture and Illumina sequencing of an approximately 20-fold coverage (Yen et al., 2013). These melanomas carried combinations of $BRAF^{V600E}$ or $NRAS^{Q61K}$ with or without mutant p53 and with additional transgenes and mutant alleles of known or suspected oncogenicity in human or fish melanoma, including ribosomal protein S29 (*rps29*), *mitf*, *krox20*, *OCT6*, *foxd3*, *pten*, and histone methyltransferase *SETDB1*. With this sequencing depth and a bait set that encompassed all annotated protein coding genes, a data quality comparable to that of the ongoing human cancer exome projects was achieved. A high heterogeneity of 991 CNAs was detected, with one recurrent (ten of 53) amplification of a region of chromosome 3 that contains the gene for one of the two catalytic subunits of protein kinase A, *prkacaa*. This suggested a role for this gene, which has already been shown to serve as a signal transducer in the melanogenesis pathway of normal melanocytes (Alizadeh, Fitch, Niswender, McKnight, & Barsh, 2008), as well as for pigmentation of the fish tumors. A total of 403 point mutations were detected in all 53 tumors. Not unexpectedly, due to their "hereditary" etiology, the mutations did not show a UV signature but were predominantly C to T substitutions; most likely due to spontaneous deamination which is also the most prominent mutation type in human tumors (Krauthammer et al., 2012). Half of the mutations were in a subset of only eight tumors, indicating that genomic instability is an acquired state affecting only some tumors, while others remain more stable. An enrichment of mutations was seen in pathways known to be critical for melanoma development, for example, in MAPkinase pathway signaling, apoptosis, vascular endothelial growth factor signaling or the p53 branch of cell-cycle regulation. However, in total only very few recurrently mutated genes were found.

The genomic approaches conducted so far to explore cancer in the different zebrafish tumor lines, besides delivering a plethora of data that contribute to a better understanding of the molecular mechanisms of tumor development, have also taught us several important general lessons. First, they confirm the general utility of aquarium fish models for validating findings from the human cancer genome-sequencing projects and for the identification of novel genetic factors that can contribute to the neoplastic phenotype of cancer cells. This should motivate the application of the same technology to other fish tumor models, in particular medaka and *Xiphophorus*, in order to provide a larger database for comparative studies and to improve our ability to distinguish driver from passenger mutations. Secondly, all studies have revealed a very pronounced genetic tumor heterogeneity. This picture is particularly clear from the many CNA studies. It is certainly consistent with the common perception that tumorigenic processes are dependent on many different mutated genes that cooperate in bringing about the full malignant phenotype. The findings from zebrafish studies clearly assign copy number variations (that may result in differences of expression levels) of cooperating genes a prominent role in addition to point mutations in protein-coding genes that have been the focus of many studies to date. On the other hand, the high level of genetic heterogeneity may not have been intuitively predicted. All tumor models studied so far are engineered transgenic fish lines, which carry highly potent driver genes expressed from strong cell type-specific promoters. Obviously, tumor development is still dependent on additional genetic changes that activate cooperating oncogenes or inactivate tumor suppressor genes.

Genomic instability appears to be a very early event and may be elicited by the transgene. This is supported by the fact that such genetic changes occur very rapidly and accompany the onset of tumor formation as most impressively documented by the high heterogeneity in embryonal rhabdomyosarcomas (ERMSs) that arise already at around 10 days of age. This is an extremely short period for somatic mutations to occur spontaneously. Activated *RAS* and *RAF* oncogenes have been shown to cause DNA damage through the production of noxious amounts of reactive oxygen species (ROS), which initially induces cellular senescence as a tumor escape mechanism (Di Micco et al., 2006). However, when senescence is disrupted and tumorigenesis is initiated by the *RAS* or *RAF* oncogenes, the hyperactivation of the RAS/RAF/MAPkinase pathway will continue to cause DNA damage. This then may be responsible for the occurrence of point mutations or larger structural changes that already become apparent as CNAs in the primary tumor at a very early stage of tumor development. Genomic instability may also be a specific feature of the genetic background in which a driver transgene acts, brought about by allelic variations for instance in DNA repair genes that would then fulfill the criterion of preexisting tumor modifier genes. Indeed, zebrafish that carry mutations in DNA mismatch repair genes display a higher susceptibility to spontaneous tumor development (Feitsma, Kuiper, Korving, Nijman, & Cuppen, 2008). In a genetic screen for mutations that cause genetic instability using a visible test for loss of wild-type gene function at a single locus, *golden*, which is an eye color phenotype, all of the mutant lines that were established also displayed an increased cancer susceptibility of about 10% even as heterozygotes (Moore, Rush, Breneman, Mohideen, & Cheng, 2006).

## 4. TRANSCRIPTOMIC APPROACHES

Cancer cells are characterized by aberrations in expression of a wide variety of genes. Besides contributing to a better understanding of the processes of tumor formation and progression, the identification of differential tumor-specific transcript levels of genes may provide information for designing novel therapeutic strategies and identifying new tumor biomarkers. The most complete information about differential expression of protein-coding genes between a tumor and what is regarded as its nonmalignant precursor lesion or corresponding normal organ can be obtained by analyzing the entire mRNA population by microarray or more recently RNA-seq technology.

## 4.1 Melanoma

In the transgenic *xmrk* medaka melanoma model (Fig. 1), due to the lack of availability of standardized genome-wide microarrays in medaka, studies on tumor transcriptomes had to wait until the arrival of the deep-sequencing technology.

RNA-seq transcriptomes from different single tumors in medaka were generated and compared with each other (Schartl et al., 2012). High expression levels of genes involved in the melanin pathway in extracutaneous melanoma (MM) and heavily melanized uveal melanoma (UM) as well as a low expression of these genes in xanthoerythrophoroma (XE) and hyperpigmented skin (HP) were found. This result was not unexpected as the black-pigmented tumors originate from melanocytes while XE and the precursor lesion HP are derived from pteridin- and carotinoid-containing pigment cells. Apart from this and more interestingly, the pigment cell tumor transcriptomes revealed a large number of tumor-specific differentially expressed genes including cell cycle-associated genes, growth factors, ROS detoxifying enzymes and inhibitors of apoptosis. *cyclinD1* (*ccnd1*), which is involved in the regulation of cell cycle progression, is upregulated in the fish tumors and is similarly overexpressed in human melanoma and other types of human cancer (Sauter et al., 2002). An upregulation of growth promoting genes such as *kit* or *fyn* was observed in tumor compared to control samples. The metabolism of ROS is associated with oncogene-induced senescence and could be involved in the transition from precursor lesions to malignant tumors (Fried & Arbiser, 2008; Leikam, Hufnagel, Schartl, & Meierjohann, 2008). ROS detoxifying enzymes, including superoxide dismutase, catalase, and glutathione peroxidase, showed an increased expression in tumors compared with the benign precursor lesion. All pigment cell tumors exhibited upregulated expression of the apoptosis regulator *bcl2* and downregulated expression of the tumor suppressor p53 and its homologs p63 and p73.

The comparison of the medaka RNA-seq data with microarray analyses from zebrafish melanoma (*mitfa:BRAF*$^{V600E}$; *p53*$^{-/-}$ transgenic line) revealed a comparable expression pattern. Several upregulated genes in zebrafish also showed enhanced expression levels in at least one of the different tumor types in medaka. For example, *dct, tyr, tuba8l3, rtn1b, uap1l1*, or *spty4* are upregulated in zebrafish as well as in medaka MM and UM. The most downregulated genes (47 of 63 genes) were similar in medaka and zebrafish tumors (Schartl et al., 2012; White et al., 2011), for example, the *desm*, *tpm1*, *pvalp3*, and *atp2a2* genes were downregulated in zebrafish and in medaka UM and XE.

Finally, the transcriptomes from the medaka pigment cell lesions were compared with microarray expression data of human melanoma (18 benign melanocytic skin nevi and 19 cutaneous primary melanoma) (Talantov et al., 2005). For this comparative analysis, the HP data were aligned with human skin nevi and the three different fish tumors (UM, MM, and XE) were aligned with cutaneous primary melanoma. The comparison showed common downregulation of 49 genes but upregulation of only one gene, *MAP3K12*, in fish and human tumors.

This study also revealed that differentially expressed genes, such as the melanosome component *SLC45A2*, which is upregulated in MM and UM of medaka as well as in the human melanoma data set, could potentially serve as melanoma biomarkers. This was substantiated by the finding that this gene was overexpressed in eight of 11 human melanoma cell lines compared to normal melanocytes (Schartl et al., 2012). In summary this study revealed that the expression signatures of pigment cell tumors are largely conserved between fish and human, confirmed, for example, by an enhanced expression of *MART1* (Nazarian, Prieto, Elder, & Duncan, 2010) and *N*-cadherin (*CDH2*) as well as a diminished *E*-cadherin (*CDH1*) (Kuphal & Bosserhoff, 2011) and *AIM1* (Ray, Wistow, Su, Meltzer, & Trent, 1997) expression.

Since tumors affect the biology of the entire organism, the investigation of the transcriptome changes that reflect the response of the whole body to malignant melanoma is a new and promising approach. To analyze these global transcriptional changes, RNA-seq data were generated from whole-body RNA preparations of four healthy wild-type and four *mitfa:xmrk* melanoma fish (Schartl et al., 2015). The animals were analyzed as juveniles when melanoma formation had just initiated (about 3-4 weeks after hatching) (Fig. 2). By comparison of the RNA-seq data from transgenic melanoma medakas with the wild-type control fish, a set of 396 significantly differentially expressed genes was identified. Interestingly, more genes were downregulated (about 70%) than upregulated in melanoma fish compared to control fish. Comparison of the expression pattern of genes associated with the development of melanocytes as well as pigmentation pathways showed a significant reduction of *shp*, *pi3k*, and *plcg2* expression as well as a significantly increased *tyrp1b* and *creb* expression in melanoma developing fish.

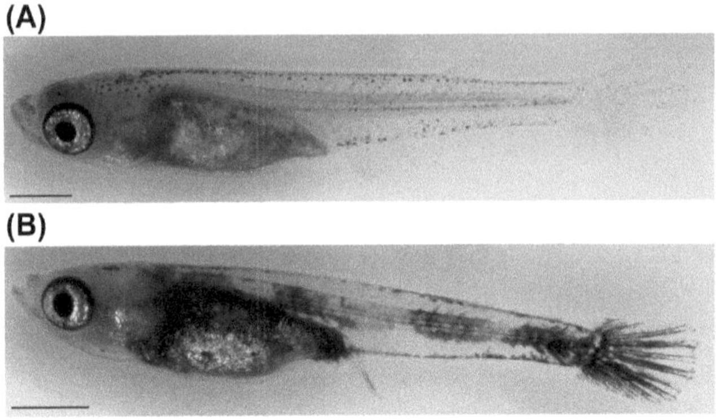


**Figure 2** Wild-type (A) and transgenic (B) medaka at 3-4 weeks of age. The fish with the *mitfa:xmrk* transgene shows onset of melanoma development in the caudal fin, inside the belly and the trunk musculature. Scale bars represent 1 mm. (See color plate)

The *mitf* gene, known as master regulator of melanocyte development and as a critical transcriptional regulator of melanoma, also tended to be upregulated in melanoma fish. Interestingly, a number of genes (eg, *irf1*, *stat4*, and *tlr2*) involved in the immune response were downregulated in the tumor-bearing medakas, suggesting a suppression of immune genes, in particular affecting the innate immune system, during the early phase (3-4 weeks) of melanoma growth in these fish. Of note, 33 genes that are known to be regulated by interferon $\gamma$ showed differential expression in the melanoma developing fish and indicated a suppression of this important component of the innate immune response. The fact that in the whole-body transcriptome, the upregulation of genes that are associated with melanocyte development and pigmentation was readily detected, demonstrates that data obtained from whole-body RNA samples are suitable for visualizing tumor-dependent gene expression changes over whole-body background expression levels.

The data from whole-body transcriptomes of juvenile fish at a time point when melanoma formation was just initiating were consistent with previously described gene expression changes in canonical pathways in human melanoma cell lines (Zhao et al., 2014). For this analysis 379 of the 396 significantly differentially expressed medaka genes could be evaluated because they showed a 1:1 match to human orthologs. Fourteen of these matched human genes show recurrent somatic mutations in melanoma (Ding et al., 2014). Another set of ten differentially expressed genes matched to human gene orthologs with already known drug-gene interactions (Griffith et al., 2013; Schartl et al., 2015). The strongest upregulated and downregulated differentially expressed genes with human orthologs were *fkbp5* (log2FC + 2.48) and *snord14c* (log2FC − 5.57), respectively. The immunophilin protein Fkbp5 is involved in basic cellular processes, immunoregulation, and melanoma dissemination (D'Angelillo, Staibano, Russo, Romano, & Romano, 2015). Furthermore, *fkbp5* is overexpressed in human melanoma tissue and associated with cell survival (D'Angelillo et al., 2015; Romano et al., 2010; Romano, Mallardo, & Romano, 2011). *snord14c* is a member of the small nucleolar RNAs and is linked with RNA methylation (Bachellerie, Cavaille, & Huttenhofer, 2002). The role of *fkpb5* had already been described in melanoma, but this represented the first time that *snord14c* was recognized in melanoma transcriptomes and at the moment there is no causal link between carcinogenesis or melanoma progression and *snord14c*. Thus, *snord14c* represents a new candidate gene for further investigation and understanding of melanoma progression on the molecular level.

Small aquarium fish provide useful disease models which are suitable for high-throughput chemical screens aiming at the identification of potential new therapeutic drugs (Giacomotto & Segalat, 2010). Generating whole-body transcriptomes from melanoma and healthy fish provides not only a new approach for the identification of previously unnoticed genes involved in human melanoma formation but also represents a tool for detecting the action of tumor inhibiting drugs at the gene expression level in high-throughput drug screens.

About 60% of human melanomas exhibit activating mutations in the *BRAF* gene and the resulting hyperactivation of the MAPkinase pathway has been associated with tumor growth (Davies et al., 2002). In zebrafish, the expression of human $BRAF^{V600E}$ under control of the pigment cell-specific *mitfa* promoter induces the development of nevus-like lesions from neural crest derived melanocytes (Patton et al., 2005), but for the development of invasive melanoma, an additional p53 inactivation is required in these fish.

To identify characteristic gene expression changes that occur in an early state of $BRAF^{V600}$ activation in the neural crest lineage, microarray-based gene expression profiles of *mitfa*:$BRAF^{V600}$; $p53^{-/-}$ transgenic embryos at 72 h postfertilization and melanomas of adult fish (from the same transgenic line) were compared and commonly deregulated genes were identified (White et al., 2011). The common signature identified was enriched for lineage-specific genes of the neural crest (eg, *sox10* and *ednrb*) as well as for several melanocyte marker genes (eg, *tyr* and *dct*), thereby supporting the concept of "lineage-addiction" in melanoma (Garraway et al., 2005).

Besides *BRAF*, *NRAS* is an important oncogene in melanoma and mutations in *NRAS* are detected in 15-30% of human melanomas (Chin, Garraway, & Fisher, 2006; Gray-Schopfer, Wellbrock, & Marais, 2007). To study the role of *NRAS* in zebrafish melanoma, transgenic strains have been engineered expressing the human oncogenic $NRAS^{Q61K}$ under the control of the *mitfa* promoter (Dovey et al., 2009). To analyze whether the microarray-based expression signatures of zebrafish and human tumors are similar to each other, Gene Set Enrichment Analysis (GSEA) was performed. The analysis disclosed a statistically significant association between the upregulated genes in zebrafish melanoma (two independent transgenic lines *NRAS1; $p53^{-/-}$* and *NRAS2; $p53^{-/-}$*) and in human melanoma. The observed overlap between the zebrafish and human gene set is significant for both transgenic zebrafish lines (*NRAS1; $p53^{-/-}$*: q-value = 0.023 and *NRAS2; $p53^{-/-}$*: q-value = 0.004). Despite the similarities in the transcriptional expression regulation, the two *NRAS; $p53^{-/-}$* tumors differed with respect to the pathways that they activate. Genes associated with Wnt/β-catenin signaling as well as melanocyte development and pigmentation signaling are highly expressed in *NRAS1; $p53^{-/-}$* tumors. This is in accordance with the strong hyperpigmented phenotype of the *NRAS1; $p53^{-/-}$* fish. Examples of statistically significantly enriched pathways identified in the *NRAS2; $p53^{-/-}$* tumors are the MAPkinase pathway, actin cytoskeleton, and calcium signaling.

All upregulated genes in *NRAS1*; *$p53^{-/-}$* and *NRAS2*; *$p53^{-/-}$* tumors are also overexpressed in human melanoma and the pathological as well as molecular characteristics are shared between human and zebrafish melanoma. This indicates a high degree of interspecies conservation in the transcriptionally activated pathways between zebrafish and human melanoma. Interestingly, conservation between zebrafish and human melanomas was only observed for the upregulated genes but not for the downregulated genes.

## 4.2 Rhabdomyosarcoma

The transcriptomic profile of embryonal rhabdomyosarcoma (ERMS) in zebrafish was studied by microarray technology and compared with human cancers (Langenau et al., 2007). Similar to the medaka *xmrk* melanoma model, this model was characterized by a very early onset of tumor formation at late or postlarval stages. In the *rag2:KRAS*$^{G12D}$-transgenic line, development of externally visible tumors begins at 10 days postfertilization.

Interestingly, diagnostic markers of human ERMS (eg, *myod* or *desmin*) were found to be expressed in 30 days old transgenic zebrafish with ERMS and furthermore morphological similarities between zebrafish and human ERMS were observed. A conserved gene signature between zebrafish and human tumors was identified. Higher expressed gene sets in zebrafish were statistically associated with the human ERMS data set, indicating a molecular similarity between zebrafish ERMS and human ERMS. *MYF5* (myogenic factor 5) is among the enriched genes in zebrafish and human ERMS.

In contrast the downregulated gene list identified in zebrafish ERMS was not conserved in human EMRS. A similar situation was also observed in a zebrafish melanoma model (Dovey et al., 2009). Why the expression profiles from upregulated genes are more conserved between zebrafish and human cancerous diseases than the downregulated genes is totally unclear and has not been discussed so far. The reason may be technical, because if the reduction in transcript levels is not considerable, then such differential expression will be more difficult to detect and might not meet the levels of significance. There may also be biological reasons, but currently no hypothesis exists. To analyze whether the upregulated gene list identified in zebrafish ERMS is similar to those of other types of cancer and thus represents a common cancer-associated molecular profile, the authors compared the fish signature with signatures of other human tumor types. Thereby a conserved signature was identified between fish ERMS and human pancreatic adenocarcinoma (PDAC), suggesting a RAS-specific signature, since the *KRAS* gene is mutated in over 90% of pancreatic carcinomas. The upregulated gene set in zebrafish EMRS contained genes known to represent direct targets of the RAS signaling pathway. G3BP, for example, is essential for RAS signaling and its expression is enhanced in EMRS as well as in human breast cancer (Barnes et al., 2002). MDM2 functions as an anti-apoptotic factor and is positioned downstream of the MAPkinase pathway (Ashcroft et al., 2002).

To validate the hypothesis of a common RAS-induced gene signature, a microarray-based expression profile of human mammary epithelial cells, infected with an adenovirus expressing activated *HRAS*, was compared with the significantly upregulated gene set of zebrafish EMRS. This revealed a significant correlation between the upregulated genes identified in zebrafish EMRS and RAS-infected human mammary epithelial cells, thereby confirming the presence of a conserved RAS signature. Moreover, this study demonstrated evolutionarily highly conserved expression signatures shared by zebrafish and human ERMS.

## 4.3 Hepatocellular Carcinoma

Another cancer in zebrafish that has been extensively studied using transcriptomic approaches is hepatocellular carcinoma (HCC). In an initial study, microarray technology was employed to identify the gene expression signature of carcinogen-induced zebrafish liver tumors compared with normal liver tissue (Lam & Gong, 2006; Lam et al., 2006). GSEA cross-species comparison of this signature with a human liver cancer microarray data set revealed a core set of 76 genes which are deregulated in both human and zebrafish HCC. Moreover, the zebrafish liver tumor signature was found to be enriched for genes of the RAS/RAF/MAPkinase, Wnt/β-catenin, p53, and RB pathways, all of which are frequently deregulated in human HCCs (Kudo, 2012; Laurent-Puig & Zucman-Rossi, 2006; Meng, Franklin, Dong, & Zhang, 2014; Waisberg & Saba, 2015). These findings strongly indicated that zebrafish liver cancers share common molecular features with human liver carcinomas. When the zebrafish HCC signature was compared with a set of human HCC microarray data, where the samples had been histopathologically graded according to clinical stages, 132 genes were identified that showed similar expression profiles in fish and human tumors and were correlated with human liver tumor aggressiveness. This study thus revealed that molecular similarities between zebrafish and human HCC even extend to tumor progression.

In all following studies, transgenic zebrafish models for HCC were used. RAS signaling is ubiquitously activated in human HCCs when compared to the non-neoplastic surrounding tissue and normal livers (Calvisi et al., 2006). Activating mutations in the *KRAS* oncogene can be found in approximately 7% of human liver tumors (Karnoub & Weinberg, 2008). To gain insight into the process of $kras^{G12V}$-driven liver tumorigenesis and cancer progression, a pathway analysis of microarray data was performed for hyperplastic lesions (HL) and HCC of $kras^{G12V}$-transgenic zebrafish (Nguyen et al., 2011). Both zebrafish tumor signatures were enriched for genes in pathways, which are also commonly deregulated in human HCC, like the RAS/RAF/MAPkinase cascade, the canonical Wnt pathway, the VEGF pathway and the complement cascade (Aravalli, Steer, & Cressman, 2008; Kudo, 2012; Waisberg & Saba, 2015). Interestingly, the Jak/Stat and the Tlr/NfκB pathways were significantly activated only in HCC, whereas the p53 pathway was activated exclusively in HL, which is regarded as the precursor lesion of HCC. As the two transcription factors NFκB and STAT3 are known to be involved in cancer-related inflammation (Mantovani, Allavena, Sica, & Balkwill, 2008), their combined upregulation hints at an important role for inflammatory processes in HCC tumor growth and progression. Moreover, as a high level of oncogene-induced senescence (OIC) was demonstrated in zebrafish liver hyperplasia compared to HCC, a role was proposed for p53 in mediating OIC in these lesions, which finally results in latent tumor development.

By comparing the zebrafish HL- and HCC-enriched genes with transcriptomic profiles of different stages of human liver tumors using cross-species GSEA, a liver cancer progression signature was identified, which comprises genes that remained upregulated throughout human and zebrafish liver cancer progression. Among the 20 genes of this signature, several known prognostic markers for human HCC were found, for example, *APOE* (Yokoyama et al., 2006) and *CCNB1* (Wurmbach et al., 2007). This further underscores the conservation of molecular processes underlying hepatocarcinogenesis in different vertebrate species. Most interestingly, the signature also comprises several genes involved in cell cycle regulation and DNA damage repair that were not associated with HCC before and hence might serve as new prognostic markers or potential therapeutic targets. One of these genes is *nibrin* (*NBN/NBS1*), a DNA double-strand break repair complex, which is overexpressed in several human tumors where it might exert a protumorigenic effect (Chen et al., 2005; Hematulin, Sagan, Eckardt-Schupp, & Moertl, 2008).

In a subsequent study, the gene expression signature of a doxycycline-inducible *Myc*-transgenic zebrafish liver tumor model was determined using an RNA-seq-based approach (Li et al., 2013). Gene Ontology (GO) term enrichment analysis performed on the list of differentially expressed genes revealed that ribosome biogenesis and translation-related genes represent the most prominent groups in the upregulated fraction. All ribosomal protein genes and several translation factor genes were significantly upregulated in zebrafish liver tumors compared to normal livers of noninduced transgenic and wild-type siblings. As tumor cells show an increased proliferation rate compared to normal liver cells, the upregulation of genes necessary for protein synthesis could simply reflect this phenomenon. However, such a strong upregulation of ribosomal genes was not observed in $kras^{G12V}$-induced (Nguyen et al., 2011) or *xmrk*-induced (Li et al., 2014) zebrafish liver tumors. Therefore, the authors of the study proposed that this upregulation is driven specifically by the *Myc* oncogene. This reasoning is in line with data showing that MYC serves as a regulator of ribosome biogenesis by directly controlling transcription of RNA and protein components of the ribosome (van Riggelen, Yetil, & Felsher, 2010).

A comparison of the zebrafish liver tumor transcriptome data with transcriptomes from different stages of human liver tumors, ranging from dysplastic lesions to advanced stages of HCC, revealed a high similarity between *Myc*-induced zebrafish liver tumors and very advanced human HCC. Furthermore, the authors identified a Myc target gene signature by comparing the significantly upregulated genes of the zebrafish liver tumors with a Myc target gene set from the Molecular Signature Database (MSigDB). The 16 genes of this signature, when used for GSEA to classify human HCC, also showed a higher correlation with more advanced stages of human HCC, thereby indicating a correlation between upregulated Myc signatures and a more advanced tumor status. In line with this, *MYC* is often amplified in human HCC and high *MYC* expression levels are correlated with later tumor stages (Gan, Gesell, Alousi, & Luk, 1993; Kaposi-Novak et al., 2009). This underscores the usefulness of this zebrafish model to study human HCCs induced by hyperactivation of the *MYC* oncogene and furthermore reveals an evolutionary conserved role for *Myc* in promoting hepatocarcinogenesis. Moreover, the 16 gene *Myc* target gene signature identified in this study could help to refine prognostic signatures for HCC patients.

Besides zebrafish $kras^{G12V}$ and mouse *Myc*, the constitutively activated EGFR variant Xmrk has been used as driver oncoprotein in an inducible zebrafish liver tumor model (Li et al., 2012). Activation of growth factor receptor signaling is commonly observed during HCC pathogenesis (Whittaker, Marais, & Zhu, 2010) and activation of EGFR signaling has been correlated with a poor prognosis of HCC cases (Foster et al., 2007). Liver-specific overexpression of *xmrk* in adult and juvenile zebrafish led to development of liver tumors, which were histopathologically classified as HCCs. Interestingly, cessation of *xmrk* expression through doxycycline withdrawal led to rapid regression of liver tumors in these fish (Li et al., 2012). A similar phenomenon was also observed in $kras^{G12V}$- and *Myc*-driven zebrafish liver tumors upon inducer withdrawal (Li et al., 2013; Nguyen et al., 2012), which suggests an "oncogene-addiction" of these zebrafish liver tumors. In humans spontaneous regression of HCCs has been observed in rare cases, but the underlying molecular mechanisms are largely unknown. However, since the identification of these mechanisms might offer new therapeutic strategies for HCC treatment, the *xmrk*-induced zebrafish liver tumor model was used in a follow-up study to determine the gene expression signature of regressing liver tumors (Li et al., 2014). For this purpose, RNA-sequencing was performed on fully developed HCC as well as early (R1) and late (R2) stages of tumor regression. These data were compared with a human transcriptome data set representing different progression stages of human liver tumors, ranging from cirrhotic liver to advanced stages of HCC. The use of a human HCC progression data set was justified by the lack of available transcriptome profiles of regressing human HCC. Similar to *kras*$^{G12V}$- and *Myc*-induced zebrafish liver tumors, *xmrk*-induced zebrafish HCCs showed the highest similarity with very advanced human HCC. The early tumor regression stage R1, in contrast, correlated with low-grade dysplastic lesions as well as cirrhotic liver, while R2 resembled cirrhotic liver. This clearly demonstrated that the expression profiles of regressing liver tumors gradually revert to a nonmalignant liver tissue signature. Consistently, GSEA pathway analysis revealed a significant downregulation of several metabolic pathways in zebrafish HCC and a gradual reactivation of these pathways during tumor regression, which probably reflects a recovery of normal liver function during this process. Among the upregulated pathways, the antigen processing and presentation pathway was commonly altered in HCC and both regression stages compared to control. Additional bioinformatics analyses revealed an enrichment of distinct immune functions in the different tumor stages, namely enriched major histocompatibility complex class I (MHC I) related genes in HCC, enriched immune cell trafficking functions in early regression and an enriched MHC II gene set in late regression. These findings confirmed earlier data from human cancer studies which pointed to an important involvement of the immune system in tumor progression (Finn, 2012) as well as tumor regression. The cessation of *xmrk* oncogene expression in the zebrafish liver tumors mimics targeted EGFR inhibition in cancer therapy. The data from the zebrafish liver tumor regression model may therefore provide new insight into the role of immune responses in targeted liver cancer therapy. Understanding the interplay of immune responses and targeted therapy is of special importance, as a combination of immunotherapy and targeted approaches has emerged as a new promising treatment strategy for human cancer (Vanneman & Dranoff, 2012).

Although distinct driver oncogenes promote hepatocarcinogenesis in *kras*$^{G12V}$-, *myc*-, and *xmrk*-transgenic zebrafish, the gene expression signatures of all three resulting zebrafish liver tumor models showed the highest similarity with advanced or very advanced human HCC when compared with the transcriptome profiles of human liver tumors of different stages (Li et al., 2013, 2014; Zheng et al., 2014). To compare the different zebrafish liver tumor models at the molecular level, a comparative analysis of their RNA profiles was performed (Zheng et al., 2014). This revealed only small overlaps of differentially expressed genes, indicating that the three driver oncogenes used in these transgenic lines regulate distinct sets of genes. To examine whether the gene expression signatures of the three zebrafish liver tumors correlate with different subsets of human HCC, the zebrafish transcriptome data were compared to microarray data sets containing more than 1200 clinical HCC samples. It is well established that human HCCs are very heterogeneous, especially at intermediate and advanced stages of the disease (Jeng, Chang, Jeng, Sheen, & Jeng, 2015; Llovet et al., 1999) and that varying mutations and altered molecular pathways are probably causative for hepatocarcinogenesis in different HCC subgroups. Each of the zebrafish signatures correlated with 25-30% of all human HCC samples. Although some clinical samples were positively correlated with more than one zebrafish signature, these data clearly demonstrated that each of the transgenic zebrafish models represents a distinct subset of human HCC cases. Moreover, similar upregulated pathways were detected among those human HCC cases that were correlated with the same zebrafish gene signature. These analyses demonstrated convincingly that the different transgenic zebrafish lines represent valuable model systems to reveal the main molecular drivers underlying hepatocarcinogenesis in different HCC subgroups, which is ultimately important for the development of targeted therapeutic approaches for each subgroup. However, as the analyzed zebrafish signatures represented only half of the human HCC cases, these three zebrafish lines are not able to reproduce the full spectrum of molecular mechanisms responsible for human hepatocarcinogenesis. Thus, the authors concluded that there is a need to develop further transgenic animal models for HCC using a different set of driver oncogenes.

By crossing the *xmrk*- and *Myc*-driven single transgenic lines used as liver cancer models in previous studies, a *Myc/xmrk* double transgenic zebrafish line was established (Li et al., 2015). Compared to single transgenic siblings, *Myc/xmrk* double transgenic fish showed an earlier onset of tumor formation and a more severe phenotype, which hints to a potential synergistic effect of *Myc* and *xmrk* on hepatocarcinogenesis in this zebrafish model. To better understand this synergy at the molecular level, a comparative RNA-seq-based approach was employed using liver tumors from single and double transgenic fish. A set of genes was identified, which was differentially expressed exclusively in the *Myc/xmrk* double transgenic liver tumors. Bioinformatics analyses revealed a significant deregulation of different metabolism-related pathways in these tumors. Interestingly, an upregulation of glucose metabolism combined with downregulated fatty acid and amino acid metabolism was encountered, which is indicative for a potential Warburg effect. This effect is most probably mediated by a synergistic interaction between *Myc* and *xmrk* in these double transgenic fish. A possible role for *MYC* (Dang, Le, & Gao, 2009; Nilsson et al., 2012) and *EGFR* (Yang et al., 2012) in promoting the Warburg effect during tumorigenesis was already proposed in previous studies; however, a synergism between these two oncogenes was not considered. This model therefore may also stimulate new research directions for studies on the Warburg effect in human tumors.

## 5. OUTLOOK

The genomic and transcriptomic approaches applied to fish cancer models have provided us with a wealth of information. First, these studies have not only confirmed previous analyses based at the single gene level but have shown on a much larger scale that medaka and zebrafish tumors closely resemble their human counterparts. It remains to be clarified whether the observed discrepancies are due to expected differences in the biological organization of fish and mammals or if they reflect so far uncharacterized differences in tumor subtypes, etiology or histopathology between the piscine tumors and the human cancer samples with which they were compared. Although such differences undeniably exist, they are obviously minor when considered in the general context of cancer initiation and progression in vertebrates.

Second, the genomic and transcriptomic approaches have disclosed genes and pathways that are conserved between fish and mammalian tumors. Cancers are usually characterized by a plethora of changes, many of which are considered to be unrelated to tumorigenesis. The evolutionary argument postulates that those changes that are found in distantly related species that diverged over 400 million years ago most likely reflect central processes that are relevant for establishing the neoplastic phenotype. Thus, comparative oncology using data from the fish models is useful to define the relevance of gene alterations and expression changes detected in cancer cells.

Third, a considerable number of genes identified in the fish studies had not been recognized previously in human tumors nor had they been intensively studied. Of course all these novel genes that may ultimately serve as biomarkers or targets need to be investigated further in the fish models because the genome-wide CNA studies and whole-transcriptome analyses, like all unbiased large-scale and high-throughput screening methods, can only identify candidates. Each one needs to be subjected to detailed analysis; importantly, functional studies have to be conducted to evaluate their importance during development of the respective cancer. Once confirmed the follow-up studies that translate the results from basic research to clinical applications are eagerly awaited and hopefully, the first such molecules that make it from the fish tank to bedside will be published soon.

New sequencing technologies have boosted the research on genomes and transcriptomes. Full genome sequences, even at low coverage, give CNA studies a greatly enhanced resolution, down to the individual gene level, and a much higher sensitivity than CGH could ever achieve. RNA-seq has many advantages over microarrays, offering increased specificity and sensitivity for the enhanced detection of low abundance transcripts and weak differential gene expression. RNA-seq technology quantifies discrete, digital sequencing read counts, thus offering a broader dynamic range. With array hybridization technology, gene expression measurement is limited by background at the low end and signal saturation at the high end. Instead, sequencing coverage depth can easily be increased to detect rare transcripts, single transcripts per cell or weakly expressed genes. Unlike arrays, RNA-seq technology does not require species- or transcript-specific probes and can detect novel transcripts, gene fusions, single nucleotide variants, indels (small insertions and deletions), and other previously unknown changes.

Besides conventional genome sequencing and RNA-seq that has been applied to study fish cancer models, deep-sequencing methodology has many modifications that certainly can also be applied. Long noncoding RNAs and the different classes of small noncoding RNAs, like miRNAs, can be easily analyzed by minor modifications of the standard RNA-seq protocols; however, at present no such studies have been reported using the fish cancer models. Single-cell transcriptomics holds great promise for a better understanding of many tumor features, including tumor stem cell function, tumor heterogeneity, and the role of the microenvironment. The genomes of single cells from tumors should enhance our understanding of genome instability, which is one of the hallmarks of many cancers and poses almost insurmountable problems for cancer prognosis determination and for successful therapies. Sequencing of chromatin immunoprecipitates (ChIP-seq) can deliver genome-wide information on the role of certain transcription factors in tumor development. Furthermore, an important role for epigenetic changes in tumor development and progression is just emerging from studies on human cancers (Shinjo & Kondo, 2015); however, deep-sequencing analysis of fish tumor methylomes has so far not been conducted. The combined use of all these new approaches for studying cancer in fish model systems has the fantastic potential to provide us not only with new insight into tumor biology of the fish models themselves but also crucially, a better understanding of human cancer.

## ACKNOWLEDGMENTS

This study was supported by the National Institutes of Health under award number R24OD018555, sub-award number 215420C. The use of deep-sequencing technology to approach a large spectrum of questions in molecular biology and genetics is an extremely fast-moving field and we apologize to all our colleagues whose work has escaped our attention and has not been mentioned appropriately.

## REFERENCES

Alizadeh, A., Fitch, K. R., Niswender, C. M., McKnight, G. S., & Barsh, G. S. (2008). Melanocyte-lineage expression of Cre recombinase using Mitf regulatory elements. *Pigment Cell & Melanoma Research, 21*(1), 63–69.

Amsterdam, A., Sadler, K. C., Lai, K., Farrington, S., Bronson, R. T., Lees, J. A., & Hopkins, N. (2004). Many ribosomal protein genes are cancer genes in zebrafish. *PLoS Biology, 2*(5), E139.

Aravalli, R. N., Steer, C. J., & Cressman, E. N. (2008). Molecular mechanisms of hepatocellular carcinoma. *Hepatology: Official Journal of the American Association for the Study of Liver Diseases, 48*(6), 2047–2063.

Arndt, C. A., & Crist, W. M. (1999). Common musculoskeletal tumors of childhood and adolescence. *The New England Journal of Medicine, 341*(5), 342–352.

Ashcroft, M., Ludwig, R. L., Woods, D. B., Copeland, T. D., Weber, H. O., MacRae, E. J., & Vousden, K. H. (2002). Phosphorylation of HDM2 by Akt. *Oncogene, 21*(13), 1955–1962.

Bachellerie, J. P., Cavaille, J., & Huttenhofer, A. (2002). The expanding snoRNA world. *Biochimie, 84*(8), 775–790.

Barnes, C. J., Li, F., Mandal, M., Yang, Z., Sahin, A. A., & Kumar, R. (2002). Heregulin induces expression, ATPase activity, and nuclear localization of G3BP, a Ras signaling component, in human breast tumors. *Cancer Research, 62*(5), 1251–1255.

Berghmans, S., Murphey, R. D., Wienholds, E., Neuberg, D., Kutok, J. L., Fletcher, C. D., … Look, A. T. (2005). tp53 mutant zebrafish develop malignant peripheral nerve sheath tumors. *Proceedings of the National Academy of Sciences of the United States of America, 102*(2), 407–412.

Box, N. F., & Terzian, T. (2008). The role of p53 in pigmentation, tanning and melanoma. *Pigment Cell & Melanoma Research, 21*(5), 525–533.

Calvisi, D. F., Ladu, S., Gorden, A., Farina, M., Conner, E. A., Lee, J. S., … Thorgeirsson, S. S. (2006). Ubiquitous activation of Ras and Jak/Stat pathways in human HCC. *Gastroenterology, 130*(4), 1117–1128.

Cancer Genome Atlas, N. (2015). Genomic classification of cutaneous melanoma. *Cell, 161*(7), 1681–1696.

Chen, E. Y., Dobrinski, K. P., Brown, K. H., Clagg, R., Edelman, E., Ignatius, M. S., … Langenau, D. M. (2013). Cross-species array comparative genomic hybridization identifies novel oncogenic events in zebrafish and human embryonal rhabdomyosarcoma. *PLoS Genetics, 9*(8), e1003727.

Chen, Y. C., Su, Y. N., Chou, P. C., Chiang, W. C., Chang, M. C., Wang, L. S., … Wu, K. J. (2005). Overexpression of NBS1 contributes to transformation through the activation of phosphatidylinositol 3-kinase/Akt. *The Journal of Biological Chemistry, 280*(37), 32505–32511.

Chin, L., Garraway, L. A., & Fisher, D. E. (2006). Malignant melanoma: genetics and therapeutics in the genomic era. *Genes & Development, 20*(16), 2149–2182.

Ciombor, K. K., Haraldsdottir, S., & Goldberg, R. M. (2014). How can next-generation sequencing (genomics) help us in treating colorectal cancer? *Current Colorectal Cancer Reports, 10*(4), 372–379.

Dang, C. V., Le, A., & Gao, P. (2009). MYC-induced cancer cell energy metabolism and therapeutic opportunities. *Clinical Cancer Research: An Official Journal of the American Association for Cancer Research, 15*(21), 6479–6483.

D'Angelillo, A., Staibano, S., Russo, M., Romano, M. F., & Romano, S. (2015). Molecular aspects of FKBP51 that enable melanoma dissemination. *Current Molecular Pharmacology, 9*(2), 141–147.

Davies, H., Bignell, G. R., Cox, C., Stephens, P., Edkins, S., Clegg, S., … Futreal, P. A. (2002). Mutations of the BRAF gene in human cancer. *Nature, 417*(6892), 949–954.

Di Micco, R., Fumagalli, M., Cicalese, A., Piccinin, S., Gasparini, P., Luise, C., … d'Adda di Fagagna, F. (2006). Oncogene-induced senescence is a DNA damage response triggered by DNA hyper-replication. *Nature, 444*(7119), 638–642.

Ding, L., Kim, M., Kanchi, K. L., Dees, N. D., Lu, C., Griffith, M., … Weber, J. S. (2014). Clonal architectures and driver mutations in metastatic melanomas. *PLoS One, 9*(11), e111153.

Dovey, M., White, R. M., & Zon, L. I. (2009). Oncogenic NRAS cooperates with p53 loss to generate melanoma in zebrafish. *Zebrafish, 6*(4), 397–404.

Fanciulli, M., Petretto, E., & Aitman, T. J. (2010). Gene copy number variation and common human disease. *Clinical Genetics, 77*(3), 201–213.

Feitsma, H., Kuiper, R. V., Korving, J., Nijman, I. J., & Cuppen, E. (2008). Zebrafish with mutations in mismatch repair genes develop neurofibromas and other tumors. *Cancer Research, 68*(13), 5059–5066.

Finn, O. J. (2012). Immuno-oncology: understanding the function and dysfunction of the immune system in cancer. *Annals of Oncology: Official Journal of the European Society for Medical Oncology/ESMO, 23*(Suppl. 8), viii6–9.

Foster, J., Black, J., LeVea, C., Khoury, T., Kuvshinoff, B., Javle, M., & Gibbs, J. F. (2007). COX-2 expression in hepatocellular carcinoma is an initiation event; while EGF receptor expression with downstream pathway activation is a prognostic predictor of survival. *Annals of Surgical Oncology, 14*(2), 752–758.

Freeman, J. L., Ceol, C., Feng, H., Langenau, D. M., Belair, C., Stern, H. M., … Lee, C. (2009). Construction and application of a zebrafish array comparative genomic hybridization platform. *Genes, Chromosomes & Cancer, 48*(2), 155–170.

Fried, L., & Arbiser, J. L. (2008). The reactive oxygen-driven tumor: relevance to melanoma. *Pigment Cell & Melanoma Research, 21*(2), 117–122.

Gan, F. Y., Gesell, M. S., Alousi, M., & Luk, G. D. (1993). Analysis of ODC and c-myc gene expression in hepatocellular carcinoma by in situ hybridization and immunohistochemistry. *The Journal of Histochemistry and Cytochemistry: Official Journal of the Histochemistry Society, 41*(8), 1185–1196.

Garraway, L. A., Widlund, H. R., Rubin, M. A., Getz, G., Berger, A. J., Ramaswamy, S., … Sellers, W. R. (2005). Integrative genomic analyses identify MITF as a lineage survival oncogene amplified in malignant melanoma. *Nature, 436*(7047), 117–122.

Giacomotto, J., & Segalat, L. (2010). High-throughput screening and small animal models, where are we? *British Journal of Pharmacology, 160*(2), 204–216.

Gray-Schopfer, V., Wellbrock, C., & Marais, R. (2007). Melanoma biology and new targeted therapy. *Nature, 445*(7130), 851–857.

Griffith, M., Griffith, O. L., Coffman, A. C., Weible, J. V., McMichael, J. F., Spies, N. C., … Wilson, R. K. (2013). DGIdb: mining the druggable genome. *Nature Methods, 10*(12), 1209–1210.

Häussler, G. (1928). Über Melanombildungen bei Bastarden von *Xiphophorus maculatus* var. rubra. *Klinische Wochenschrift, 7*, 1561–1562.

Hawryluk, E. B., & Tsao, H. (2014). Melanoma: clinical features and genomic insights. *Cold Spring Harbor Perspectives in Medicine, 4*(9), a015388.

Hematulin, A., Sagan, D., Eckardt-Schupp, F., & Moertl, S. (2008). NBS1 is required for IGF-1 induced cellular proliferation through the Ras/Raf/MEK/ERK cascade. *Cellular Signalling, 20*(12), 2276–2285.

Howe, K., Clark, M. D., Torroja, C. F., Torrance, J., Berthelot, C., Muffato, M., … Stemple, D. L. (2013). The zebrafish reference genome sequence and its relationship to the human genome. *Nature, 496*(7446), 498–503.

Jeng, K. S., Chang, C. F., Jeng, W. J., Sheen, I. S., & Jeng, C. J. (2015). Heterogeneity of hepatocellular carcinoma contributes to cancer progression. *Critical Reviews in Oncology/Hematology, 94*(3), 337–347.

Kaposi-Novak, P., Libbrecht, L., Woo, H. G., Lee, Y. H., Sears, N. C., Coulouarn, C., … Thorgeirsson, S. S. (2009). Central role of c-Myc during malignant conversion in human hepatocarcinogenesis. *Cancer Research, 69*(7), 2775–2782.

Karnoub, A. E., & Weinberg, R. A. (2008). Ras oncogenes: split personalities. *Nature Reviews Molecular Cell Biology, 9*(7), 517–531.

Kasahara, M., Naruse, K., Sasaki, S., Nakatani, Y., Qu, W., Ahsan, B., … Kohara, Y. (2007). The medaka draft genome and insights into vertebrate genome evolution. *Nature, 447*(7145), 714–719.

Kaufmann, W. K., Carson, C. C., Omolo, B., Filgo, A. J., Sambade, M. J., Simpson, D. A., … Thomas, N. E. (2014). Mechanisms of chromosomal instability in melanoma. *Environmental and Molecular Mutagenesis, 55*(6), 457–471.

Kosswig, C. (1928). Über Kreuzungen zwischen den Teleostiern *Xiphophorus helleri* und *Platypoecilus maculatus. Z Indukt Abstammungs-Vererbungsl, 47*, 150–158.

Krauthammer, M., Kong, Y., Ha, B. H., Evans, P., Bacchiocchi, A., McCusker, J. P., … Halaban, R. (2012). Exome sequencing identifies recurrent somatic RAC1 mutations in melanoma. *Nature Genetics, 44*(9), 1006–1014.

Kudo, M. (2012). Signaling pathway/molecular targets and new targeted agents under development in hepatocellular carcinoma. *World Journal of Gastroenterology: WJG, 18*(42), 6005–6017.

Kuphal, S., & Bosserhoff, A. K. (2011). Phosphorylation of beta-catenin results in lack of beta-catenin signaling in melanoma. *International Journal of Oncology, 39*(1), 235–243.

Lam, S. H., & Gong, Z. (2006). Modeling liver cancer using zebrafish: a comparative oncogenomics approach. *Cell Cycle, 5*(6), 573–577.

Lam, S. H., Wu, Y. L., Vega, V. B., Miller, L. D., Spitsbergen, J., Tong, Y., … Gong, Z. (2006). Conservation of gene expression signatures between zebrafish and human liver tumors and tumor progression. *Nature Biotechnology, 24*(1), 73–75.

Langenau, D. M., Keefe, M. D., Storer, N. Y., Guyon, J. R., Kutok, J. L., Le, X., … Zon, L. I. (2007). Effects of RAS on the genesis of embryonal rhabdomyosarcoma. *Genes & Development, 21*(11), 1382–1395.

Langenau, D. M., Traver, D., Ferrando, A. A., Kutok, J. L., Aster, J. C., Kanki, J. P., … Look, A. T. (2003). Myc-induced T cell leukemia in transgenic zebrafish. *Science, 299*(5608), 887–890.

Laurent-Puig, P., & Zucman-Rossi, J. (2006). Genetics of hepatocellular tumors. *Oncogene, 25*(27), 3778–3786.

Leikam, C., Hufnagel, A., Schartl, M., & Meierjohann, S. (2008). Oncogene activation in melanocytes links reactive oxygen to multinucleated phenotype and senescence. *Oncogene, 27*(56), 7070–7082.

Lengauer, C., Kinzler, K. W., & Vogelstein, B. (1998). Genetic instabilities in human cancers. *Nature, 396*(6712), 643–649.

Li, Z., Huang, X., Zhan, H., Zeng, Z., Li, C., Spitsbergen, J. M., … Gong, Z. (2012). Inducible and repressable oncogene-addicted hepatocellular carcinoma in Tet-on xmrk transgenic zebrafish. *Journal of Hepatology, 56*(2), 419–425.

Li, Z., Luo, H., Li, C., Huo, X., Yan, C., Huang, X., … Gong, Z. (2014). Transcriptomic analysis of a transgenic zebrafish hepatocellular carcinoma model reveals a prominent role of immune responses in tumour progression and regression. *International Journal of Cancer, 135*(7), 1564–1573.

Li, Z., Zheng, W., Li, H., Li, C., & Gong, Z. (2015). Synergistic induction of potential Warburg effect in zebrafish hepatocellular carcinoma by co-transgenic expression of myc and xmrk oncogenes. *PLoS One, 10*(7), e0132319.

Li, Z., Zheng, W., Wang, Z., Zeng, Z., Zhan, H., Li, C., … Gong, Z. (2013). A transgenic zebrafish liver tumor model with inducible Myc expression reveals conserved Myc signatures with mammalian liver tumors. *Disease Models & Mechanisms, 6*(2), 414–423.

Llovet, J. M., Bustamante, J., Castells, A., Vilana, R., Ayuso Mdel, C., Sala, M., … Bruix, J. (1999). Natural history of untreated nonsurgical hepatocellular carcinoma: rationale for the design and evaluation of therapeutic trials. *Hepatology: Official Journal of the American Association for the Study of Liver Diseases, 29*(1), 62–67.

Mantovani, A., Allavena, P., Sica, A., & Balkwill, F. (2008). Cancer-related inflammation. *Nature, 454*(7203), 436–444.

Meierjohann, S., & Schartl, M. (2006). From Mendelian to molecular genetics: the *Xiphophorus* melanoma model. *Trends in Genetics: TIG, 22*(12), 654–661.

Meng, X., Franklin, D. A., Dong, J., & Zhang, Y. (2014). MDM2-p53 pathway in hepatocellular carcinoma. *Cancer Research, 74*(24), 7161–7167.

Meyer, N., & Penn, L. Z. (2008). Reflecting on 25 years with MYC. *Nature Reviews Cancer, 8*(12), 976–990.

Moore, J. L., Rush, L. M., Breneman, C., Mohideen, M. A., & Cheng, K. C. (2006). Zebrafish genomic instability mutants and cancer susceptibility. *Genetics, 174*(2), 585–600.

Nazarian, R. M., Prieto, V. G., Elder, D. E., & Duncan, L. M. (2010). Melanoma biomarker expression in melanocytic tumor progression: a tissue microarray study. *Journal of Cutaneous Pathology, 37*(Suppl. 1), 41–47.

Nguyen, A. T., Emelyanov, A., Koh, C. H., Spitsbergen, J. M., Lam, S. H., Mathavan, S., … Gong, Z. (2011). A high level of liver-specific expression of oncogenic Kras(V12) drives robust liver tumorigenesis in transgenic zebrafish. *Disease Models & Mechanisms, 4*(6), 801–813.

Nguyen, A. T., Emelyanov, A., Koh, C. H., Spitsbergen, J. M., Parinov, S., & Gong, Z. (2012). An inducible kras(V12) transgenic zebrafish model for liver tumorigenesis and chemical drug screening. *Disease Models & Mechanisms, 5*(1), 63–72.

Nilsson, L. M., Forshell, T. Z., Rimpi, S., Kreutzer, C., Pretsch, W., Bornkamm, G. W., & Nilsson, J. A. (2012). Mouse genetics suggests cell-context dependency for Myc-regulated metabolic enzymes during tumorigenesis. *PLoS Genetics, 8*(3), e1002573.

Patton, E. E., Mathers, M. E., & Schartl, M. (2011). Generating and analyzing fish models of melanoma. *Methods in Cell Biology, 105*, 339–366.

Patton, E. E., Widlund, H. R., Kutok, J. L., Kopani, K. R., Amatruda, J. F., Murphey, R. D., … Zon, L. I. (2005). BRAF mutations are sufficient to promote nevi formation and cooperate with p53 in the genesis of melanoma. *Current Biology: CB, 15*(3), 249–254.

Pollock, P. M., Harper, U. L., Hansen, K. S., Yudt, L. M., Stark, M., Robbins, C. M., … Meltzer, P. S. (2003). High frequency of BRAF mutations in nevi. *Nature Genetics, 33*(1), 19–20.

Ray, M. E., Wistow, G., Su, Y. A., Meltzer, P. S., & Trent, J. M. (1997). AIM1, a novel non-lens member of the betagamma-crystallin superfamily, is associated with the control of tumorigenicity in human malignant melanoma. *Proceedings of the National Academy of Sciences of the United States of America, 94*(7), 3229–3234.

Remke, M., Pfister, S., Kox, C., Toedt, G., Becker, N., Benner, A., … Kulozik, A. E. (2009). High-resolution genomic profiling of childhood T-ALL reveals frequent copy-number alterations affecting the TGF-beta and PI3K-AKT pathways and deletions at 6q15-16.1 as a genomic marker for unfavorable early treatment response. *Blood, 114*(5), 1053–1062.

van Riggelen, J., Yetil, A., & Felsher, D. W. (2010). MYC as a regulator of ribosome biogenesis and protein synthesis. *Nature Reviews Cancer, 10*(4), 301–309.

Romano, S., D'Angelillo, A., Pacelli, R., Staibano, S., De Luna, E., Bisogni, R., … Romano, M. F. (2010). Role of FK506-binding protein 51 in the control of apoptosis of irradiated melanoma cells. *Cell Death and Differentiation, 17*(1), 145–157.

Romano, S., Mallardo, M., & Romano, M. F. (2011). FKBP51 and the NF-kappaB regulatory pathway in cancer. *Current Opinion in Pharmacology, 11*(4), 288–293.

Rudner, L. A., Brown, K. H., Dobrinski, K. P., Bradley, D. F., Garcia, M. I., Smith, A. C., … Frazer, J. K. (2011). Shared acquired genomic changes in zebrafish and human T-ALL. *Oncogene, 30*(41), 4289–4296.

Sauter, E. R., Yeo, U. C., von Stemm, A., Zhu, W., Litwin, S., Tichansky, D. S., … Bastian, B. C. (2002). Cyclin D1 is a candidate oncogene in cutaneous melanoma. *Cancer Research, 62*(11), 3200–3206.

Schartl, M., Kneitz, S., Wilde, B., Wagner, T., Henkel, C. V., Spaink, H. P., & Meierjohann, S. (2012). Conserved expression signatures between medaka and human pigment cell tumors. *PLoS One, 7*(5), e37880.

Schartl, M., Shen, Y., Maurus, K., Walter, R., Tomlinson, C., Wilson, R. K., … Warren, W. C. (2015). Whole body melanoma transcriptome response in medaka. *PLoS One, 10*(12), e0143057.

Schartl, M., Wilde, B., Laisney, J. A., Taniguchi, Y., Takeda, S., & Meierjohann, S. (2010). A mutated EGFR is sufficient to induce malignant melanoma with genetic background-dependent histopathologies. *The Journal of Investigative Dermatology, 130*(1), 249–258.

Schilsky, R. L. (2014). Implementing personalized cancer care. *Nature Reviews Clinical Oncology, 11*(7), 432–438.

Shinjo, K., & Kondo, Y. (2015). Targeting cancer epigenetics: linking basic biology to clinical medicine. *Advanced Drug Delivery Reviews, 95*, 56–64.

Simon, R., & Roychowdhury, S. (2013). Implementing personalized cancer genomics in clinical trials. *Nature Reviews Drug Discovery, 12*(5), 358–369.

Stephens, P. J., Greenman, C. D., Fu, B., Yang, F., Bignell, G. R., Mudie, L. J., … Campbell, P. J. (2011). Massive genomic rearrangement acquired in a single catastrophic event during cancer development. *Cell, 144*(1), 27–40.

Talantov, D., Mazumder, A., Yu, J. X., Briggs, T., Jiang, Y., Backus, J., … Wang, Y. (2005). Novel genes associated with malignant melanoma but not benign melanocytic lesions. *Clinical Cancer Research: An Official Journal of the American Association for Cancer Research, 11*(20), 7234–7242.

Van Raamsdonk, C. D., & Deo, M. (2013). Links between Schwann cells and melanocytes in development and disease. *Pigment Cell & Melanoma Research, 26*(5), 634–645.

Vanneman, M., & Dranoff, G. (2012). Combining immunotherapy and targeted therapies in cancer treatment. *Nature Reviews Cancer, 12*(4), 237–251.

Waisberg, J., & Saba, G. T. (2015). Wnt-/-beta-catenin pathway signaling in human hepatocellular carcinoma. *World Journal of Hepatology, 7*(26), 2631–2635.

Wellbrock, C., & Arozarena, I. (2015). Microphthalmia-associated transcription factor in melanoma development and MAP-kinase pathway targeted therapy. *Pigment Cell & Melanoma Research, 28*(4), 390–406.

White, R. M., Cech, J., Ratanasirintrawoot, S., Lin, C. Y., Rahl, P. B., Burke, C. J., … Zon, L. I. (2011). DHODH modulates transcriptional elongation in the neural crest and melanoma. *Nature, 471*(7339), 518–522.

Whittaker, S., Marais, R., & Zhu, A. X. (2010). The role of signaling pathways in the development and treatment of hepatocellular carcinoma. *Oncogene, 29*(36), 4989–5005.

Wurmbach, E., Chen, Y. B., Khitrov, G., Zhang, W., Roayaie, S., Schwartz, M., … Llovet, J. M. (2007). Genome-wide molecular profiles of HCV-induced dysplasia and hepatocellular carcinoma. *Hepatology: Official Journal of the American Association for the Study of Liver Diseases, 45*(4), 938–947.

Yang, W., Zheng, Y., Xia, Y., Ji, H., Chen, X., Guo, F., … Lu, Z. (2012). ERK1/2-dependent phosphorylation and nuclear translocation of PKM2 promotes the Warburg effect. *Nature Cell Biology, 14*(12), 1295–1304.

Yen, J., White, R. M., Wedge, D. C., Van Loo, P., de Ridder, J., Capper, A., … Futreal, P. A. (2013). The genetic heterogeneity and mutational burden of engineered melanomas in zebrafish models. *Genome Biology, 14*(10), R113.

Yokoyama, Y., Kuramitsu, Y., Takashima, M., Iizuka, N., Terai, S., Oka, M., … Sakaida, I. (2006). Protein level of apolipoprotein E increased in human hepatocellular carcinoma. *International Journal of Oncology, 28*(3), 625–631.

Zhang, G., Hoersch, S., Amsterdam, A., Whittaker, C. A., Beert, E., Catchen, J. M., … Lees, J. A. (2013). Comparative oncogenomic analysis of copy number alterations in human and zebrafish tumors enables cancer driver discovery. *PLoS Genetics, 9*(8), e1003734.

Zhang, G., Hoersch, S., Amsterdam, A., Whittaker, C. A., Lees, J. A., & Hopkins, N. (2010). Highly aneuploid zebrafish malignant peripheral nerve sheath tumors have genetic alterations similar to human cancers. *Proceedings of the National Academy of Sciences of the United States of America, 107*(39), 16940–16945.

Zhang, X., Jia, H., Lu, Y., Dong, C., Hou, J., Wang, Z., … Wang, K. (2014). Exome sequencing on malignant meningiomas identified mutations in neurofibromatosis type 2 (NF2) and meningioma 1 (MN1) genes. *Discovery Medicine, 18*(101), 301–311.

Zhao, H., Li, Y., Wang, S., Yang, Y., Wang, J., Ruan, X., ... Fang, X. (2014). Whole transcriptome RNA-seq analysis: tumorigenesis and metastasis of melanoma. *Gene, 548*(2), 234–243.

Zheng, W., Li, Z., Nguyen, A. T., Li, C., Emelyanov, A., & Gong, Z. (2014). Xmrk, kras and myc transgenic zebrafish liver cancer models share molecular signatures with subsets of human hepatocellular carcinoma. *PLoS One, 9*(3), e91179.

CHAPTER THREE

# Gal4 Driver Transgenic Zebrafish: Powerful Tools to Study Developmental Biology, Organogenesis, and Neuroscience

**K. Kawakami*[,1], K. Asakawa*, M. Hibi[§], M. Itoh[¶], A. Muto* and H. Wada***

*National Institute of Genetics and SOKENDAI (The Graduate University for Advanced Studies), Mishima, Japan
[§]Nagoya University, Nagoya, Japan
[¶]Chiba University, Chiba, Japan
[1]Corresponding author: E-mail: kokawaka@nig.ac.jp

## Contents



## Abstract

Targeted expression by the Gal4-UAS system is a powerful genetic method to analyze the functions of genes and cells in vivo. Although the Gal4-UAS system has been extensively used in genetic studies in *Drosophila*, it had not been applied to genetic studies in vertebrates until the mid-2000s. This was mainly due to the lack of an efficient

*Advances in Genetics*, Volume 95
ISSN 0065-2660
http://dx.doi.org/10.1016/bs.adgen.2016.04.002

transgenesis tool in model vertebrates, such as the *P*-transposable element of *Drosophila*, that can create hundreds or thousands of transgene insertions in different loci on the genome and thereby enables the generation of transgenic lines expressing Gal4 in various tissues and cells via enhancer trapping. This situation was revolutionized when a highly efficient transgenesis method using the *Tol2* transposable element was developed in the model vertebrate zebrafish. By using the *Tol2* transposon system, we and other labs successfully performed gene trap and enhancer trap screens in combination with the Gal4-UAS system. To date, numerous transgenic fish lines that express engineered versions of Gal4 in specific cells, organs, and tissues have been generated and used for various aspects of biological studies. By constructing transgenic fish lines harboring genes of interest downstream of UAS, the Gal4-expressing cells and tissues in those transgenic fish have been visualized and manipulated via the Gal4-UAS system. In this review, we describe how the Gal4-UAS system works in zebrafish and how transgenic zebrafish that express Gal4 in specific cells, tissues, and organs have been used for the study of developmental biology, organogenesis, and neuroscience.

## 1. INTRODUCTION

Gal4 is a yeast transcription factor that contains a DNA-binding domain and a transcription activation domain at its N-terminus and C-terminus, respectively (Keegan, Gill, & Ptashne, 1986; Ma & Ptashne, 1987). Gal4 binds to a specific sequence, $UAS_G$ (UAS stands for upstream activating sequence), and activates transcription from a basal promoter (ie, TATA sequence) placed downstream of UAS in various animal cells (Fischer, Giniger, Maniatis, & Ptashne, 1988; Webster, Jin, Green, Hollis, & Chambon, 1988). The Gal4-UAS system was initially employed as a two-component gene expression system in *Drosophila* (Brand & Perrimon, 1993). Specifically, fly lines expressing Gal4 in specific tissues and cells and a fly line carrying the *lacZ* gene downstream of UAS were created and crossed. In the double transgenic progeny, LacZ was expressed in regions where Gal4 was expressed. A major advantage of this method is that any genes can be expressed in a desired place and time by crossing specific Gal4-expressing lines with UAS-reporter and UAS-effector lines. For instance, when a fly line harboring the tetanus toxin light chain gene, that inhibits neuronal function, downstream of UAS was crossed with a fly line expressing Gal4 in a subpopulation of neurons, behavioral phenotypes due to inhibition of those neuronal activities were observed (Sweeney, Broadie, Keane, Niemann, & O'Kane, 1995). The Gal4-UAS approach has been extensively used in flies since the *P*-transposable element–mediated enhancer trapping can create a number of transgenic fly lines that express Gal4 in desired cell types very efficiently.

It had been desired for a long time that the Gal4-UAS system can be applied to genetic studies in vertebrates, especially in zebrafish in which a large-scale forward genetic approach is practically possible even in small- or middle-scale labs. However, such a method had not been developed in zebrafish for a long time mainly because of the lack of an efficient transposon system such as the *P* transposon system in *Drosophila*, by which a large number of insertions of a Gal4 enhancer trap construct in different loci on the genome can be efficiently generated. In this review, we will describe how we and other labs have successfully developed the Gal4-UAS system in zebrafish and how transgenic fish that express Gal4 in specific cells, tissues, and organs have been employed for the study of developmental biology, organogenesis, and neuroscience.

## 2. *Tol2*-MEDIATED TRANSGENESIS IN ZEBRAFISH

The *Tol2* transposable element was identified from the genome of the Japanese medaka fish that shares sequence similarities with transposons of the hAT family (Koga, Suzuki, Inagaki, Bessho, & Hori, 1996). It was shown that the *Tol2* element contains a gene encoding a fully functional transposase (Kawakami, Koga, Hori, & Shima, 1998; Kawakami & Shima, 1999) (Fig. 1A). Thus, the *Tol2* element represents the first active transposon to be identified from a vertebrate genome. The *Tol2* element also contains DNA sequences that are recognized by the transposase. The minimal *cis*-sequences essential for transposition were analyzed, and it was shown that 200-bp from the left end and 150-bp DNA from the right end of the *Tol2* element are necessary and sufficient for transposition (Balciunas et al., 2006; Urasaki, Morvan, & Kawakami, 2006) (Fig. 1A). Therefore, any DNA fragment can be cloned between these *cis*-sequences.

The *Tol2* transposition system consists of two components, a transposon-donor plasmid carrying a *Tol2* construct and the transposase activity supplied in the form of mRNA synthesized by using the transposase cDNA as a template or an expression plasmid carrying that cDNA under the control of an appropriate promoter. It has been shown that the *Tol2* system is active in all vertebrate cells tested so far (Kawakami, 2007). In zebrafish, a transposon-donor plasmid and mRNA synthesized in vitro are injected into fertilized eggs. The *Tol2* construct is excised from the donor-plasmid and integrated into the genome of the germ cell lineage during embryonic development, and the transposon insertions are transmitted to the next generation through germ cells (Kawakami, Shima, & Kawakami, 2000; Kawakami et al., 2004)

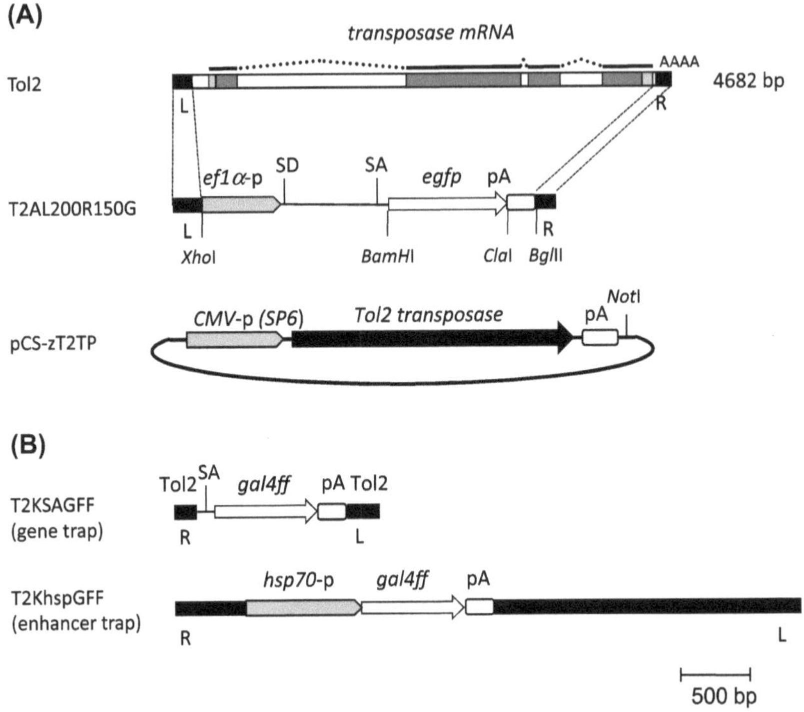


**Figure 1** ***Structures of the*** **Tol2** ***transposable element, gene trap, and enhancer trap constructs.*** (A) The full-length *Tol2* element, minimal *Tol2* vector T2AL200R150G, and pCS-zT2TP. *Tol2* is 4682 bp in length and encodes mRNA for the transposase (*dotted lines* indicate introns). T2AL200R150G contains 200-bp and 150-bp DNA from the left (L) and right (R) terminals of *Tol2* that are essential for transposition, the *Xenopus* EF1α promoter (*ef1α*-p), the rabbit β-globin intron (from SD to SA), the *egfp* gene, and the SV40 polyA signal (pA). pCS-zT2TP contains the codon-optimized transposase coding sequence downstream of the CMV and SP6 promoters. Unique restriction enzyme sites are indicated. (B) Gene and enhancer trap constructs. T2KSAGFF contains the splice acceptor (SA) and the *gal4ff* gene. T2KhspGFF contains the *hsp70* promoter and the *gal4ff* gene.

(Fig. 1A). Transgenesis using the *Tol2* transposon system is highly efficient. Namely, 50–70% of fish injected with the *Tol2* system at the one-cell stage and grown up to adulthood become germ line–transmitting founder fish that transmit transgenes to their offspring (Kawakami et al., 2004; Urasaki et al., 2006). Thus, the *Tol2*-mediated transgenesis became a standard method to create transgenic zebrafish. *Tol2* transposon–mediated transgenesis has the following merits. First, since a transposon construct is integrated as a single copy, the expression of the transgene on the construct is less sensitive to silencing in comparison to multimeric or concatemeric transgene integrations. Second, since a transposon vector functions as a cassette, end-to-end integration of a transgene is guaranteed. Third, the transposon insertion does not cause unwanted rearrangements at the integration locus. Fourth, the *Tol2* vector has a fairly large cargo capacity. 10-kb DNA can be cloned without reducing the transpositional activity (Balciunas et al., 2006; Urasaki et al., 2006), and, furthermore, even a BAC-size piece of DNA, namely 100–200 kb in length, can be cloned into the *Tol2* vector (Suster, Sumiyama, & Kawakami, 2009).

It was demonstrated that six to seven insertions on average are transmitted from single germ line–transmitting founder fish to the next generation (Kawakami et al., 2004; Urasaki et al., 2006). This advantage, together with the high germ line transmission frequencies, enables the generation of thousands of transposon insertions in a small- to mid-scale laboratory. Therefore, we and other labs have successfully developed important forward genetic methodologies, such as enhancer trapping, gene trapping (exon trapping), and protein trapping by using the *Tol2* system (Clark et al., 2011; Kawakami et al., 2004; Nagayoshi et al., 2008; Parinov, Kondrichin, Korzh, & Emelyanov, 2004; Trinh le et al., 2011). In these methods, *Tol2*-based gene trap and enhancer trap constructs are integrated into the genome nearly randomly, and reporter genes on the constructs are expressed under the control of promoters or enhancers located close to the integration sites. The transgenic zebrafish thus generated express reporter genes in spatially and temporally restricted fashions and have been used as valuable resources to visualize specific cell types during embryonic development and organogenesis.

## 3. THE Gal4-UAS SYSTEM IN ZEBRAFISH

### 3.1 The Emergence of the Gal4-UAS System in Zebrafish

The Gal4-UAS system was first applied to genetic studies in zebrafish by Scheer and Campos-Ortega (1999). Transgenic fish expressing the full-length yeast *GAL4* gene under the control of ubiquitous promoters, the *hsp* promoter or the *deltaD* promoter, and transgenic fish carrying a constitutively active form of the Notch receptor downstream of UAS (UAS:*myc-notch1a-intra*) were constructed by microinjection of the plasmid DNAs into fertilized eggs. These fish were crossed, and it was demonstrated that the effector gene was ectopically expressed in the Gal4-expressing cells in the double transgenic offspring (Scheer, Groth, Hans, & Campos-Ortega, 2001). Koster and Fraser (2001) employed Gal4-VP16 instead of the full-length Gal4 to overcome the weak expression activity of the full-length form. Gal4-VP16 is a fusion of the Gal4 DNA-binding domain and the transcriptional activation domain from the herpes simplex virus VP16 protein and has stronger transcriptional activity (Sadowski, Ma, Triezenberg, & Ptashne, 1988). Plasmid constructs containing both Gal4-VP16 downstream of a tissue-specific promoter and a reporter gene downstream of UAS in tandem were injected into zebrafish embryos, and high levels of reporter gene expression were observed in the injected embryos transiently in regions where the tissue-specific promoter was active. Sagasti et al. constructed a stable transgenic line harboring the construct containing the Gal4-VP16 gene under the control of an enhancer from the *islet-1* gene and the UAS:GFP gene via the meganuclease (I-SceI) mediated transgenesis method (Sagasti, Guido, Raible, & Schier, 2005; Thermes et al., 2002). In the transgenic embryos, GFP was expressed in sensory neurons, which allowed them to investigate the shapes and sizes of individual sensory arbors.

These works demonstrated that the Gal4-UAS system can enhance expression of a reporter or an effector gene in tissue-specific and cell type–specific manners in both transient and stable transgenic fish. Despite this promising work, only a handful of research projects that employed transgenic Gal4 lines and UAS-transgene lines had been published by the mid-2000s. There were two main reasons. First, the number of enhancers and promoters available to achieve tissue-specific and cell type–specific expression had been limited. Second, transgenesis efficiencies using conventional methodology, either microinjection of plasmid DNA or the I-SceI–mediated method, were not high enough to create many different transgenic lines.

## 3.2 *Tol2*-Mediated Gal4 Gene Trapping and Enhancer Trapping

To apply the Gal4-UAS system to genetic studies in zebrafish, we employed Gal4FF, a modified version of the Gal4 yeast transcription activator, that has the Gal4 DNA-binding domain and two short transcription activator segments from the herpes simplex viral protein VP16 (Asakawa & Kawakami, 2008). It was shown that high levels of expression of Gal4-VP16 inhibited transcription independently of the UAS sequence. This phenomenon was thought to be caused by titration of endogenous transcription machineries by the strong activation domain, the so-called "squelching" (Sadowski et al., 1988). The lethality observed in zebrafish in association with Gal4-VP16 expression (Argenton, Arava, Aronheim, & Walker, 1996; Koster & Fraser, 2001; Scott et al., 2007) may be partly explained by such a mechanism. Gal4FF was shown to be better tolerated than Gal4-VP16 in vertebrate cells (Baron, Gossen, & Bujard, 1997; Seipel, Georgiev, & Schaffner, 1992).

To generate transgenic fish with various different patterns of Gal4FF expression, we constructed a gene trap construct T2KSAGFF that contains a splice acceptor from the rabbit β-globin gene and the *gal4ff* gene and an enhancer trap construct T2KhspGFF that contains 650-bp DNA of the zebrafish *hsp70* promoter and the *gal4ff* gene (Fig. 1B). At normal temperatures, the activity of the *hsp70* promoter is repressed, and it works as a minimal promoter. When the gene trap construct is integrated within a host gene and the splice acceptor has "trapped" its endogenous transcript, the *gal4ff* gene is expressed under the control of the promoter activity. Then, when the enhancer trap construct is integrated in the genome and the *hsp70* promoter is influenced by a nearby enhancer, the *gal4ff* gene is expressed in a pattern dictated by the trapped enhancer (Asakawa & Kawakami, 2008).

To visualize Gal4FF expression, we created the UAS:GFP fish that contained the *egfp* gene downstream of five repeats of UAS (5xUAS) by *Tol2*-mediated transgenesis. Detection of GFP fluorescence and GFP mRNA by whole-mount in situ hybridization is much more sensitive than detection of Gal4 mRNA by in situ hybridization, indicating that the signal was amplified through the Gal4-UAS system. The following two things should be noted. First, there should be a time lag between the Gal4-dependent transcription of the reporter gene and maturation of the reporter protein. Due to this "transcript to activity lag," the reporter expression always temporally follows the expression of Gal4 (Phelps & Brand, 1998). Second, turnover of the reporter protein needs to be considered. Specifically, GFP is a fairly stable protein, and even after Gal4 expression has been terminated and the residual activity of Gal4 has disappeared, the original Gal4-expressing cells may still contain the GFP protein (Asakawa & Kawakami, 2008). For this reason, reporter expression is not always consistent with Gal4 expression and therefore with endogenous enhancer/promoter activities. We injected these Gal4FF trap constructs into fertilized eggs, crossed the injected fish with the UAS:GFP reporter fish, and demonstrated that various different patterns of GFP expression can be identified in the F1 embryos (Fig. 2). Notably, obtaining 129 patterns from 250 injected fish represents a high

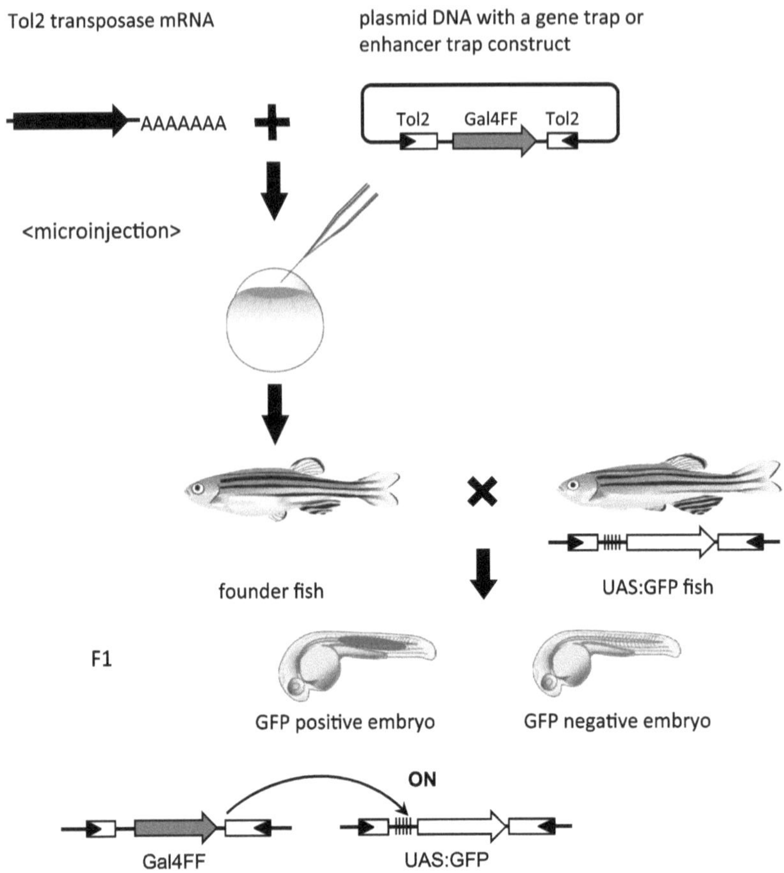


**Figure 2** ***Gene trapping and enhancer trapping with the Gal4-UAS system.*** A trap construct containing *gal4ff* is injected into fertilized eggs with the transposase mRNA. Injected fish are raised and mated with homozygous UAS:GFP reporter fish. Double transgenic F1 embryos expressing GFP in specific regions should be obtained. (See color plate)

enough frequency to permit a small laboratory to efficiently collect hundreds of different Gal4FF expression patterns (Asakawa & Kawakami, 2008).

Scott et al. (2007) constructed *Tol2* enhancer trap constructs carrying the Gal4-VP16 gene downstream of long (1.5 kb) and short (600 bp) versions of the zebrafish *hsp70* promoter. Then, transgenic fish carrying random integrations of these constructs in the genome were crossed with the UAS: Kaede transgenic fish line that contained a gene for a photoconvertible fluorescent protein, Kaede, cloned downstream of 14xUAS. In the double transgenic embryos, various patterns of Kaede expression were observed. Davison et al. (2007) constructed SAGVG that contained the Gal4-VP16 gene downstream of a splice acceptor from the rabbit β-globin gene, 14xUAS, and the EGFP gene. Thus, SAGVG is a self-reporting construct which can express GFP when the splice acceptor has trapped an endogenous transcript and Gal4-VP16 is expressed. Random integrations of the SAGVG construct in the genome were created, and transgenic fish expressing GFP in various patterns were generated. These transgenic fish were crossed with the UAS:Kaede reporter fish, and it was shown that Gal4-VP16 expression can also activate the Kaede reporter. An advantage of SAGVG is that one can collect Gal4-expressing lines without using UAS reporter fish. However, this may also be a disadvantage when the Gal4-expressing lines are crossed with other UAS reporter or effector fish containing green fluorescent markers.

### 3.3 Additional Improvements in the Gal4-UAS System

Goll et al. observed variegated expression of GFP in transgenic fish that contained the self-reporting SAGVG construct with 14xUAS, from GFP-weak to GFP-high, and showed that the 14xUAS sequence was heavily methylated in the GFP-weak lines. Interestingly, the GFP expression was reactivated in the *dnmt1* mutant that had a mutation in a gene encoding DNA methyltransferase-1 (Goll, Anderson, Stainier, Spradling, & Halpern, 2009). To circumvent this problem, reducing the number of UAS repeats is one solution. Indeed, Gal4FF-dependent GFP expression using our UAS:GFP fish line using 5xUAS has been reliable and reproducible for more than 15 generations (Asakawa & Kawakami, 2009; unpublished observation). Akitake, Macurak, Halpern, and Goll (2011) also reported reduced silencing and methylation by using four repeats of nonrepetitive UAS sequences. Distel, Wullimann, and Koster (2009) extensively investigated the effects of the number of UAS repeats on the strength of gene expression, finding that 14xUAS was less active than 5xUAS and comparable to 4xUAS, and so used 4xUAS.

Strong Gal4 expression is the key to achieving reliable expression of UAS effector genes in Gal4-expressing cells. Therefore, researchers need to select transgenic fish that express Gal4 not only specifically but also strongly for their experiments. When Gal4 expression is not high enough, effector genes will not be expressed to the levels which are necessary to obtain reliable results. To circumvent this problem, Distel et al. (2009) have described a system called "Kaloop". Namely, transgenic fish containing a gene for a variant of Gal4 (KalTA4) downstream of 4xUAS was created, in which KalTA4 can self-activate its own expression through 4xUAS. Such a method may be helpful to boost Gal4 expression when the original is not sufficient.

## 4. APPLICATIONS OF Gal4 DRIVERS

The power of the Gal4-UAS approach is that any genes can be expressed in Gal4-expressing cells. For this purpose, a collection of transgenic fish that express Gal4 in various cells, tissues, and organs is necessary. Since the mid-2000s, my lab has been performing a genetic screen to generate such Gal4 fish. All of the transgenic fish have been characterized at the molecular level (Abe, Suster, & Kawakami, 2011), and, for most, fish lines with single transposon insertions have been established and the transposon integration sites have been cloned and determined. In addition, to make the best use of those transgenic lines, we have constructed a database named *z*Trap by which researchers can search transgenic fish based on expression patterns and transposon integration sites (Kawakami et al., 2010). Here, we describe how the Gal4-UAS approach has been applied to the study of developmental biology, organogenesis, and neuroscience. Transgenic zebrafish lines described are summarized in Tables 1 and 2.

### 4.1 Inhibition of Neuronal Activities via Targeted Expression of a Neurotoxin Gene

We constructed UAS:TeTxLC:CFP transgenic lines (Asakawa et al., 2008) containing a gene encoding the tetanus toxin light chain fused to the CFP gene downstream of UAS. Tetanus toxin light chain (TeTxLC) blocks synaptic transmission by cleaving a synaptic vesicle-associated membrane protein VAMP-2 to block neurotransmitter release (Schiavo et al., 1992). By performing the Gal4FF gene trapping and enhancer trapping, we generated more than 100 transgenic fish lines that expressed Gal4FF in various types of neurons. We crossed those transgenic fish with the UAS:TeTxLC:CFP effector line and analyzed the double transgenic larvae for touch response behavior (Fig. 3A). A wild-type larva at 2 dpf swims rapidly away from the stimulus when it is gently touched on the tail with a needle. In contrast, when the SAGFF36B gene trap fish, that expressed Gal4FF in the Rohon-Beard sensory neurons, was crossed with the UAS:TeTxLC:CFP fish, the SAGFF36B;UAS:TeTxLC:CFP double transgenic larva did not respond to touch. When the SAGFF31B gene trap fish, that expressed Gal4FF in the subsets of interneurons in the spinal cord, was crossed with the UAS: TeTxLC:CFP fish, the SAGFF31B;UAS:TeTxLC:CFP double transgenic

**Table 1** Gal4 driver transgenic zebrafish lines

| Line name | Expression pattern (gene) | References |
|---|---|---|
| ***The motor system in the spinal cord and hindbrain*** | | |
| SAGFF31B | Ventral spinal cord (moto- and interneurons) | Asakawa et al. (2008) |
| SAGFF36B | RB sensory neurons | Asakawa et al. (2008) |
| HGj4A | Ventral spinal cord (motoneurons) (*mnr2b*) | Asakawa et al. (2012) |
| mnGFF7 | Ventral spinal cord (motoneurons) | Asakawae et al. (2013) |
| hspGFF62A | Mauthner cells | Yamanaka et al. (2013) |
| ***Calcium imaging of neuronal activities*** | | |
| SAIGFF213A | CaP neurons (*prdm14*) | Muto et al. (2011) |
| gSA2AzGFF49A | Optic tectum (*disc large homolog 2*) | Muto et al. (2013) |
| ***The cerebellar neural circuits*** | | |
| gSA2AzGFF152B | Granule cells | Takeuchi et al. (2015) |
| hspzGFFgDMC156A | Eurydendroid cells | Takeuchi et al. (2015) |
| hspGFFDMC28C | Inferior olive nuclei | Takeuchi et al. (2015) |
| ***The lateral line system*** | | |
| HGn39D | Lateral line sensory neurons (*cntnap2a*) | Pujol-Marti et al. (2012) |
| hspGFFDMC131A | Lateral line sensory neurons (*rspo2*) | Wada, Dambly-Chaudiere, et al. (2013) |
| gSAzGFF15A | Subpopulation of hair cells | Wada et al. (2014) |
| gSAGFF202A | Schwann cells (*erbB2*) | Xiao et al. (2015) |
| Notch signaling | | |
| hspGFFDMC72A | Primary neurons (*deltaA*) | Okigawa et al. (2014) |
| SAGFF214A | Vacuolated cells in the notochord (*cyb5r2*) | Yamamoto et al. (2010) |

**Table 2** UAS reporter and effector transgenic zebrafish lines

| Line name | Reporter or effector gene | References |
|---|---|---|
| UAS:GFP | EGFP | Asakawa et al. (2008) |
| UAS:TeTxLC:CFP | The tetanus toxin light chain fused to the CFP gene | Asakawa et al. (2008) |
| UAS:GCaMP-HS | Calcium indicator GCaMP-HS | Muto et al. (2011) |
| UAS:GCaMP7a | Calcium indicator GCaMP7a | Muto et al. (2013) |
| UAS:WGA | Wheat germ agglutinin | Takeuchi et al. (2015) |

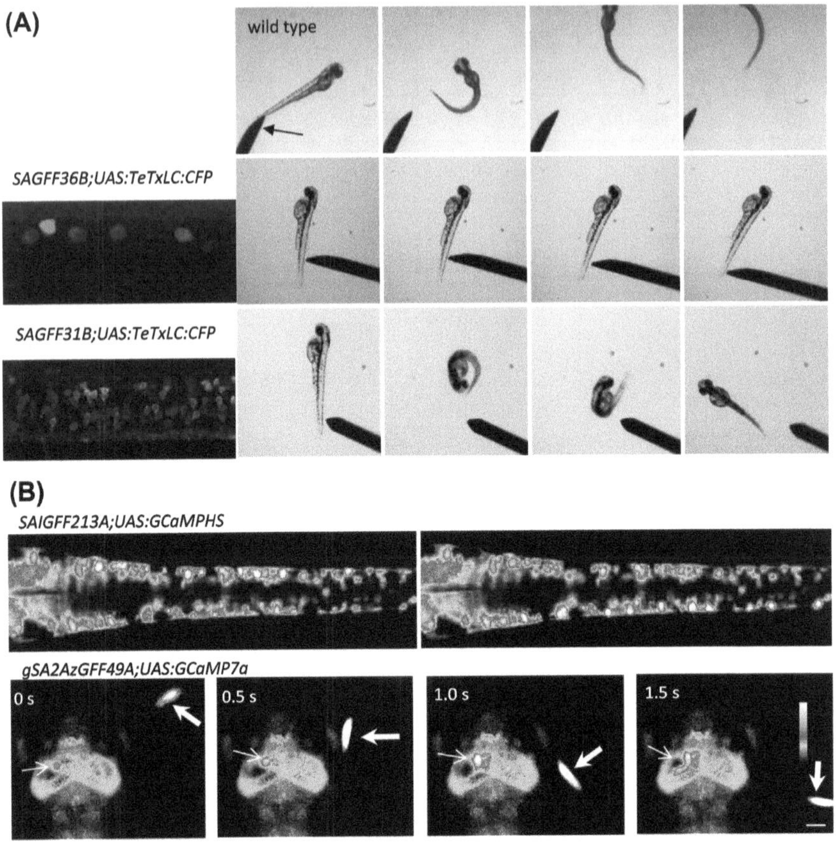


**Figure 3** ***Gal4FF driver transgenic fish and targeted expression via the Gal4FF-UAS system.*** (A) Inhibition of neuronal activity via targeted expression of a neurotoxin gene. A wild-type embryo at 2 dpf rapidly escapes from a gentle touch to the tail. The SAGFF36B;UAS:TeTxLC:CFP fish, which expresses TeTxLC:CFP in the sensory neurons, does not respond to touch. The SAGFF31B;UAS:TeTxLC:CFP fish, which expresses TeTxLC:CFP in subsets of interneurons and motor neurons, responds to the touch but shows abnormal escape swimming. (B) Imaging of the activity of CaP motor neurons in the SAIGFF213A;UAS:GCaMPHS double transgenic embryo. A top view of an immobilized embryo at 24 hpf. Anterior to the left. Simultaneous calcium signals on one side are detected alternately (pseudocolored). Imaging of the tectal neurons in the gSA2AzGFF49A;UAS:GCaMP7a double transgenic larva. Top view of an immobilized embryo at 6 dpf. A paramecium is moving from anterior to posterior on the right-hand side in a space around the larva, and calcium signals on the left tectum move from anterior to posterior (pseudocolored). (C) Genetic dissection of the motor system. HGj4A and mnGFF7;UAS:GFP at 48 hpf express GFP and Gal4FF in the ventral spinal cord, respectively. Single neurons are visualized in the mnGFF7 larva injected with the UAS:GFP plasmid. The hspGFF62A transgenic fish expresses Gal4FF in the Mauthner cells. (D) The architecture of cerebellar neural circuits. Top view of the cerebellum. *CCe*, corpus cerebelli; *EG*, eminentia granularis; *GC*, granule cells; *LCa*, lobus caudalis cerebelli; *PC*, Purkinje ells; *PF*, parallel fibers; *Va*, valvula cerebelli. The gSA2AzGFF152B, hspzGFFgDMC156A, and hspGFFDMC28C transgenic fish express Gal4FF in the granule cells, eurydendroid cells, and IO (inferior olive) neurons, respectively. An *arrow* indicates contralateral projections of the climbing fibers. (E) Transgenic fish that label the lateral line system. The HGn39D transgenic fish expresses Gal4FF in sensory axons (green) and

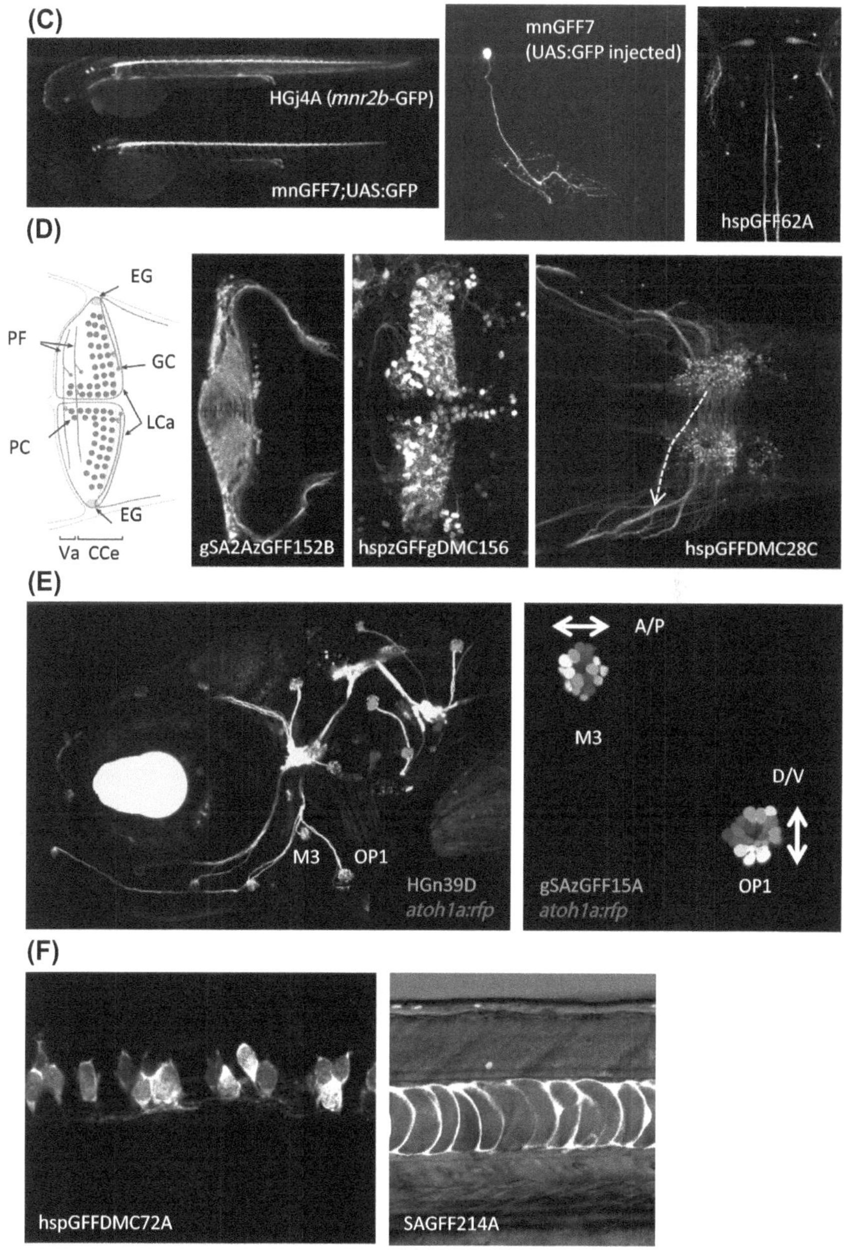


**Figure 3** (*continued*).

*atoh1a:rfp* labels hair cells (red) in the cranial region of an embryo at 5 dpf. Anterior to the left. gSAzGFF15A transgenic fish express Gal4FF in older hair cells (green) and *atoh1a:rfp* labels younger hair cells (red). *Arrows* indicate the direction of hair cell migration in each neuromast. (F) Transgenic fish used to study spatiotemporal functions of Notch signaling. The hspGFFDMC72A transgenic fish expresses Gal4FF in early differentiating primary neurons. The SAGFF214A transgenic fish expresses Gal4FF in vacuolated cells in the notochord. (See color plate) larva show abnormal escape behavior. Thus, we achieved functional dissection of the neuronal circuits by application of the Gal4FF-UAS system. The UAS:TeTxLC:CFP fish line has been used to analyze other neuronal circuitries that regulate vertebrate behavior. For example, it was used to inhibit the activities of Kolmer—Agduhr cells in the spinal cord, contacting the cerebrospinal fluid (Wyart et al., 2009), and of the lateral subnucleus of the dorsal habenula in the diencephalon (Agetsuma et al., 2010).

## 4.2 Visualization of Neuronal Activities by Calcium Imaging

The Gal4-UAS system has also been applied to imaging of neuronal activity with the genetically encoded calcium indicator GCaMP in zebrafish larvae. The GCaMP protein is a fusion of the calmodulin-binding domain from the myosin light chain kinase (which is called an M13 peptide), permutated EGFP, and the calmodulin and increases its intensity of green fluorescence upon binding to calcium ions (Nakai, Ohkura, & Imoto, 2001). Using GCaMP, generation of action potentials in neurons can be indirectly monitored by detecting calcium influx via voltage-gated calcium channels. The sensitivity of GCaMP and its high expression levels are critical to obtain an optimal signal-to-noise ratio and therefore for successful detection of neuronal activity. After the first version of GCaMP, efforts have been made to generate brighter and more sensitive versions of GCaMP by introducing amino acid substitutions. Thus, we generated the UAS:GCaMP-HS (Muto et al., 2011) and UAS:GCaMP7a transgenic fish (Muto, Ohkura, Abe, Nakai, & Kawakami, 2013) that contained improved versions of GCaMP downstream of 5xUAS.

Zebrafish larvae exhibit spontaneous body contraction in the embryonic stages. Specifically, they spontaneously contract their left and right trunk muscles alternately without an external stimulus, starting after 17—19 hpf. We generated the SAIGFF213A gene trap line that expressed Gal4FF in a subset of primary motoneurons (CaP neurons) in the spinal cord, which extend their axons to the ventral muscles. The SAIGFF213A fish were crossed with UAS:GCaMPHS and the SAIGFF213A;UAS:GCaMPHS double transgenic larvae were used for calcium imaging. We observed periodic synchronized fluorescence changes of GCaMP-HS in the ipsilateral CaP neurons in embryos embedded in agarose. We also observed that the synchronized activation of the CaP neurons on the right and left side occurred alternately with regular intervals, concomitantly with the alternating muscle contractions (Muto et al., 2011) (Fig. 3B).

Zebrafish larvae start to eat paramecia after 4–5 dpf, exhibiting a behavior called prey capture. The prey capture behavior comprises a sequence of stereotyped behaviors, ie, perception of prey, eye convergence, and approach swimming. As a first step toward understanding the functional neural circuits that control the prey capture behavior, we visualized how neurons in the brain respond when a zebrafish larva perceives a natural object, a swimming paramecium. We generated the gSA2AzGFF49A gene trap line that expresses Gal4FF strongly in the optic tectum, and the gSA2AzGFF49A;UAS:GCaMP7a double transgenic larva was used for imaging. Firstly, spontaneous neuronal activities in the tectum could be detected at a single cell resolution. Next, we analyzed the neuronal activity in the tectum using 5–6 dpf larvae embedded in agarose. We demonstrated clear $Ca^{2+}$ signals evoked by the paramecium swimming around the immobilized larva in the neuropil area and the cell bodies of the tectum. The functional visuotopic map revealed by imaging was consistent with the retinotopic map created by anatomical studies; namely, when a paramecium moved from the right to left hemifield, from ventral to dorsal, or from anterior to posterior, $Ca^{2+}$ transients moved from the left to right tectum, from the margin (ventral) to center (dorsal), or from anterior to posterior in the neuropil area of the contralateral tectum, respectively (Fig. 3B). Finally, we imaged a freely swimming gSA2AzGFF49A;UAS:GCaMP7a larvae and showed that activation of the anterior tectum may be functionally connected to a subsequent visuomotor pathway that induces the prey capture behavior (Muto et al., 2013). In these imaging experiments, while the brain at larval stages is fairly transparent, we use the *nacre* mutant strain that completely lacks melanophores (O'Malley et al., 2004). The *nacre* mutant is especially useful for the study of the visual system because it retains the intact retinal pigment epithelium.

## 4.3 Genetic Dissection of the Motor System

The vertebrate motor system consists of various cell types. We generated an enhancer trap line HGj4A that expresses EGFP in the spinal motor column (Asakawa, Higashijima, & Kawakami, 2012). We analyzed transposon integration site and found that, in the HGj4A line, the enhancer trap construct with GFP trapped the enhancer activity of a gene encoding the Mnx-class homeobox *mnr2b/mnx2b* (hereafter referred to as *mnr2b*), which is expressed in the ventral spinal cord, including the spinal motor neurons. To reveal the cellular architecture of the spinal cord motor column, we exploited the Gal4-UAS system. First, we introduced the Gal4FF gene into the bacterial artificial chromosome (BAC) containing the *mnr2b* locus by recombineering, injected the BAC construct into fertilized eggs, and generated a Gal4 driver line mnGFF7 (Asakawa, Abe, & Kawakami, 2013) (Fig. 3C). Then, we injected the UAS:GFP plasmid into mnGFF7 embryos at the one-cell or blastula stages. By this approach, we stochastically labeled single Gal4-expressing cells in the spinal motor column (Fig. 3C). From 103 single cells labeled with EGFP, we identified at least 11 different types of spinal motor neurons based on their axon trajectories and pattern of muscle innervation at 4–5 dpf. Thus, this approach identified previously uncharacterized spinal motor neurons and revealed their topographic relationships with the skeletal muscle.

We generated the hspGFF62A enhancer trap transgenic fish that expresses Gal4FF in the Mauthner cells (M-cells), a pair of giant reticulospinal neurons located in rhombomere 4 of the hindbrain (Fig. 3C). The hspGFF62A was used for targeted expression of autocamtide-2-related inhibitory peptide II that inhibited CaMKII activation, to show that postsynaptic activation is required for glycine receptor (GlyR) clustering in M-cells (Yamanaka et al., 2013), and for targeted expression of GCaMP6s to observe M-cell activation following acoustic stimulation (Marsden & Granato, 2015).

## 4.4 The Architecture of Cerebellar Neural Circuits

The cerebellum is involved in some forms of motor coordination and motor learning. The major excitatory glutamatergic and inhibitory GABAergic neurons in the cerebellum are granule cells and Purkinje cells, respectively. Purkinje cells receive inputs from the climbing fibers (CFs), which are neuronal axons emanating from the inferior olive nuclei located in the ventroposterior hindbrain. Granule cells receive inputs from the mossy fibers (MFs), which are neuronal axons projected from precerebellar nuclei, located in various regions of the brain. The signals carried by the MFs are transmitted to the Purkinje cell dendrites by granule cell axons. Thus, signals carried by both the MFs and the CFs are integrated by the Purkinje cells. In teleosts, neurons known as eurydendroid cells receive signals from both Purkinje cell axons and granule cell axons; eurydendroid cells further integrate the information and project their axons to targets outside the cerebellum. In addition to cerebellar neurons, Bergmann glial cells are involved in the development and function of the cerebellum.

We generated transgenic fish that express Gal4FF in the granule cells (gSA2AzGFF152B and other lines), Purkinje cells (*aldoca:gal4ff* containing the *gal4ff* gene downstream of the promoter of the *aldolase Ca* gene), eurydendroid cells (hspzGFFgDMC156A), Bergmann glia (SAGFF251A, SAGFF226B), or IO neurons (hspGFFDMC28C). We crossed these Gal4 lines with UAS reporter lines and analyzed the anatomy and developmental processes of the cerebellar neural circuitry (Fig. 3D). The hspGFFDMC28C (IO-specific Gal4 driver) was crossed with a UAS: photoconvertible-Kaede transgenic line, and the contralateral projections of CFs were clearly visualized. The gSA2AzGFF152B (granule cell–specific Gal4 driver) was crossed with the UAS:WGA transgenic fish containing a gene encoding the *trans*-synaptic protein wheat germ agglutinin (WGA) downstream of UAS. Thereby, it was revealed that the granule cells are directly or indirectly connected with Purkinje cells, eurydendroid cells, and IO neurons. The gSA2AzGFF152B (granule cell–specific Gal4 driver) was crossed with the UAS:GFP line and used for time-lapse imaging. We demonstrated initial random movements and ventral migration of granule cell nuclei. These demonstrate the utility of cerebellar Gal4 transgenic lines for studying the development and function of cerebellar neural circuits (Takeuchi et al., 2015).

## 4.5 Proliferation and Differentiation of the Lateral Line System

The lateral line is a mechanosensory system for fish to detect water movements. The system comprises sensory organs, neuromasts, distributed over the body surface. Each neuromast is composed of a core of sensory hair cells surrounded by nonsensory support cells, and the hair cells are innervated by sensory neurons located in anterior and posterior sensory ganglia (ALL and PLL). Since the lateral line is located in the skin, enhancer trap lines have been successfully used to visualize each cell type. Here we describe four lines that label specific cell types in the lateral line.

The HGn39D enhancer trap line expresses GFP in the lateral line sensory neurons and lens (Fig. 3E). This enhancer trap construct was integrated within the contactin-associated protein-like2/Caspr2 (*cntnap2a*) gene. Hence, the promoter region of *cntnap2a* was cloned and the *cntnap2a*:Gal4 fish was generated. The lateral line nerves consist of afferent (sensory) and efferent components, forming a single fascicle. In HGn39D, the afferent (sensory) neurons were exclusively labeled. Thus, the HGn39D line can be used to visualize the central projection of the lateral line nerves within the brain (Pujol-Marti et al., 2012).

The hspGFFDMC131A enhancer trap line expresses Gal4FF in the lateral line sensory neurons. Here, the enhancer trap construct was integrated close to the *r-spondin2* (*rspo2*) gene, and we thereby discovered that the *rspo2* gene is expressed in the lateral line ganglion. Rspo family proteins are secreted factors that activate Wnt signaling. We demonstrated that innervation of the sensory axons is required for neuromast proliferation and that Wnt signaling activity is essential for proliferation of hair cell precursors. Thus, we hypothesize that Rspo2 emanating from the axon terminals may regulate neuromast growth (Wada, Dambly-Chaudiere, Kawakami, & Ghysen, 2013; Wada, Ghysen, et al., 2013).

The gSAzGFF15A gene trap line expresses Gal4FF in a subpopulation of sensory hair cells (Fig. 3E). Hair cells are continuously generated from progenitor cells, migrate away from the center of a neuromast along the A/P or D/V axis of the body depending on a polarized organization of neuromasts, and then eventually die. The *atoh1a:rfp* transgenic line containing the RFP gene downstream of the *atoh1a* promoter labels hair cells and their precursors at early stages. In contrast, gSAzGFF15A labels hair cells at later stages. Therefore, the gSAzGFF15A;UAS:GFP;*atoh1a:rfp* triple transgenic line can visualize a turnover of the hair cells along the A/P or the D/V axis of a neuromast (Wada, Iwasaki, & Kawakami, 2014).

The gSAGFF202A gene trap line expresses Gal4FF in Schwann cells that are intimately associated with the lateral line axons. The enhancer trap construct is integrated within and disrupts the function of the *erbB2* gene, which is essential for the migration of Schwann cells along the axons. Homozygous gSAGFF202A fish show nerve defasciculation due to loss of myelination. Schwann cells also play important roles in repair of the nervous system through interactions with axons. We performed in vivo live imaging of regeneration of the lateral line nerve using the gSAGFF202A fish and demonstrated that Schwann cells extend bridging processes to close the injury gap, and subsequently, regenerating axons grow faster after Schwann cells clear distal debris (Xiao et al., 2015).

## 4.6 Spatiotemporal Functions of Notch Signaling

The Notch signaling pathway is highly evolutionarily conserved and is involved in a variety of cell fate decisions during development. Notch receptors are activated by their ligands, the Delta and Jagged family of proteins, whose functions are modulated by the ubiquitin ligase Mib1 (Itoh et al., 2003). Zebrafish is a highly suitable model animal to explore the spatiotemporal functions of Notch signaling in vivo.

The hspGFFDMC72A enhancer trap line expresses Gal4FF in early differentiating primary neurons (Fig. 3F). In hspGFFDMC72A, the enhancer trap construct is inserted into the first exon of the *deltaA* gene which leads to loss of DeltaA function in the allele. Using the *deltaA* mutant, we investigated the functions of Notch signaling in sensory organ and interneuron development (Mizoguchi, Togawa, Kawakami, & Itoh, 2011; Okigawa et al., 2014). We investigated how neurons and the sensory epithelium develop separately from the same preplacodal progenitors using posterior lateral line (PLL) as a model organ (Mizoguchi et al., 2011). We found that the number of posterior lateral line ganglion (PLLG) neurons increases in embryos with reduced Notch activities, suggesting that Notch-mediated lateral inhibition regulates the PLLG cell fate. Fate-mapping analysis revealed that cells adjacent to the PLLG neurons in the pre-PLL placodal region give rise to the anterior part of the PLL primordium (ie, sensory epithelial progenitor cells). Notch signaling determines whether putative precursor cells become PLLG neurons or PLL primordia. We also investigated the differentiation of interneurons in the spinal cord, namely V2-interneuron (V2-IN) progenitor proliferation and V2a/V2b cell fate determination that occur concurrently during development (Okigawa et al., 2014). In the spinal cord, V2-IN progenitors (p2) differentiate into excitatory V2a-INs and inhibitory V2b-INs. Two ligands, DeltaA and DeltaD, and three receptors, Notch1a, Notch1b, and Notch3, redundantly contribute to p2 progenitor maintenance. On the other hand, DeltaA, DeltaC, and Notch1a mainly contribute to V2a/V2b cell fate determination.

The SAGFF214A gene trap line expresses Gal4FF in vacuolated cells in the notochord (Fig. 3F). We crossed SAGFF214A with UAS:GFP fish and monitored the development of vacuolated cells in live embryos. We found that the Mib1-mediated activation of Notch signaling through Jag1 regulates cell fate determination in the notochord-lineage cells, namely vacuolated cells versus nonvacuolated epithelial-like cells (Yamamoto et al., 2010).

## 5. CONCLUSION

Transgenic fish expressing Gal4 in specific cells, tissues, and organs can be generated by standard *Tol2*-transgenesis or *Tol2*-mediated BAC transgenesis methods using constructs containing the gal4 gene downstream of a cell type–specific promoter (Asakawa et al., 2012; Suster et al., 2009), and by gene trap and enhancer trap methods (Asakawa et al., 2008; Davison et al., 2007; Scott et al., 2007). In addition, gal4 transgenic fish can be created by the CRISPR/Cas9-mediated knock-in method (Auer, Duroure, De Cian, Concordet, & Del Bene, 2014). In either case, it is important to create transgenic zebrafish with specific and strong Gal4 expression. For UAS reporter and effector fish to function successfully, it is important to generate lines carrying at least 10 insertions of the UAS-reporter or UAS-effector construct and select the one that works best with the Gal4 transgenic fish of interest. We and other labs have demonstrated that the Gal4 driver fish are applicable to various biological studies. Thus, the Gal4-UAS approach represents a useful and powerful genetic methodology in zebrafish.

## ACKNOWLEDGMENTS

This work was partly supported by National BioResource Project (KK), Sumitomo Foundation (grant No. 11061) (MI), Takeda Science Foundation (MI), Program for Improvement of Research Environment for Young Researchers from SCF commissioned by the Ministry of Education, Culture, Sports, Science, and Technology of Japan (MEXT) (MI), PRESTO program of the Japan Science and Technology Agency (JST) (HW), and Grant-in-Aids (15H02370 to KK; 25830020 to KA; 22657048 and 24370080 to MI; 22500299 and 24120521 to AM; 19770204 and 25440118 to HW) from MEXT.

## REFERENCES

Abe, G., Suster, M. L., & Kawakami, K. (2011). *Tol2*-mediated transgenesis, gene trapping, enhancer trapping, and the Gal4-UAS system. *Methods in Cell Biology, 104*, 23–49.

Agetsuma, M., Aizawa, H., Aoki, T., Nakayama, R., Takahoko, M., Goto, M., … Okamoto, H. (2010). The habenula is crucial for experience-dependent modification of fear responses in zebrafish. *Nature Neuroscience, 13*(11), 1354–1356.

Akitake, C. M., Macurak, M., Halpern, M. E., & Goll, M. G. (2011). Transgenerational analysis of transcriptional silencing in zebrafish. *Developmental Biology, 352*(2), 191–201.

Argenton, F., Arava, Y., Aronheim, A., & Walker, M. D. (1996). An activation domain of the helix-loop-helix transcription factor E2A shows cell type preference in vivo in microinjected zebra fish embryos. *Molecular and Cellular Biology, 16*(4), 1714–1721.

Asakawa, K., Abe, G., & Kawakami, K. (2013). Cellular dissection of the spinal cord motor column by BAC transgenesis and gene trapping in zebrafish. *Frontiers in Neural Circuits, 7*, 100.

Asakawa, K., Higashijima, S., & Kawakami, K. (2012). An mnr2b/hlxb9lb enhancer trap line that labels spinal and abducens motor neurons in zebrafish. *Developmental Dynamics, 241*(2), 327–332.

Asakawa, K., & Kawakami, K. (2008). Targeted gene expression by the Gal4-UAS system in zebrafish. *Development, Growth and Differentiation, 50*(6), 391–399.

Asakawa, K., & Kawakami, K. (2009). The *Tol2*-mediated Gal4-UAS method for gene and enhancer trapping in zebrafish. *Methods, 49*(3), 275–281.

Asakawa, K., Suster, M. L., Mizusawa, K., Nagayoshi, S., Kotani, T., Urasaki, A., … Kawakami, K. (2008). Genetic dissection of neural circuits by *Tol2* transposon-mediated Gal4 gene and enhancer trapping in zebrafish. *Proceedings of the National Academy of Sciences of the United States of America, 105*(4), 1255–1260.

Auer, T. O., Duroure, K., De Cian, A., Concordet, J. P., & Del Bene, F. (2014). Highly efficient CRISPR/Cas9-mediated knock-in in zebrafish by homology-independent DNA repair. *Genome Research, 24*(1), 142–153.

Balciunas, D., Wangensteen, K. J., Wilber, A., Bell, J., Geurts, A., Sivasubbu, S., … Ekker, S. C. (2006). Harnessing a high cargo-capacity transposon for genetic applications in vertebrates. *PLoS Genetics, 2*(11), e169.

Baron, U., Gossen, M., & Bujard, H. (1997). Tetracycline-controlled transcription in eukaryotes: novel transactivators with graded transactivation potential. *Nucleic Acids Research, 25*(14), 2723–2729.

Brand, A. H., & Perrimon, N. (1993). Targeted gene expression as a means of altering cell fates and generating dominant phenotypes. *Development, 118*(2), 401–415.

Clark, K. J., Balciunas, D., Pogoda, H. M., Ding, Y., Westcot, S. E., Bedell, V. M., … Ekker, S. C. (2011). In vivo protein trapping produces a functional expression codex of the vertebrate proteome. *Nature Methods, 8*(6), 506–515.

Davison, J. M., Akitake, C. M., Goll, M. G., Rhee, J. M., Gosse, N., Baier, H., … Parsons, M. J. (2007). Transactivation from Gal4-VP16 transgenic insertions for tissue-specific cell labeling and ablation in zebrafish. *Developmental Biology, 304*(2), 811–824.

Distel, M., Wullimann, M. F., & Koster, R. W. (2009). Optimized Gal4 genetics for permanent gene expression mapping in zebrafish. *Proceedings of the National Academy of Sciences of the United States of America, 106*(32), 13365–13370.

Fischer, J. A., Giniger, E., Maniatis, T., & Ptashne, M. (1988). GAL4 activates transcription in *Drosophila*. *Nature, 332*(6167), 853–856.

Goll, M. G., Anderson, R., Stainier, D. Y., Spradling, A. C., & Halpern, M. E. (2009). Transcriptional silencing and reactivation in transgenic zebrafish. *Genetics, 182*(3), 747–755.

Itoh, M., Kim, C. H., Palardy, G., Oda, T., Jiang, Y. J., Maust, D., … Chitnis, A. B. (2003). Mind bomb is a ubiquitin ligase that is essential for efficient activation of Notch signaling by Delta. *Developmental Cell, 4*(1), 67–82.

Kawakami, K. (2007). *Tol2*: a versatile gene transfer vector in vertebrates. *Genome Biology, 8*(Suppl. 1), S7.

Kawakami, K., Abe, G., Asada, T., Asakawa, K., Fukuda, R., Ito, A., … Yoshida, M. (2010). zTrap: zebrafish gene trap and enhancer trap database. *BMC Developmental Biology, 10*, 105.

Kawakami, K., Koga, A., Hori, H., & Shima, A. (1998). Excision of the *Tol2* transposable element of the medaka fish, *Oryzias latipes*, in zebrafish, *Danio rerio*. *Gene, 225*(1–2), 17–22.

Kawakami, K., & Shima, A. (1999). Identification of the *Tol2* transposase of the medaka fish *Oryzias latipes* that catalyzes excision of a nonautonomous *Tol2* element in zebrafish *Danio rerio*. *Gene, 240*(1), 239–244.

Kawakami, K., Shima, A., & Kawakami, N. (2000). Identification of a functional transposase of the *Tol2* element, an Ac-like element from the Japanese medaka fish, and its transposition in the zebrafish germ lineage. *Proceedings of the National Academy of Sciences of the United States of America, 97*(21), 11403–11408.

Kawakami, K., Takeda, H., Kawakami, N., Kobayashi, M., Matsuda, N., & Mishina, M. (2004). A transposon-mediated gene trap approach identifies developmentally regulated genes in zebrafish. *Developmental Cell, 7*(1), 133–144.

Keegan, L., Gill, G., & Ptashne, M. (1986). Separation of DNA binding from the transcription-activating function of a eukaryotic regulatory protein. *Science, 231*(4739), 699–704.

Koga, A., Suzuki, M., Inagaki, H., Bessho, Y., & Hori, H. (1996). Transposable element in fish. *Nature, 383*(6595), 30.

Koster, R. W., & Fraser, S. E. (2001). Tracing transgene expression in living zebrafish embryos. *Developmental Biology, 233*(2), 329–346.

Ma, J., & Ptashne, M. (1987). Deletion analysis of GAL4 defines two transcriptional activating segments. *Cell, 48*(5), 847–853.

Marsden, K. C., & Granato, M. (2015). In vivo Ca(2+) imaging reveals that decreased dendritic excitability drives startle habituation. *Cell Reports, 13*(9), 1733–1740.

Mizoguchi, T., Togawa, S., Kawakami, K., & Itoh, M. (2011). Neuron and sensory epithelial cell fate is sequentially determined by Notch signaling in zebrafish lateral line development. *Journal of Neuroscience, 31*(43), 15522–15530.

Muto, A., Ohkura, M., Abe, G., Nakai, J., & Kawakami, K. (2013). Real-time visualization of neuronal activity during perception. *Current Biology, 23*(4), 307–311.

Muto, A., Ohkura, M., Kotani, T., Higashijima, S., Nakai, J., & Kawakami, K. (2011). Genetic visualization with an improved GCaMP calcium indicator reveals spatiotemporal activation of the spinal motor neurons in zebrafish. *Proceedings of the National Academy of Sciences of the United States of America, 108*(13), 5425–5430.

Nagayoshi, S., Hayashi, E., Abe, G., Osato, N., Asakawa, K., Urasaki, A., … Kawakami, K. (2008). Insertional mutagenesis by the *Tol2* transposon-mediated enhancer trap approach generated mutations in two developmental genes: tcf7 and synembryn-like. *Development, 135*(1), 159–169.

Nakai, J., Ohkura, M., & Imoto, K. (2001). A high signal-to-noise Ca(2+) probe composed of a single green fluorescent protein. *Nature Biotechnology, 19*(2), 137–141.

Okigawa, S., Mizoguchi, T., Okano, M., Tanaka, H., Isoda, M., Jiang, Y. J., … Itoh, M. (2014). Different combinations of Notch ligands and receptors regulate V2 interneuron progenitor proliferation and V2a/V2b cell fate determination. *Developmental Biology, 391*(2), 196–206.

O'Malley, D. M., Sankrithi, N. S., Borla, M. A., Parker, S., Banden, S., Gahtan, E., & Detrich, H. W., 3rd (2004). Optical physiology and locomotor behaviors of wild-type and nacre zebrafish. *Methods in Cell Biology, 76*, 261–284.

Parinov, S., Kondrichin, I., Korzh, V., & Emelyanov, A. (2004). *Tol2* transposon-mediated enhancer trap to identify developmentally regulated zebrafish genes in vivo. *Developmental Dynamics, 231*(2), 449–459.

Phelps, C. B., & Brand, A. H. (1998). Ectopic gene expression in *Drosophila* using GAL4 system. *Methods, 14*(4), 367–379.

Pujol-Marti, J., Zecca, A., Baudoin, J. P., Faucherre, A., Asakawa, K., Kawakami, K., & Lopez-Schier, H. (2012). Neuronal birth order identifies a dimorphic sensorineural map. *Journal of Neuroscience, 32*(9), 2976–2987.

Sadowski, I., Ma, J., Triezenberg, S., & Ptashne, M. (1988). GAL4-VP16 is an unusually potent transcriptional activator. *Nature, 335*(6190), 563–564.

Sagasti, A., Guido, M. R., Raible, D. W., & Schier, A. F. (2005). Repulsive interactions shape the morphologies and functional arrangement of zebrafish peripheral sensory arbors. *Current Biology, 15*(9), 804–814.

Scheer, N., & Campos-Ortega, J. A. (1999). Use of the Gal4-UAS technique for targeted gene expression in the zebrafish. *Mechanisms of Development, 80*(2), 153–158.

Scheer, N., Groth, A., Hans, S., & Campos-Ortega, J. A. (2001). An instructive function for Notch in promoting gliogenesis in the zebrafish retina. *Development, 128*(7), 1099–1107.

Schiavo, G., Benfenati, F., Poulain, B., Rossetto, O., Polverino de Laureto, P., DasGupta, B. R., & Montecucco, C. (1992). Tetanus and botulinum-B neurotoxins block neurotransmitter release by proteolytic cleavage of synaptobrevin. *Nature, 359*(6398), 832–835.

Scott, E. K., Mason, L., Arrenberg, A. B., Ziv, L., Gosse, N. J., Xiao, T., … Baier, H. (2007). Targeting neural circuitry in zebrafish using GAL4 enhancer trapping. *Nature Methods, 4*(4), 323–326.

Seipel, K., Georgiev, O., & Schaffner, W. (1992). Different activation domains stimulate transcription from remote ('enhancer') and proximal ('promoter') positions. *EMBO Journal, 11*(13), 4961–4968.

Suster, M. L., Sumiyama, K., & Kawakami, K. (2009). Transposon-mediated BAC transgenesis in zebrafish and mice. *BMC Genomics, 10*, 477.

Sweeney, S. T., Broadie, K., Keane, J., Niemann, H., & O'Kane, C. J. (1995). Targeted expression of tetanus toxin light chain in *Drosophila* specifically eliminates synaptic transmission and causes behavioral defects. *Neuron, 14*(2), 341–351.

Takeuchi, M., Matsuda, K., Yamaguchi, S., Asakawa, K., Miyasaka, N., Lal, P., … Hibi, M. (2015). Establishment of Gal4 transgenic zebrafish lines for analysis of development of cerebellar neural circuitry. *Developmental Biology, 397*(1), 1–17.

Thermes, V., Grabher, C., Ristoratore, F., Bourrat, F., Choulika, A., Wittbrodt, J., & Joly, J. S. (2002). I-SceI meganuclease mediates highly efficient transgenesis in fish. *Mechanisms of Development, 118*(1–2), 91–98.

Trinh le, A., Hochgreb, T., Graham, M., Wu, D., Ruf-Zamojski, F., Jayasena, C. S., … Fraser, S. E. (2011). A versatile gene trap to visualize and interrogate the function of the vertebrate proteome. *Genes and Development, 25*(21), 2306–2320.

Urasaki, A., Morvan, G., & Kawakami, K. (2006). Functional dissection of the *Tol2* transposable element identified the minimal cis-sequence and a highly repetitive sequence in the subterminal region essential for transposition. *Genetics, 174*(2), 639–649.

Wada, H., Dambly-Chaudiere, C., Kawakami, K., & Ghysen, A. (2013). Innervation is required for sense organ development in the lateral line system of adult zebrafish. *Proceedings of the National Academy of Sciences of the United States of America, 110*(14), 5659–5664.

Wada, H., Ghysen, A., Asakawa, K., Abe, G., Ishitani, T., & Kawakami, K. (2013). Wnt/Dkk negative feedback regulates sensory organ size in zebrafish. *Current Biology, 23*(16), 1559–1565.

Wada, H., Iwasaki, M., & Kawakami, K. (2014). Development of the lateral line canal system through a bone remodeling process in zebrafish. *Developmental Biology, 392*(1), 1–14.

Webster, N., Jin, J. R., Green, S., Hollis, M., & Chambon, P. (1988). The yeast UASG is a transcriptional enhancer in human HeLa cells in the presence of the GAL4 trans-activator. *Cell, 52*(2), 169–178.

Wyart, C., Del Bene, F., Warp, E., Scott, E. K., Trauner, D., Baier, H., & Isacoff, E. Y. (2009). Optogenetic dissection of a behavioural module in the vertebrate spinal cord. *Nature, 461*(7262), 407–410.

Xiao, Y., Faucherre, A., Pola-Morell, L., Heddleston, J. M., Liu, T. L., Chew, T. L., … Lopez-Schier, H. (2015). High-resolution live imaging reveals axon-glia interactions during peripheral nerve injury and repair in zebrafish. *Disease Models and Mechanisms, 8*(6), 553–564.

Yamamoto, M., Morita, R., Mizoguchi, T., Matsuo, H., Isoda, M., Ishitani, T., … Itoh, M. (2010). Mib-Jag1-Notch signalling regulates patterning and structural roles of the notochord by controlling cell-fate decisions. *Development, 137*(15), 2527–2537.

Yamanaka, I., Miki, M., Asakawa, K., Kawakami, K., Oda, Y., & Hirata, H. (2013). Glycinergic transmission and postsynaptic activation of CaMKII are required for glycine receptor clustering in vivo. *Genes to Cells: Devoted to Molecular and Cellular Mechanisms, 18*(3), 211–224.

CHAPTER FOUR

# Whole-Organism Cellular Pathology: A Systems Approach to Phenomics

**K.C. Cheng[1], S.R. Katz, A.Y. Lin, X. Xin and Y. Ding**
The Pennsylvania State University College of Medicine, Hershey, PA, United States
[1]Corresponding author: E-mail: kcheng76@gmail.com

## Contents



## Abstract

Phenotype is defined as the state of an organism resulting from interactions between genes, environment, disease, molecular mechanisms, and chance. The purpose of the emerging field of phenomics is to systematically determine and measure phenotypes across biology for the sake of understanding. Phenotypes can affect more than one cell type and life stage, so ideal phenotyping would include the state of every cell type within the context of both tissue architecture and the whole organism at each life

*Advances in Genetics*, Volume 95
ISSN 0065-2660
http://dx.doi.org/10.1016/bs.adgen.2016.05.003

stage. In medicine, high-resolution anatomic assessment of phenotype is obtained from histology. Histology's interpretative power, codified by Virchow as *cellular pathology*, is derived from its ability to discern diagnostic and characteristic cellular changes in *diseased* tissues. Cellular pathology is observed in every major human disease and relies on the ability of histology to detect cellular change in any cell type due to unbiased pan-cellular staining, even in optically opaque tissues. Our laboratory has shown that histology is far more sensitive than stereomicroscopy for detecting phenotypes in zebrafish mutants. Those studies have also shown that more complete sampling, greater consistency in sample orientation, and the inclusion of phenotypes extending over longer length scales would provide greater coverage of common phenotypes. We are developing technical approaches to achieve an ideal detection of cellular pathology using an improved form of X-ray microtomography that retains the strengths and addresses the weaknesses of histology as a screening tool. We are using zebrafish as a vertebrate model based on the overlaps between zebrafish and mammalian tissue architecture, and a body size small enough to allow whole-organism, volumetric imaging at cellular resolution. Automation of whole-organism phenotyping would greatly increase the value of phenomics. Potential societal benefits would include reduction in the cost of drug development, a reduction in the incidence of unexpected severe drug and environmental toxicity, and more rapid elucidation of the contributions of genes and the environment to phenotypes, including the validation of candidate disease alleles identified in population and personal genetics.

## 1. INTRODUCTION

In this chapter, we have taken a systems approach to thinking about phenomics, focusing on what is ideally needed for an *anatomical phenotyping* that involves the evaluation of tissue architecture. After defining relevant terms, we begin with a consideration of what we have learned from the study of normal and diseased human and zebrafish tissues using histology, and consider potential approaches to an ideal higher-throughput, whole-organism, cellular phenotyping. We consider the history of phenotyping and phenomics, and review how light microscopy caused a revolution in biology and medicine. We then discuss limitations of histology learned from our experience with anatomic pathology and the histological phenotyping of zebrafish mutants. We conclude that whole-animal cellular pathology with computational phenotyping of all cell types and developmental stages will enable a phenomics that provides the greatest benefit to society.

### 1.1 Phenotype

Phenotypes are observable traits of organisms that represent the consequences of unique interactions between genetic background and environmental factors. Phenotyping involves recording the observed features of an organism and is of central importance in biology and medicine. High-throughput phenotyping, a primary focus in the emerging field of phenomics, has the potential to anchor the study of tens of thousands of genetic and chemical variables at a time and to connect genetic and environmental influences that share molecular pathways. However, advances in technology are necessary to facilitate large-scale phenotyping, both in terms of the number of observed individuals (so-called *extensive phenotyping*) and the depth of detail obtained per individual (*intensive phenotyping*) (Houle, Govindaraju, & Omholt, 2010; Mahner & Kary, 1997).

## 1.2 Phenomics and Phenome Projects

*Phenomics* is the large-scale study of phenotypes (reviewed by Cheng, Xin, Clark, & La Riviere, 2011; Hancock, 2014; Houle et al., 2010; Lussier & Liu, 2007). Since a phenotype is the product of interactions between genes and environment, and those variables are typically studied one class at a time, phenomics can be separately considered in the contexts of *genetic phenomics* and *environmental phenomics*. The study of phenotypes caused by genetic manipulations or polymorphisms falls within the domain of genetic phenomics, while environmental influences on phenotype, including those associated with drug candidates and potential environmental toxins, can be studied within the context of environmental phenomics. Within environmental phenomics, the study of phenotypes caused by chemicals and potential toxins can be called *chemical phenomics* and *toxicological phenomics*, respectively.

Phenomics is commonly discussed in the context of gene function at a genome-wide level (Mahner & Kary, 1997), and has been applied to the study of metabolism (Schilling, Edwards, & Palsson, 1999), plant biology (Yang, Duan, Chen, Xiong, & Liu, 2013), evolutionary biology (Houle, 2010), and human health and disease (Lussier & Liu, 2007), including cancer (Zbuk & Eng, 2007), heart development (Li et al., 2015) and neurological disorders (Gerlai, 2002; Bilder et al., 2009). *Clinical phenomics* (Lussier & Liu, 2007) in the form of medical informatics is playing an increasingly important role in evidence-based medicine and may become a focal point for *comparative phenomics* across model systems. If we take the hint from sets of phenotypes serving as the starting point for *differential diagnosis* in medicine, with some phenotypes being "pathognomonic" (unique to a given disease), it follows that molecular pathways may be associated with sets of phenotypes in one or more model systems that comprise *phenotypic signatures* (discussed in Canada, 2010). Phenotypic signatures have been applied to screens for drugs that may be useful in neuropsychiatry. The behavioral screens used take advantage of the ability to test multiple zebrafish larvae per well in multiwell plates. Their behavioral phenotype-driven chemical screens are based on computer analysis of movement patterns induced by stimuli such as flashing lights and sound, and have yielded new biologically active compounds that fit into existing drug classes with shared phenotypic signatures (Rennekamp & Peterson, 2015).

Model system phenome projects are becoming critical tools for understanding disease, discovering new drugs, assessing potential environmental toxicants, and improving genotype-phenotype correlations. Vertebrate model organisms for which large-scale, collaborative genetic phenomics has been pursued include the mouse (Austin et al., 2004; Brown & Moore, 2012; Grubb, Maddatu, Bult, & Bogue, 2009; Mouse Phenome Database, http://phenome.jax.org; Knockout Mouse Project, http://www.mousephenotype.org), zebrafish (Zebrafish Phenome Project, and Zebrafish Mutation Project, http://www.sanger.ac.uk/resources/zebrafish/zmp; Cheng et al., 2011), and dog (Canine Phenome Project, http://www.caninephenome.org). Since some orthologous genes may show species-specific effects, comparative phenomics (Kahraman et al., 2005) will play an important role in making model systems phenomics relevant to human biology and health. Since human disease is often defined by its histological, physiological, and neurological correlates, the detection of similar phenotypes in model systems helps to define their human relevance.

## 1.3 Phenotype Omission Error

The omission of phenotypes is a characteristic of present-day phenomics, and causes misleading uncertainties in both the presence and absence of a reported phenotype. Studies involving positional cloning, and now in personalized medicine, commonly remind us that our knowledge of the connections between genes and phenotypes is far from complete. A classic example of this problem was encountered during the positional cloning of the zebrafish *golden* pigmentation gene. Genetic mapping reduced the number of candidate genes to five within one genomic region, in the same gene order in both zebrafish and humans. At the time, however, none of these genes had any known connection with pigmentation. Only by testing each candidate by morpholino knockdown (Nasevicius & Ekker, 2000) was it possible to associate a pigmentation phenotype with one of those genes; this association was subsequently supported by numerous genetic, genomic, and molecular tests in zebrafish and humans, and by correlation between pigmentation and genotype in African Americans (Lamason et al., 2005). Some of the deficiencies in existing phenotypic databases are a consequence of variations in phenotypic analysis between different laboratories, but most are likely due to the incomplete nature of phenome projects and current phenotyping methodologies.

The assignment of a specific phenotype to a single gene ideally means that all phenotypes have been tested, and that only one phenotype was found. However, a single phenotype more commonly means that only one phenotype was investigated, that the assay used did not have the sensitivity to detect changes in other organs, or alternatively, that the study was focused on one organ system due to the demands and constraints of publication and research funding. Likewise, unless otherwise specified, absence of any phenotype in a given organ or cell type may be the result of any of a number of possibilities: that no other phenotype has been detected during a detailed study, no test was performed, that the assay had insufficient sensitivity to detect other phenotypes that are present, or that other phenotypes were ignored. One frequent incomplete phenotype that was common in early mouse phenomics is "embryonic lethal," which means that the embryos were not examined in detail. Such issues may be collectively referred to as *phenotype omission error*. When trying to decide whether or not a drug is safe, it is of course crucial that a negative result is truly negative. Lack of a stated phenotype may otherwise create a false sense of confidence that the chemical in question is benign. The morbidity associated with the use of thalidomide is a striking example of a tragedy that potentially could have been avoided by the use of more complete phenotypic assays during drug testing (Gordon, Spielberg, Blake, & Balasubramanian, 1981). Thus, it is in everyone's interest to progressively improve our methods to eliminate phenotype omission error by improving our technologies and assays until they are capable of providing complete sets of meaningful phenotypic measurements.

One approach to addressing phenotype omission error originates from the practice of human autopsy pathology (as observed during residency training by KCC), in which every organ in the body is studied grossly and sampled for anatomic and cellular disease phenotypes. The ideal complete autopsy effectively represents whole-organism phenotyping to the extent that every organ is examined. In reality, however, only a small fraction of each organ is examined histologically during autopsy. We are unaware of any multicellular organism for which every cell type, in the context of the entire organism, has been characterized in histological detail in a large-scale phenotypic screen. We are working to address this important scientific deficit.

## 1.4 Stereomicroscopy

Stereomicroscopy, which utilizes "dissecting microscopes" with typical resolutions in the range of 5 μm, has been proven to be a practical and rapid tool for identifying developmental mutants in model organisms, and is commonly used to study the toxicological effects of chemicals because it provides a practical balance of large working distance and resolution. Surface features and organismal and organ shape may be readily scorable. However, functionally important details of internal tissue architecture are not readily discernable in unstained internal tissues, particularly if the organism is only translucent (as opposed to transparent), or is optically opaque for reasons such as tissue differentiation and pigmentation. Furthermore, higher-resolution studies of differentiated tissues are required to provide insight into cellular change.

## 1.5 Molecular Imaging Versus Pan-Cellular Phenotyping

Gene- and protein-specific stains and fluorophores are popular, powerful tools for studying patterns of expression and mutant phenotypes, but are subject to the problem of phenotype omission error because they do not provide histology-level data about potential phenotypes in the nonfluorescent cells, even for clearing-based methodologies (Chung & Diesseroth, 2013). *Pan-cellular phenotyping*, or the unbiased study of all cell types in an affected tissue, is required to minimize this type of omission error. For example, epithelial cancers (carcinomas) are so named for their effects on epithelia, but are typically associated with clinically relevant changes to stromal tissues within which the tumor grows (Fig. 1). In fact, the hardness of a palpated mass in breast cancer is due to the collagenous stromal response of the body to the tumor rather than the tumor cells themselves (Fig. 1, Cotran, Kumar, Collins, & Robbins, 1999). Studies of epithelial cancers (carcinomas) that focus exclusively on cancer cells and preneoplastic epithelial tissues are likely to miss evidence that may indicate that the stroma contributes to the cancer phenotype (eg, Bhowmick, Neilson, & Moses, 2004; Kalluri & Zeisberg, 2006).

## 1.6 Why Histological Phenotyping?

In medicine, the morbidities and mortality of every major disease, from cardiovascular disease and diabetes to cancer, are closely associated with micron to millimeter-scale cellular and tissue architectural changes that are both typical and informative of pathophysiological mechanisms; most of these microanatomical changes are evident by histology (Cotran et al., 1999). This knowledge motivates our present work to expand phenotyping to

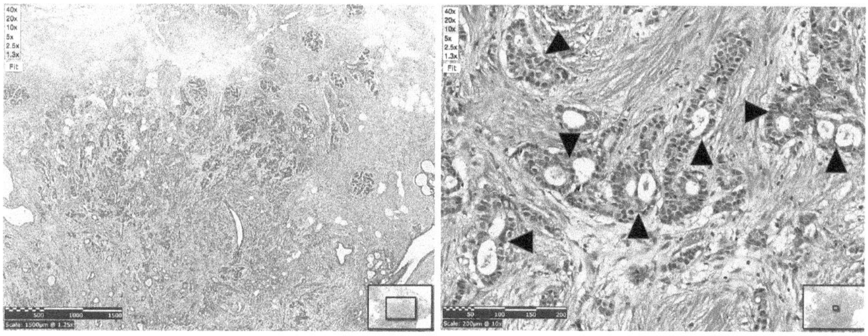

**Figure 1** ***Infiltrating ductal carcinoma illustrating what is visible at different imaging resolutions.*** The nuclei of nests of invading tumor cells are evident at 1.3× magnification, at which pixels are ~7.7 × 7.7 μm, which makes cell types are difficult to discern. What is evident at this power, however, is that the cancer cells, generally represented by the purple cells, are surrounded by dense collections of collagen (stained pink) that cause the palpable hardness of tumors in the clinic and represents a reaction to the tumor. Virtually all cell types are distinguishable at 10× magnification, at which pixels are about 1 × 1 μm. *Arrowheads*: Invasive nests of cancer cells with multiple, "back-to-back" glandular spaces, typical of ductal adenocarcinoma. *Both panels are from the Penn State Histopathology Collection at bio-atlas.psu.edu, at http://bio-atlas.psu.edu/view.php?s=1817&atlas=107.* (See color plate)

include microanatomy for studying the effects of disease, genetic, and/or environmental variables on organismal phenotype. The fascinating history of microscopy and its role in science and medicine makes clear the motivation underlying our current focus on the challenging problem of microanatomical phenotyping.

# 2. LESSONS FROM LIGHT MICROSCOPY

## 2.1 Early Tissue Phenotyping

Tissue phenotyping was first made possible by the light microscopic study of stained tissue sections, which enabled the startling and unprecedented ability to assess both tissue architecture and the physiological state of component cells at the same time. A striking and historic confluence of multiple scientific advances made the development of the compound microscope possible. Those advances included the discoveries of tissue clearing, hardening, and embedding to allow nearly transparent sections to be cut using microtomes (Hill, 1770; reviewed by Bracegirdle, 1987), the invention of the microtome for slicing tissue thin enough to distinguish individual cells in tissue sections (Bracegirdle, 1987; Davis, 1979), the production of glass of sufficient uniformity to produce lenses with cellular (submicron) resolution (reviewed by Vogel, 2012), development of the mathematics of optics (Abbe, 1873, 1881, 1883), and advances in dye chemistry leading to differential stains for cellular components (Beer, 1959; Travis, Hornix, Bud, & Homburg, 1992). Combined with a flourishing academic scientific environment at major universities, including the first medical schools based on science, these advances helped to create a market for microscope manufacture that could finance further improvements. From the microscopic study of tissue sections, we were able to learn, for the first time, that all living things are made of cells (Du Trochet, 1824; Schleiden, 1838; Schwann, 1839), that all cells come from other cells (*Omnis cellula e cellula*; reviewed by Wright & Poulsom, 2012), and that the histological study of diseased tissues can provide insight into the cellular mechanisms of disease (Virchow, 1855); the latter work was published in book form by Virchow under the title, *Cellular Pathology* (1858). The concepts and activities that yielded *Cellular Pathology* represent the founding of the field of anatomic pathology, in which cellular disease phenotypes anchor our study of molecular mechanisms. The completeness of linkages between cellular pathology and molecular mechanisms (Virchow, 1855; Vallee, Warren, & Wacker, 1963) would be greatly improved by the development of a more quantitative cellular pathology.

## 2.2 Histological Phenotyping in Medicine

Light microscopic diagnosis by specialists in anatomic pathology remains today the clinical standard for making tissue diagnoses of diseases including cancer (Cotran et al., 1999). The purview of cellular pathology includes both the gross and microscopic changes associated with disease, which are *physical phenotypes*. A particularly important property that has made histology so powerful is the ability to survey all cell types at the same time through pan-cellular staining. The basis for pan-cellular recognition is the use of a combination of stains for different cellular components: one for nucleic acids (hematoxylin, which is purple) and one for protein (eosin, which is pink; Poole, 1875, Figs. 1 and 2). Each cell type has a characteristic pattern of both nucleic acid and macromolecular protein and organellar structure that allows the recognition of different cell types as well as a readout of the physiological state of the cell. Digitization of glass slides is now common practice (Pantanowitz et al., 2011), but appears to be poorly suited for high-throughput phenotyping (see below). A long-term goal of our work is to find a practical way to apply histopathological principles to large-scale phenotypic screens in drug development, toxicology, and genetics.

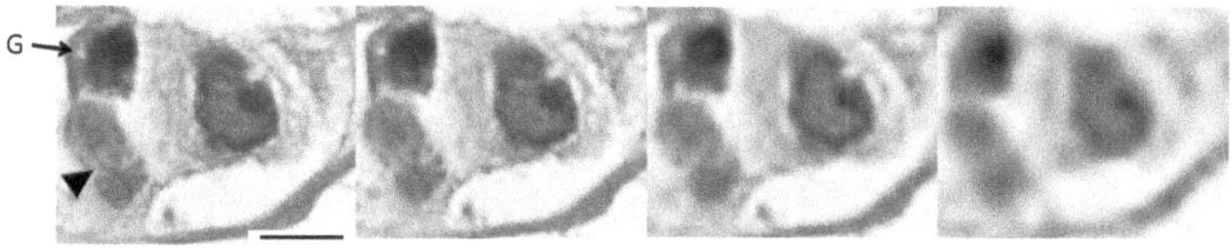


| Objective | 40x | 20x | 10x | 5x |
|---|---|---|---|---|
| Pixel (μ x μ) | 0.25 | 0.5 | 1 | 2 |
| Chromosomes visible | Yes | Barely | No | No |
| Golgi visible | Yes | Yes | Barely | No |

**Figure 2** ***Cellular features, including chromosomes and Golgi body at different imaging resolutions.*** Cells from the same infiltrating ductal carcinoma as in Fig. 1, pixels are about 1 × 1 μm. The cell at the *arrowhead* contains two clusters of chromosomes, whose threadlike appearance is visible at 40× (see http://bio-atlas.psu.edu/view.php?s=1817&atlas=107&z=0&c=59056,34025), particularly when focusing through optical sections on the microscope, and are barely discernable by individuals experienced in pathology. Individual chromosomes are not apparent when pixels reach about 1 μm, and very few cytological features are visible when pixels reach about 2 μm in dimension (about the size applicable to stereomicroscopy). *G*, Golgi apparatus, is the pale area next to the nucleus. Size bar, 20 μm. (See color plate)

## 2.3 Histological Phenotyping in Zebrafish

The limitations of stereomicroscopy in terms of resolution and the detection of internal tissue architectural phenotypes predict that stereomicroscopy tends to underestimate the number of phenotypes in a sample compared with methods such as histology that have a higher spatial resolution. To enable the use of histology for genetic screens, we first had to develop tools for the histological study of zebrafish.

### *2.3.1 Method and Tool Development for Zebrafish Histology*

In the early 1990s, a number of laboratories, beginning with the founder of the modern age of zebrafish genetics, George Streisinger (1984), began zebrafish work that was guided by an interest in cancer. Our first work was based on the idea that genetic screens could be pursued that are inspired by biological processes involved in human cancer. One of the projects involved a genetic screen for mutants showing histological features of cancer. Basic protocols for fixation, decalcification, cutting, and staining of fish were based on established practice in both human and mouse pathology performed in our department. Our earliest work showed that there is a tendency for fixed embryos to become bent or curled, and for collections of embryos to be badly misaligned. Since a screen required uniform histological sections of families of zebrafish larvae, we needed to develop a way to cut consistent sections of arrays of fish. We decided to try using agarose as a medium to hold families of fish in a uniform position within agarose wells—essentially a three-dimensional (3D) version of gel combs used for gel electrophoresis. Penn State was well known for its development of artificial hearts, so we collaborated with a machinist who was familiar with machining parts for those hearts. We created molds to create uniform wells for aligning arrays of fixed fish in agarose blocks (Moore, Aros, Steudel, & Cheng, 2002; Moore, Rush, Breneman, & Cheng, 2006; Sabaliauskas et al., 2006; Tsao-Wu, Weber, Budgeon, & Cheng, 1998). This work was performed in parallel with the development of methods for creating orthogonal sections of adults with neoplasms (Beckwith, Moore, Tsao-Wu, Harshbarger, & Cheng, 2000; Moore et al., 2006).

### 2.3.2 Lessons From Creating a Web-Based Histology Atlas

Since an understanding of normal histology is required to interpret histopathology, there was also a need for an online atlas of microanatomy for the zebrafish community. Our laboratory led this task (Cheng, 2004; see http://www.bio-atlas.psu.edu, referred to henceforth as "zebrafish atlas") based on our fortunate proximity to a histotechnologist (Lynn Budgeon) with great talent and a vast experience in processing and sectioning different animal tissues who works in the laboratory of a generous papilloma virologist and immunologist (Neil Christensen). Creation of this atlas required the generation of zebrafish at different ages, the setting of standards of fixation, decalcification (of adults), and the generation of slices oriented in the three standard orientations for histology (coronal, sagittal, and transverse). To increase the throughput and consistency of orientation of histological sections of zebrafish, we used the plastic molds cited above. The mold was useful, but creating sections of well-oriented and straight fish was nontrivial. As a result, many more fish and sections were cut than could be used for the atlas. The atlas also includes scans of slides of human, mouse, and other animals. As described below, we used scans of a human cancer to show how resolution determines our ability to visualize structures.

Resolution (explained at https://micro.magnet.fsu.edu/primer/anatomy/numaperture.html) determines our ability to visualize structures and patterns associated with phenotypes that are relevant to cellular pathology. A brief review of digital optics related to the zebrafish atlas will allow us to take advantage of the digital features of our zebrafish atlas to illustrate how the physical distance represented by a pixel determines resolution. Imaging chips are comprised of arrays of light-sensitive squares called pixels. Images that correspond to physical space are projected onto these chips. The physical dimension *corresponding* to each pixel is calculated from the pixel number across the chip and the physical dimension of the subject whose image is optically projected onto those pixels. For example, if a 1 mm (1000 μm) wide area of a slide were to be projected across 1000 pixels, each pixel would represent 1 μm. This spatial representation per pixel for digital scans is commonly set by the scanning company to take maximum advantage of lens resolution. The Aperio slide scanner we used achieves 40× magnification using a lens with a numerical aperture of 0.75 (which corresponds to an optical resolution of 0.37 μm at a wavelength of 550 nm). Available magnifications represented at www.bio-atlas.psu.edu correspond to each of the commonly used objective powers used in the practice of anatomic pathology. The highest, 40×, magnification utilizes the full pixel resolution. The progressively lower resolutions presented on our atlas were not created from scans using lower power lenses, but rather from a series of 2-fold reductions in resolution. Thus, simulations of 20×, 10×, and 5× objective images on the atlas are derived from a series of images in which the 40× scans was reduced in resolution and pixel dimensions by 2, 4, and 8-fold, respectively. In sum, the resulting pixel dimensions corresponding to slide scans at 5×, 10×, 20×, and 40× magnifications are about 2 × 2 μm, 1 × 1 μm, 0.5 × 0.5 μm, and 0.25 × 0.25 μm, respectively; these latter numbers can also be referred to as *pixel resolution*.

A rule of thumb in terms of resolution in the digital world is that a structure can be resolved in digital images when the pixel dimensions are no larger than half of the structure's dimensions (which we refer to below as the "two pixels per structure rule"). We can use a series of images at progressively lower resolutions to illustrate this point. To illustrate the effect of pixel resolution on the recognition of cellular structures, we show in Fig. 2 screen captures of an area from a human breast cancer scanned for the atlas that contains a cancer cell of ~40 μm maximum dimension and two of its neighboring cells containing small structures, a Golgi body (~2 μm in minimum dimension) and chromosomes (~0.5 μm in minimum dimension). Note increasing blurriness at progressively lower pixel resolutions corresponding to progressively lower simulated objective magnifications (Fig. 2). In accord with the two pixels per structure rule cited above, the Golgi apparatus is discernable at a pixel resolution of 1 μm (10× objective) or smaller; similarly, chromosomes begin to be discernable when pixels are 0.5 μm (20× objective) or smaller.

To show more broadly how the sizes of specific anatomic structures determine the pixel sizes required to visualize them for common structures, we tabulated the sizes of structures as measured at bio-atlas.psu.edu using a ruler tool (different forms of which are activated by shift-R) in various cell types of the zebrafish and whether or not they are visible at each of several powers represented at www.bio-atlas.psu.edu (Table 1). At least 20 measurements were made of each structure from different areas on the same slide to provide the sizes listed, and at least 20 structures were examined in the

**Table 1** Structures are distinguishable when the pixel size is half of the structure's dimension

| | Microscope Objective | 40 × | 20 × | 10 × | 5 × |
|---|---|---|---|---|---|
| | Scanned pixel dimensions (μm) | 0.25 | 0.5 | 1 | 2 |
| Structure | Skeletal muscle striations (dark 1–2 μm; light 1.5–4 μm) | Yes | Barely | Light and some dark | No |
| | Golgi (~2 × 3 μm) | Yes | Yes | Yes | No |
| | RBC (~2 × 3 μm) | Yes | Yes | Yes | No |
| | Zymogen granules (0.5 –2.2 μm diameter) | Yes | Most | Some | No |
| | Chromosomes (~0.5 μm dimeter) | Some | Some | Rare | No |
| | Space between RBCs (~1 μm) | Yes | Yes | No | No |
| | Nuclear Fragments of 0.5 μm diameter | Yes | Most | Some | No |

way illustrated in Fig. 2, at each power, to confirm the tabulated result. As predicted, each structure became visible when the pixel dimension is about half the size of the structure. Experience and random samplings of the same structures on our zebrafish slides were concordant with this conclusion. Our general conclusion from this exercise is that a pixel resolution of about 1 μm is required to distinguish different cell types.

We also used a model cell of similar dimensions to the cancer cell in Fig. 2 to illustrate how histological slices can misrepresent microanatomy (Fig. 3). Adjacent histological slices (histology slices are actually slabs of about 5 μm thickness) in the plane perpendicular to that shown in Fig. 3 are represented by rectangles labeled 1, 2, 3, and 4. These sections seriously misrepresent the cell, particularly when the diameter of the cell is greater than the slice thickness. Slice 1 would include a nucleus of the correct diameter, though in only one plane. Slice 2 contains only an edge of the nucleus, making it look smaller and less darkly stained than a full-thickness slice of nucleus. Slices 3 and 4 are misleading with respect to the anatomy of the cell because slice 3 shows no nucleus, and slice 4 shows only a fraction of the cytoplasm of the cell, again with no nucleus.

Sectioning artifacts, represented by large linear spaces between tissue planes, are another misrepresentation of anatomy associated with histology. These

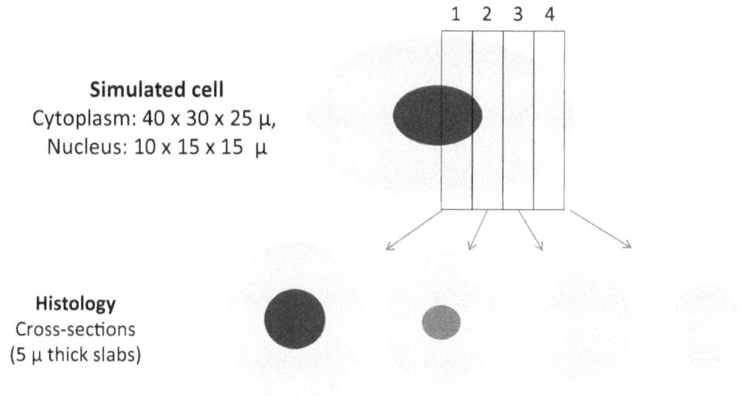


| Histology cross-section | 1 | 2 | 3 | 4 |
|---|---|---|---|---|
| Nuclear diameter (μ) | 15 | 5 | None | None |
| Nuclear stain | Dark | Light | None | None |
| Cytoplasm area (μ) | 30 x 25 | 25 x 20 | 20 x 25 | 10 x 8 |

**Figure 3** ***Histology misrepresents the microanatomy of large cells.*** The simulated large cell is modeled after that shown in Fig. 2. 5-μm thick sections typical of histology yield strikingly different cell representations, including dimensions of cytoplasm and nucleus, and DNA staining, depending upon the plane of section. (See color plate)

artifacts are generally worse with edematous and diseased tissues, and result from the need to float tissue sections on water prior to mounting them on glass slides. During the floatation step, different tissue types absorb water at different rates. Weak connections between different parts of the tissue, as encountered in damaged or edematous tissue, cause parts of the section to float away from each other, resulting in spaces between structures that are larger than in the original tissue. This artifact can be minimized by study of intact, rather than sectioned samples, as is done in tomographic imaging. Other artifacts associated with histology include the floating of tissue fragments over each other that are not really associated with each other and the loss of tissue fragments that float away from the specimen prior to mounting.

For several years we attempted to generate 3D reconstructions from serial sections. We utilized 1 μm thick plastic sections rather than the traditional 5 μm paraffin sections due to their lower sample distortion and greater z-axis resolution. Unfortunately, due to a combination of sample distortion associated with floating of slices on water, and slice thicknesses that differed from optical 2D resolution, we were unable to align slices with a precision adequate for the creation of useful 3D volumes. These experiences with zebrafish histology left us with a strong desire to find a way to create 3D images of whole fish that we could digitally slice and otherwise manipulate and study from different orientations. While digital cutting of sections in any plane would be ideal, no existing 3D imaging methods (surveyed by Cheng et al., 2011) were noted to have the combination of pan-cellular staining and cellular resolution that is required for assessment of cellular pathology.

Overall, what we have learned from generating a zebrafish atlas of histology is that sectioning of zebrafish for histology requires special tools for orienting them well enough for reproducible characterization, that a digital way to create 3D representations of fish of cellular resolution would be potentially far more flexible and powerful than histology due to the ability to create virtual slices and views computationally, and that pixel resolutions of 1 μm or less (corresponding to magnifications of 10× or higher at www.bio-atlas.psu.edu) are required to identify specific cell types and to detect and distinguish pathological processes.

### 2.3.3 Lessons From Phenotyping Zebrafish Using Histology

#### 2.3.3.1 From a Histological Mutagenesis Screen

Since histological features are so informative about disease state, we pursued a histological screen for zebrafish mutants. Based on the concept that mutations specific to tissue differentiation may only be apparent beyond early embryonic stages, ie, during larval development, we pursued a mutagenesis screen based on histological examination of about 12 interspersed sections of families of zebrafish larvae (intervening tissue sections were saved but unstained, and every third or fourth slice was stained for histological study; the staining and examination of sections at such intervals is sometimes called "levels" in the practice of pathology). At least that many sections were studied for families of 7-day-old larval zebrafish (Mohideen et al., 2003). The screen yielded six mutants; more may have been found with a faster and more consistent method of embedding, processing, and cutting groups of larvae.

#### 2.3.3.2 The Potential Utility of Volumetric Data

Volumetric data provides useful insights that histology cannot provide by itself. To illustrate this idea, consider the example of the "small eye" phenotype of one of the mutants, *hht*, identified by our histology screen (Mohideen et al., 2003). The small eye phenotype is evident from day 2 until the time of death between day 5 and 7, but the cellular mechanisms underlying this phenotype are not apparent by stereomicroscopy (Fig. 4). In contrast, histology reveals the underlying cellular basis for the small eyes (Fig. 5): At day 3, the retinal cells show extensive cell death, evident by the presence of degenerating nuclei (karyorrhexis), confirmed by both TUNEL and acridine orange staining (Lin & Cheng, unpublished; Mohideen et al., 2003). By day 5, the process of cell death is less active, and the eyes are more notable for disorganized neuropil and a paucity of

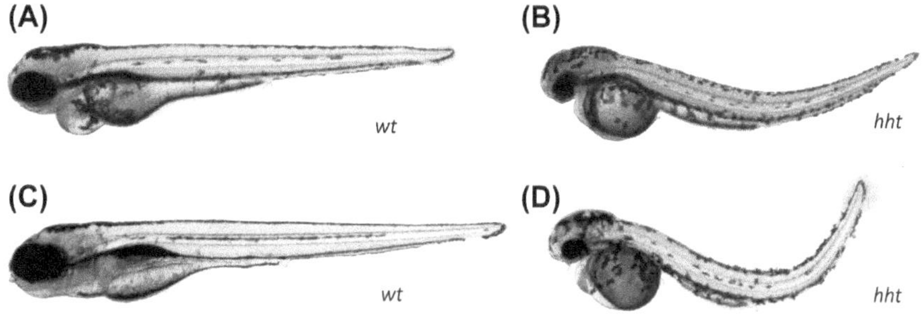


**Figure 4** ***Small-eye phenotype in 3 dpf and 5 dpf hht by stereomicroscope.*** 3 dpf (B) and 5 dpf (D) *hht* mutants both have "small eyes" compared to their 3 dpf (A) and 5 dpf (C) wild-type siblings. These views say nothing about the cellular mechanism underlying the small eye phenotype. (See color plate)

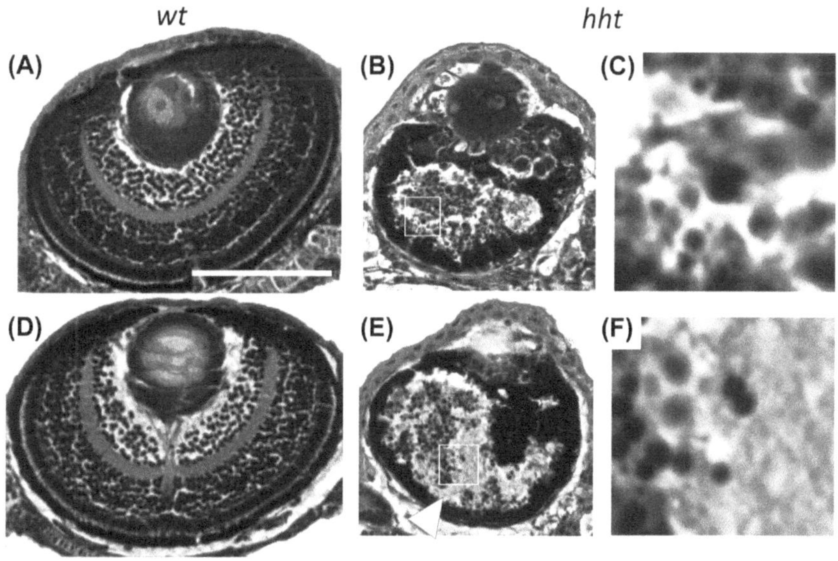


**Figure 5** ***Histology distinguishes between dying and degenerated cells as causative of "small eye" phenotype.*** Histologically, 3 dpf *hht* eyes (B) contain fewer frequently apoptotic cells (*smaller dark dots*, B, C) compared with wild-type (A). By day 5, the retinal layering of wild-type eyes (D) is still clear, but the *hht* retina is replaced by disorganized neuropil (*arrowhead*) and nuclear debris (E, F). Size bar, 100 μm. *Boxed areas*, ~25 × 25 μm.

mature retinal cells. Quantitative insight into the extent of cell death at different life stages would be gained with a 3D imaging method that would allow us to count the number of mature nuclei or nuclear fragments per eye. We note that the small eyes of *hht* are characterized by a high density of melanin pigmentation in melanophores and a retinal pigmented epithelium that has collapsed into the outer parts of the eye, greatly limiting the use of fluorescence for gaining information about the inside of the eye.

#### 2.3.3.3 From a Histological Screen of Existing Mutants Found by Stereomicroscopy

In a simultaneous study, we tested the notion that histological detail can detect additional phenotypes in mutants already identified and characterized by stereomicroscopy by performing a histological screen of 97 mutants from the Hopkins retroviral insertional mutagenesis screen at MIT (Amsterdam et al., 2004). Mutants were chosen on the basis of survival until at least 5 days of development to allow for temporal phenotypic scoring and analysis. Families of mutants were studied histologically, and semiquantitatively scored over four days by one observer (Thomas, 2009).

The comparative efficiency of detecting phenotypes in different organs by stereomicroscopy versus histology is summarized in Table 2. The mutant phenotypes of some organ systems, including kidney, pancreas, integument, fin, and brain, were almost undetectable by stereomicroscopy, but readily detected by histology. Large numbers of phenotypes were completely missed by stereomicroscopy, including 81, 48, and 66 phenotypes for kidney, integument, and pancreas, respectively. The relative insensitivity of stereomicroscopy as compared with histology was not as severe in other organs. The most commonly affected organ was the eye, where 93/97, or nearly 97% of mutants were affected, for which 21/93, or nearly a quarter (23%) of phenotypes detected using histology were not detected by stereomicroscopy. It was striking that 86% of the mutations were pleiotropic, with only 11/97, or 11%, of mutants showing a phenotype in only one organ system. This high frequency of pleiotropy suggests that comprehensive phenotyping will be required to accurately determine the range of functional effects of any given experimental variable.

Histology allowed us to achieve a greater granularity of phenotyping than was possible by stereomicroscopy, since we were able to catalog ontologically defined structures as small as cells that cannot be visualized by stereomicroscopy, along with type and degree of change (see Thomas, 2009 for details). Obtaining these data took one student two full years, long enough to make it necessary to find ways to speed up the process. One possibility is to develop imaging and image analysis tools to accomplish some of this phenotyping computationally (Canada, 2010; Canada, Cheng, & Wang, 2006; Canada, Thomas, Cheng, & Wang, 2007, 2011; Canada, Thomas, Cheng, Wang, & Liu, 2008a,b). A summary of some of the issues faced during histological phenotyping with suggested solutions is provided in Table 3.

In summary, histological studies of zebrafish showed that it is challenging but possible to screen for mutants using histology, that histology provides greater sensitivity for detection of phenotypes than stereomicroscopy, and that weaknesses to address include histology's poor control over sample

**Table 2** Phenotyping using histology is ~3-fold more sensitive than stereomicroscopy.

| | Mutants with phenotypes of organ system[a] | | | | | | | | | |
|---|---|---|---|---|---|---|---|---|---|---|
| | **Kidney** | **Integument** | **Pancreas** | **Fin** | **Brain** | **Pharynx** | **Eye** | **Liver** | **Gut** | **Total** |
| Stereomicroscopy (~5 μ pixels) | 0 | 0 | 1 | 4 | 8 | 21 | 72 | 52 | 58 | 216 |
| Histology (0.25–0.5 μ pixels) | 81 | 48 | 67 | 81 | 87 | 77 | 93 | 79 | 91 | 704 |
| Percent of histological phenotypes detected by stereomicroscopy | 0 | 0 | ~1 | 5 | 59 | 27 | 77 | 66 | 64 | 31 |

[a]Among 97 mutants at 5 dpf.

**Table 3** Observations from histological studies of zebrafish mutants and suggested next steps for microanatomical phenotyping

| Observation or Problem | Solution for microanatomical phenomics |
|---|---|
| Histology is >threefold more sensitive for detecting phenotypes than those stereomicroscopy | Retain cellular resolution for phenotyping, using pixel dimensions of 1 μm or less |
| Most developmental mutations are pleiotropic, affecting multiple cell types and organs | Develop whole-body, pan-cellular phenotyping |
| Patterns of tissue pleiotropy detected by histology are characteristic of sets of functionally related genes and classes of drugs | Utilize tissue phenotyping with the sensitivity of histology |
| Quantitative analysis adds power to phenotypic analysis | Develop and utilize quantitative analysis of phenotypes that are presently descriptive |
| Time course adds a useful dimension to phenotype | Score mutants and treated animals at more than one time point |
| Histology is impractical for large-scale microanatomic phenotyping | Develop a more practical form of microanatomic phenotyping |
| Some structures such as nuclei and large cells are frequently cut, leaving an unknown proportion of the structure in the section | Develop a 3D form of histology that has cellular resolutions so that the full volume of structures can be determined |
| Many microanatomical phenotypes are volumetric, and therefore only indirectly measurable by histology | Measure phenotypes from 3D image files with histological resolution |
| The study of tissue scans at various microscope objective resolutions, such as those at bio-atlas.psu.edu, allow the identification of some cell types at 5× and most cell types at about 10× magnification or better (corresponding to pixel sizes of 2 μm and 1 μm, respectively) | Perform 3D scans at voxel resolutions of 1 μm or better |
| Floating of sections on water prior to mounting on the slide causes distortion or loss of fragments of tissue sections | Image intact samples in 3D after they are mounted in solid media such as paraffin or plastic so that there is no physical cutting of the tissue |
| Sampling error and limited control over sample orientation and distortion contribute to difficulties in phenotyping | Utilize full-volume, whole-body imaging, and registration of individual organisms to standard anatomy |
| Histopathological studies in high throughput will require automation to achieve scale | Utilize automation such as robotic loading and automated sample tracking for high throughput |
| Small round zebrafish nuclei are about 4 μm in diameter; some elongated nuclei are only about 2 μm in diameter; many vacuoles and chromosomes are about 1 μm in diameter | Aim for 3D imaging that achieves voxel dimensions of 1 μm or less; 0.5 μm voxels may be an ideal balance of resolution and file size |
| Interpretative variation between users limits the large-scale use of manual phenotyping | Develop and utilize computational tools for phenotyping to minimize manual phenotyping |
| Computational tools for microanatomic phenotyping are currently very limited | Develop computational tools for phenotyping and utilize machine learning |
| Accessibility to data is required for validation and reinterpretation | Implement web-based sharing of whole-organism, raw image data |

orientation, insensitivity to macroscopic changes such as misshapen skeletons, and an inability to provide accurate volumetric data about cellular phenotype or tissue structure. Three-dimensional imaging may address some of these problems, and we expect it to make increasingly important contributions to phenome projects as technologies (reviewed by Cheng et al., 2011) improve beyond the already powerful methods in existence today (eg, Badea, Gewalt, Avants, Cook, & Johnson, 2012; Johnson, Cofer, Gewalt, & Hedlund, 2002; Johnson et al., 2006).

# 3. WHOLE-ORGANISM CELLULAR PATHOLOGY AS A SYSTEMS APPROACH TO PHENOMICS

## 3.1 The Need for Whole-Organism Cellular Pathology

Our consideration of the benefits of incorporating cellular pathology into phenotypic screens points to a need for whole-organism 3D imaging at a high resolution. A list of features for ideal whole-animal cellular pathology is suggested in Table 4. A survey of available technologies suggests that X-ray micro-computed tomography (microCT) has the potential to approach such a 3D imaging ideal for small model organisms (reviewed by Cheng et al., 2011). Both cone-beam and parallel-beam microCT work by Metscher has shown great promise in the use of fixed, metal-stained samples for visualizing the cellular structure of vertebrate and invertebrate species (Metscher, 2009a, 2009b; Metscher & Müller, 2011). Cytological resolutions that are sufficient for pathology have yet to be achieved by commercial machines, most of which use cone-beam geometry. Some examples of what can currently be achieved using cone-beam geometry are shown in Figs. 6–8. Early images of a juvenile zebrafish using a Skyscan machine produced images in which organ systems could be readily identified, but in which few individual cell types could be distinguished (Fig. 6). Imaging of a larva at nominal voxel resolutions of 0.87 μm on the Xradia XRM500 yielded images in which individual cell types can be distinguished, but cellular state cannot be readily assessed (Fig. 7). These images require long (~10 h or more) scans and cannot produce high-resolution images of the full organism in one operation. Individual nuclei are visible in these images, but the polychromatic X-rays from the tube source blur the edges enough that cellular phenotype is difficult to ascertain. Synchrotron-based microCT can achieve greater effective resolution than commercial microCT due to the nature of the tube source and system geometry, and has potential for high throughput as a result of its greater flux. Despite some of our improvements to synchrotron microCT (La Riviere et al., 2010), existing implementations are not yet suitable for higher-throughput whole-animal imaging due to a relatively small field-of-view, a need to retain

**Table 4** Ideal whole-animal cellular pathology

| |
|---|
| Quantitation of every phenotype |
| Identification of every cell type (pan-cellular imaging) |
| Three-dimensional imaging with isotropic voxels |
| Detection of organism-scale phenotypes |
| Size, shape, and volume |
| Appendage characteristics |
| Organ statistics |
| Skeletal or physical defects |
| Detection of cytological abnormalities |
| Nuclear/cytoplasmic ratio |
| Swelling (lower density) |
| Changes in chromatin pattern in the nucleus |
| Intracellular inclusions (cytoplasmic or nuclear) |
| Detection of changes in cellular pattern |
| Position |
| Proportionality |
| Volume |
| Circularity |
| Spacing |
| Invasion |
| Number |
| Detection of abnormal structures |
| Abscesses or cysts |
| Bacteria or other pathogens |
| Parasites |
| Intracellular or extracellular deposits |

the flexibility to image a wide array of samples including rocks, soil samples, and materials, and a lack of capability to efficiently image whole, long specimens in a single operation. Fusing blocks of images is achievable but micron-scale misalignments during sample movement occur, which distort the "joints" between images. Those distortions may not be as critical for sample types such as soil, but for biological organisms, distortions coinciding with small anatomical structures may significantly affect phenotyping.

Overall, existing work indicates that neither commercial nor synchrotron sources are yet practical for whole-organism cellular pathology, but adjustments in the travel distance and precision of actuators, field-of-view, other aspects of X-ray optics, sample handling, and digital workflow will yield increases in both resolution and throughput that may make it possible for either geometry to contribute to large-scale phenomics. We are pursuing the use of such improvements.

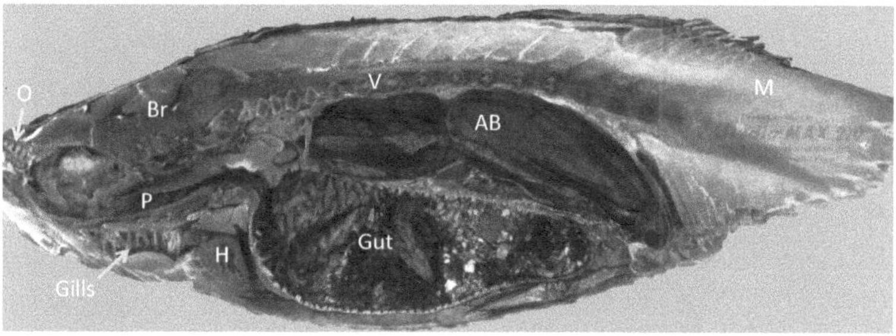


**Figure 6** ***Commercial microCT of a wild-type 35 dpf juvenile zebrafish.*** 3D rendering of cone-beam microCT image scanned in a Skyscan 1172 from VGStudioMax. This near mid-sagittal visualization mimics what one would see if the physical fish were cut in this plane. The mucosal folds of the inner surface of the right side of the air bladder (AB), gut, and pharynx (P) are visible in this view, as are a number of organs including the olfactory epithelium (O), brain (Br), gills, heart (H), vertebrae (V), gut, and muscle (M), but only rare nuclei and individual cells are discernable.

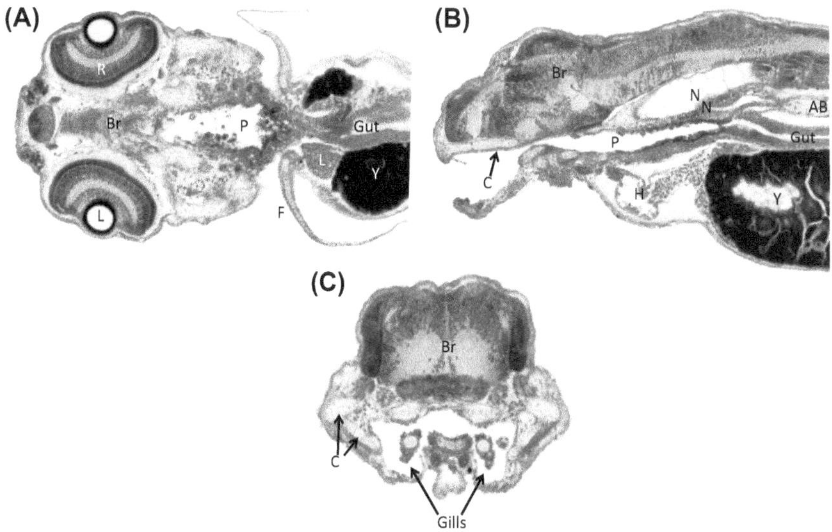


**Figure 7** ***Nominal 0.87 μm voxel resolution, cone-beam microCT slices of a wild-type 4 dpf larval zebrafish in the coronal (A), sagittal (B), and transverse (C) planes showing nearly histological resolution.*** In these Xradia VersaXRM500 microCT images created using VGStudioMax, essentially all tissues and nuclei are visible, though nuclear and other cellular detail is not. Discernable structures include the lens (L), retina (R), brain (Br), pharynx (P), gut, liver (L), fin (F), yolk (Y), notocord (N), heart (H), air bladder (AB), and cartilage (C). Notably, cellular boundaries between the cells of the notocord are apparent. The quality of these images allows the identification of most organs, but a histological level of characterization of cellular state will require greater usable resolution.

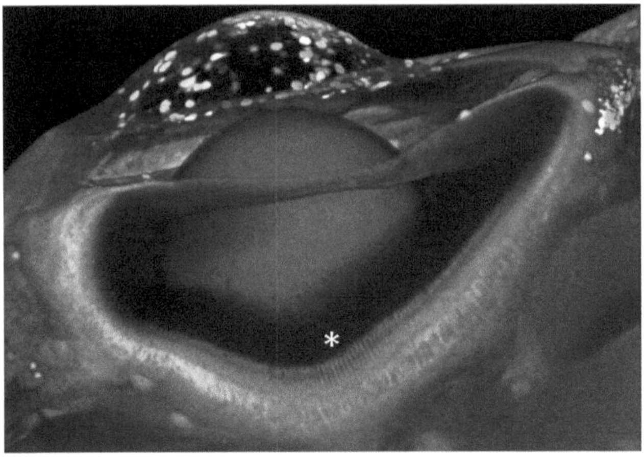


**Figure 8** ***Slice of a phosphotungstic acid-stained wild-type juvenile larval zebrafish eye imaged in a VersaXRM500 at a nominal voxel resolution of 1.15 μm voxel resolution shows nearly histological resolution.*** Patterns of the innermost layer of retinal cells are apparent (*), but photoreceptors, which are 4 μm in width, are not readily discerned. 3D rendering created using VGStudioMax.

## 3.2 Whole-Organism Cellular Pathology Using Zebrafish

Multiple factors indicate that the zebrafish is a good model for developing whole-organism cellular pathology. An image representing an organism's width is distributed across a chip of fixed pixel dimension during imaging, determining resolution. For example, a 2-cm sample imaged across a 2k × 2k chip yields a 10-μm pixel; in contrast, a 2-mm sample with the same chip yields a 1-μm pixel. The small size of zebrafish larvae (<2 mm diameter) theoretically enables this resolution. The use of a chip with more pixels can increase resolution and/or field-of-view, but smaller pixel size in larger arrays are associated with lower quantum efficiency, lower signal to noise ratio, and slower data output. The zebrafish is a well-characterized vertebrate model; about 70% of human genes have at least one obvious zebrafish orthologue (Howe et al., 2013), suggesting great potential to study vertebrate problems related to human biology. Other advantages of the zebrafish include its well-established genome editing tools, including knock-in technology (among the latest in this hot field is Hoshijima, Jurynec, & Grunwald, 2016), and for chemical phenomics, the ability to do rapid chemical screening using multiwell plates. These considerations led to our choice of zebrafish, and similar arguments can be made for the other popular model fish, the medaka.

## 4. CONCLUSIONS

Cellular pathology is of critical value in diagnostic anatomic pathology, which tells us that it would be highly beneficial to incorporate its value into large-scale phenotypic analysis. There has unfortunately been no practical way to assess cellular pathology in large-scale, whole-animal screens. We have noted that the range of cell types and organ systems that may be affected by any given variable cannot be predicted, which ideally necessitates the screening of every cell within its tissue context in whole, intact organisms. Furthermore, despite histology's status as the gold standard today for tissue phenotyping of human disease, and large superiority over stereomicroscopy in terms of ability to detect mutant microanatomical phenotypes in the zebrafish, its technical demands and limitations make it suboptimal for higher-throughput analysis. Its poor ability to represent 3D phenotypes such as skeletal or vascular abnormalities, and poor ability to make accurate volumetric measurements at the cellular level, motivate the development of a tool that has histology's power and adds the ability to perform volumetric analyses at cellular resolution.

The number of experimental variables number in the tens of thousands for both genes and chemicals, making rapid, quantitative analysis critical for comparing phenotypes. Considering the high resolution and whole-animal scale we desire, high-throughput imaging and image processing should be expected to be a form of "big data" analysis. The work described herein has allowed us to specify some of the ideal features that would be required for a 3D microanatomical phenotyping. As described, these needs can potentially be addressed using X-ray based, micron-scale tomography, or microCT, using fixed and stained whole organisms. We have noted how the similarity of its genes to humans and its diminutive size have made the zebrafish an ideal vertebrate model for customizing microCT's for higher-throughput whole-body cellular pathology for phenomics.

We expect the proposed microCT tools to provide knowledge important to genetic, toxicological, and disease phenomics. As reviewed here and by others, a majority of the discussion around phenomics seems to be currently motivated by the study of gene function and drug screening. It is therefore worth noting the importance of applying cellular pathology to environmental toxicology. While most toxicological studies study the effect of one potential toxicant at a time, it is worth noting that the growing abundance of chemicals and nanoproducts that are useful or potentially useful for societal functions such as agriculture, energy production, and manufacture has created a growing risk of combinatorial toxicity. For example, a significant, but unknown proportion of the global collapse in bee populations, upon which much of agriculture is dependent, is associated with multiple, rather than single environmental toxicants (Krupke, Hunt, Eitzer, Andino, & Given, 2012; Mullin et al., 2010; Pettis, Lichtenberg, Andree, Stitzinger, & Rose, 2013). Since healthy colonies exist with single toxicants, the combination of toxicants is now thought to weaken the ability of bee colonies to resist naturally occurring pathogens. We know that histology-based cellular pathology is necessary for identifying infectious agents. Infectious diseases, whether associated with global climate change or combinatorial environmental toxicity, can therefore be expected to be associated with patterns of cellular and tissue architectural change that are directly involved in disease pathogenesis.

In sum, we have described the vision of three-dimensional, cellular pathology-based, computational phenomics. We expect this approach to contribute towards a better understanding of biology, safer and more economical drug development, and the prevention and elucidation of mechanisms of new chemical and infectious threats to global societies. Widespread understanding and acceptance of the proposed infrastructures proposed here will require broader discussion, but with recent advancements in the relevant technologies, we have every reason to believe that credible infrastructures for a phenomics based on whole-organism cellular pathology can be created within a decade.

## ACKNOWLEDGMENTS

This work was supported by 2R24 RR017441-10 to KCC (NIH), 1R24OD18559-01-A2 to KCC (NIH), 1R01HD069321 from the NIH to M. Mullins and KCC, the Jake Gittlen Laboratories for Cancer Research, and the Department of Pathology at the Penn State College of Medicine. We also are grateful for the generosity of Drs. Nancy Hopkins and Adam Amsterdam of MIT for allowing us to collect mutants and to Drs. G. Thomas, B. Canada, and other members of the Cheng lab for contributing towards the bio-atlas (formerly zfatlas.psu.edu) and the histological study of mutants from the Hopkins lab.

## REFERENCES

Abbe, E. (1873). Beiträge zur Theorie des Mikroskops und der mikroskopischen Wahrnehmung. *Archiv für mikroskopische Anatomie, 9*(1), 413–418.

Abbe, E. (1881). VII.—On the estimation of aperture in the microscope. *Journal of the Royal Microscopical Society, 1*(3), 388–423.

Abbe, E. (1883). XV.—The relation of aperture and power in the microscope (continued). *Journal of the Royal Microscopical Society, 3*(6), 790–812.

Amsterdam, A., Nissen, R. M., Sun, Z., Swindell, E. C., Farrington, S., & Hopkins, N. (2004). Identification of 315 genes essential for early zebrafish development. *Proceedings of the National Academy of Sciences of the United States of America, 101*(35), 12792–12797.

Austin, C. P., Battey, J. F., Bradley, A., Bucan, M., Capecchi, M., Collins, F. S., … Zambrowitz, B. (2004). The knockout mouse project. *Nature Genetics, 36*(9), 921–924.

Badea, A., Gewalt, S., Avants, B. B., Cook, J. J., & Johnson, G. A. (November 15, 2012). Quantitative mouse brain phenotyping based on single and multispectral MR protocols. *NeuroImage, 63*(3), 1633–1645.

Beckwith, L. G., Moore, J. L., Tsao-Wu, G. S., Harshbarger, J. C., & Cheng, K. C. (2000). Ethylnitrosourea induces neoplasms in zebrafish (*Danio rerio*). *Laboratory Investigation; a Journal of Technical Methods and Pathology, 80*, 379–385.

Beer, J. J. (1959). *The emergence of the German dye industry*.

Bhowmick, N. A., Neilson, E. G., & Moses, H. L. (2004). Stromal fibroblasts in cancer initiation and progression. *Nature, 432*(7015), 332–337.

Bilder, R. M., Sabb, F. W., Cannon, T. D., London, E. D., Jentsch, J. D., Parker, D. S., … Freimer, N. B. (2009). Phenomics: the systematic study of phenotypes on a genome-wide scale. *Neuroscience, 164*(1), 30–42.

Bracegirdle, B. (1987). *A history of microtechnique: The evolution of the microtome and the development of tissue preparation* (2nd ed.). Lincolnwood, IL: Science Heritage, Ltd.

Brown, S. D., & Moore, M. W. (2012). The International Mouse Phenotyping Consortium: past and future perspectives on mouse phenotyping. *Mammalian Genome, 23*(9–10), 632–640.

Canada, B. A. (2010). *Augmentation of the zebrafish phenomic landscape by histological analysis and content-based image retrieval (Doctoral dissertation)* (Retrieved from the Penn State Electronic Theses and Dissertations Database).

Canada, B. A., Cheng, K. C., & Wang, J. Z. (2006). QCHARM: a novel computational and scientific visualization framework for facilitating discovery and improving diagnostic reliability in medicine. In *Proceedings of the American medical informatics association Annual Symposium* (p. 870).

Canada, B. A., Thomas, G. K., Cheng, K. C.*, & Wang, J. Z.* (2007). Automated segmentation and classification of zebrafish histology images for high-throughput phenotyping. *Proceedings of the third IEEE-NIH life science systems and applications workshop (LISSA 2007)* (pp. 245–248). (*equal contributors).

Canada, B. A., Thomas, G. K., Cheng, K. C.*, & Wang, J. Z.* (2011). SHIRAZ: an automated histology image annotation system for zebrafish phenomics. *International Journal of Multimedia Tools and Applications (Special Issue on Hot Topics in Multimedia), 51*, 401–440. (*equal contributors).

Canada, B. A., Thomas, G. K., Cheng, K. C., Wang, J. Z., & Liu, Y. (2008a). Towards efficient automated characterization of irregular histology images via transformation to Frieze-like patterns. In *Proceedings of the ACM international conference on image and video retrieval* (pp. 581–590).

Canada, B. A., Thomas, G. K., Cheng, K. C., Wang, J. Z., & Liu, Y. (2008b). Automatic lattice detection in near-regular histology array images. In *15th IEEE international conference on image processing* (pp. 1452–1455).

Cheng, K. C. (2004). A life-span atlas for the zebrafish. *Zebrafish, 1*(2), 69.

Cheng, K. C., Xin, X., Clark, D. P., & La Riviere, P. (2011). Whole-animal imaging, gene function, and the zebrafish phenome project. *Current Opinion in Genetics & Development, 21*, 620–629.

Chung, K., & Diesseroth, K. (2013). CLARITY for mapping the nervous system. *Nature Methods, 10*(6), 508–513.

Cotran, R. S., Kumar, V., Collins, T., & Robbins, S. L. (1999). *Robbins pathologic basis of disease*. Philadelphia: Saunders Co.

Davis, A. B. (1979). Book review: slides and microtomes. *Science, 204*(4394), 748–749. http://dx.doi.org/10.1126/science.204.4394.748.

Du Trochet, M. H. (1824). *Recherches anatomiques et physiologiques sur la structure intime des animaux et des végétaux, et sur leur motilité* (JB Baillière).

Gerlai, R. (2002). Phenomics: fiction or the future? *Trends in Neurosciences, 25*, 506–509.

Gordon, G. B., Spielberg, S. P., Blake, D. A., & Balasubramanian, V. (1981). Thalidomide teratogenesis: evidence for a toxic arene oxide metabolite. *Proceedings of the National Academy of Sciences of the United States of America, 78*(4), 2545–2548.

Grubb, S. C., Maddatu, T. P., Bult, C. J., & Bogue, M. A. (January 1, 2009). Mouse phenome database. *Nucleic Acids Research, 37*(Suppl. 1), D720–D730.

Hancock, J. M. (2014). *Phenomics*. CRC Press.

Hill, J. (1770). *The construction of timber, from its early growth; explained by the microscope, and proved from experiments, in a great variety of kinds, London* (Printed for the author).
Hoshijima, K., Jurynec, M. J., & Grunwald, D. J. (2016). Precise editing of the zebrafish genome made simple and efficient. *Developmental Cell, 36*(6), 654–667.
Houle, D. (2010). Numbering the hairs on our heads: the shared challenge and promise of phenomics. *Proceedings of the National Academy of Sciences of the United States of America, 107*, 1793–1799.
Houle, D., Govindaraju, D. R., & Omholt, S. (2010). Phenomics: the next challenge. *Nature Reviews Genetics, 11*, 855–866.
Howe, K., Clark, M. D., Torroja, C. F., Torrance, J., Berthelot, C., Muffato, M., … Stemple, D. L. (2013). The zebrafish reference genome sequence and its relationship to the human genome. *Nature, 496*(7446), 498–503.
Johnson, G. A., Cofer, G. P., Gewalt, S. L., & Hedlund, L. W. (2002). Morphologic phenotyping with MR microscopy: the visible mouse. *Radiology, 222*(3), 789–793.
Johnson, J. T., Hansen, M. S., Wu, I., Healy, L. J., Johnson, C. R., Jones, G. M., … Keller, C. (2006). Virtual histology of transgenic mouse embryos for high-throughput phenotyping. *PLoS Genetics, 2*(4), e61.
Kahraman, A., Avramov, A., Nashev, L. G., Popov, D., Ternes, R., Pohlenz, H. D., & Weiss, B. (2005). PhenomicDB: a multi-species genotype/phenotype database for comparative phenomics. *Bioinformatics, 21*(3), 418–420.
Kalluri, R., & Zeisberg, M. (2006). Fibroblasts in cancer. *Nature Reviews Cancer, 6*(5), 392–401.
Krupke, C. H., Hunt, G. J., Eitzer, B. D., Andino, G., & Given, K. (2012). Multiple routes of pesticide exposure for honey bees living near agricultural fields. *PLoS One, 7*(1), e29268.
Lamason, R. L., Mohideen, M.-A. P., Mest, J. R., Wong, A. C., Norton, H. L., Aros, M. C., … Cheng, K. C. (2005). SLC24A5, a putative cation exchanger, affects pigmentation in zebrafish and humans. *Science, 310*, 1782–1786.
La Riviere, P. J., Clark, D., Rojek, A., Vargas, P., Xiao, X., DeCarlo, F., … Cheng, K. C. (2010). Optimizing synchrotron micro-CT for high-throughput phenotyping of zebrafish. In S. R. Stock (Ed.), *Proceedings of SPIE: Vol. 7804M. Developments in x-ray tomography VII* (pp. 1–8).
Li, Y., Klena, N. T., Gabriel, G. C., Liu, X., Kim, A. J., Lemke, K., … Lo, C. W. (2015). Global genetic analysis in mice unveils central role for cilia in congenital heart disease. *Nature*. http://dx.doi.org/10.1038/nature14269.
Lussier, Y. A., & Liu, Y. (2007). Computational approaches to phenotyping. *Proceedings of the American Thoracic Society, 4*(1), 18–25.
Mahner, M., & Kary, M. (1997). What exactly are genomes, genotypes and phenotypes? And what about phenomes? *Journal of Theoretical Biology, 186*, 55–63.
Metscher, B. D. (2009a). MicroCT for comparative morphology: simple staining methods allow high-contrast 3D imaging of diverse non-mineralized animal tissues. *BMC Physiology, 9*(1), 1.
Metscher, B. D. (2009b). MicroCT for developmental biology: a versatile tool for high-contrast 3D imaging at histological resolutions. *Developmental Dynamics, 238*(3), 632–640.
Metscher, B. D., & Müller, G. B. (2011). MicroCT for molecular imaging: quantitative visualization of complete three-dimensional distributions of gene products in embryonic limbs. *Developmental Dynamics, 240*(10), 2301–2308.
Mohideen, M.-A. P. K., Beckwith, L. G., Tsao-Wu, G. S., Moore, J. L., Wong, A. C. C., Chinoy, M. R., & Cheng, K. C. (2003). Histology-based screen for zebrafish mutants with abnormal cell differentiation. *Developmental Dynamics, 228*, 414–423.
Moore, J. L., Aros, M., Steudel, K. G., & Cheng, K. C. (2002). Fixation and decalcification of adult zebrafish for histological, immunocytochemical, and genotypic analysis. *BioTechniques, 32*, 296–298.

Moore, J. L., Rush, L. M., Breneman, C., & Cheng, K. C. (2006). Zebrafish genomic instability mutants and cancer susceptibility. *Genetics, 174*, 585–600.

Mullin, C. A., Frazier, M., Frazier, J. L., Ashcraft, S., Simonds, R., & Pettis, J. S. (2010). High levels of miticides and agrochemicals in North American apiaries: implications for honey bee health. *PLoS One, 5*(3), e9754.

Nasevicius, A., & Ekker, S. C. (2000). Effective targeted gene 'knockdown'in zebrafish. *Nature Genetics, 26*(2), 216–220.

Pantanowitz, L., Valenstein, P. N., Evans, A. J., Kaplan, K. J., Pfeifer, J. D., Wilbur, D. C., … Colgan, T. J. (2011). Review of the current state of whole slide imaging in pathology. *Journal of Pathology Informatics, 2*, 36. http://www.jpathinformatics.org/text.asp?2011/2/1/36/83746.

Pettis, J. S., Lichtenberg, E. M., Andree, M., Stitzinger, J., & Rose, R. (2013). Crop pollination exposes honey bees to pesticides which alters their susceptibility to the gut pathogen *Nosema ceranae*. *PLoS One, 8*(7), e70182.

Poole, W. H. (1875). A double staining with haematoxylin and aniline. *Quarterly Journal of Microscopical Science NS, 15*, 375–377.

Rennekamp, A. J., & Peterson, R. T. (2015). 15 years of zebrafish chemical screening. *Current Opinion in Chemical Biology, 24*, 58–70.

Sabaliauskas, N. A., Foutz, C. A., Mest, J. R., Budgeon, L. R., Sidor, A., Gershenson, J., … Cheng, K. C. (2006). High-throughput zebrafish histology. *Methods: A Companion To Methods in Enzymology, 39*, 246–254.

Schilling, C. H., Edwards, J. S., & Palsson, B. O. (1999). Toward metabolic phenomics: analysis of genomic data using flux balances. *Biotechnology Progress, 15*, 288–295.

Schleiden, M. J. (1838). Beitrage zue Phytogenesis. *Archiv für Anatomie, Physiologie und Wissenschaftliche Medicin, 13*, 137–176.

Schwann, T. (1839). Mikroskopische Untersuchungen über die Übereinstimmung in der Struktur und dem Wachstum der Tiere und Pflanzen. In I. Jahn (Ed.), *Klassische Schriften zur Zellenlehre (2003)*. Frankfurt am Main: Wissenschaftlicher Verlag Harri Deutschland GmbH.

Streisinger, G. (1984). Attainment of minimal biological variability and measurements of genotoxicity: production of homozygous diploid zebrafish. *National Cancer Institute Monographs, 65*, 53–58.

Thomas, G. K. (2009). *A molecular and systems biology analysis of pleiotropic vertebrate phenotypes (Doctoral dissertation)* (Retrieved from the Penn State Electronic Theses and Dissertations Database).

Travis, A. S., Hornix, W. J., Bud, R., & Homburg, E. (1992). The emergence of research laboratories in the dyestuffs industry, 1870–1900. *British Journal for the History of Science, 25*(01), 91–111.

Tsao-Wu, G. S., Weber, C. H., Budgeon, L. R., & Cheng, K. C. (1998). Agarose embedded tissue arrays for histological and genetic analysis. *BioTechniques, 25*, 614–618.

Vallee, B. L., Warren, E. C., & Wacker, W. E. (1963). Medical biology: a perspective. *JAMA, 184*(6), 485–489.

Virchow, R. (1855). Cellular-pathologie. *Virchows Archiv, 8*(1), 3–39.

Virchow, R. (1858). *Die cellular pathologie*. Berlin: August Hirschwald.

Vogel, W. (2012). *Glass chemistry*. Springer Science & Business Media.

Wright, N. A., & Poulsom, R. (2012). Omnis cellula e cellula revisited: cell biology as the foundation of pathology. *Journal of Pathology, 226*, 145–147.

Yang, W., Duan, L., Chen, G., Xiong, L., & Liu, Q. (2013). Plant phenomics and high-throughput phenotyping: accelerating rice functional genomics using multidisciplinary technologies. *Current Opinion in Plant Biology, 16*, 180–187.

Zbuk, K. M., & Eng, C. (2007). Cancer phenomics: RET and PTEN as illustrative models. *Nature Reviews Cancer, 7*, 35–45.

CHAPTER FIVE

# Evolutionary Genetics of the Cavefish *Astyanax mexicanus*

**D. Casane*,§,1 and S. Rétaux¶,1**
*Laboratory EGCE, CNRS and University of Paris-Sud, Gif-sur-Yvette, France
§Paris Diderot University, Sorbonne Paris Cité, France
¶Paris-Saclay Institute of Neuroscience, CNRS and University Paris-Sud, Gif-sur-Yvette, France
[1]Corresponding authors: E-mail: didier.casane@egce.cnrs-gif.fr; retaux@inaf.cnrs-gif.fr

## Contents



*Advances in Genetics*, Volume 95
ISSN 0065-2660
http://dx.doi.org/10.1016/bs.adgen.2016.03.001

## Abstract

Blind and depigmented fish belonging to the species *Astyanax mexicanus* are outstanding models for evolutionary genetics. During their evolution in the darkness of caves, they have undergone a number of changes at the morphological, physiological, and behavioral levels, but they can still breed with their river-dwelling conspecifics. The fertile hybrids between these two morphotypes allow forward genetic approaches, from the search of quantitative trait loci to the identification of the mutations underlying the evolution of troglomorphism. We review here the past 30 years of evolutionary genetics on *Astyanax*: from the first crosses and the discovery of convergent evolution of different *Astyanax* cavefish populations to the most recent evolutionary transcriptomics and genomics studies that have provided researchers with potential candidate genes to be tested using functional genetic approaches. Although significant progress has been made and some genes have been identified, cavefish have not yet fully revealed the secret of their adaptation to the absence of light. In particular, the genetic determinism of their loss of eyes seems complex and still puzzles researchers. We also discuss future research directions, including searches for the origin of cave alleles and searches for selection genome-wide, as well as the necessary but missing information on the timing of cave colonization by surface fish.

## 1. INTRODUCTION

Recently, a book entitled "Biology and Evolution of the Mexican Cavefish" has been published, reporting specialized views on topics covering ecology and evolution, genetic diversity and quantitative genetics, morphology and development, as well as behavior (Keene, Yoshizawa, & McGaugh, 2015). Here, we would like to give a more general and historical perspective on the current knowledge in evolutionary genetics and genomics on the fish *Astyanax mexicanus,* as a follow-up and update of Wilkens contribution back in 1988 (Wilkens, 1988) (see also chapter: Studying the evolution of the vertebrate circadian clock: The power of fish as comparative models by Foulkes et al., 2016). First and as an introduction, we discuss the interest of studying teleost fishes to address questions related to evolutionary biology.

Among more than 66,000 extant species of vertebrates currently described, about 60 are jawless vertebrates (lampreys and hagfishes), about 900 are cartilaginous fish (sharks, rays, and chimeras), and all the others species are bony fish (Osteichthyes). Bony fishes are divided into two groups of about the same size (~33,000 species), lobe-finned fishes (Sarcopterygii) and ray-finned fishes (Actinopterygii). Most extant lobe-finned fishes are tetrapods; they are air-breathing terrestrial animals that show a huge diversity (Lecointre & Le Guyader, 2006) and some species such as mouse and humans have been extensively studied. Extant ray-finned fish comprise on the one hand about 50 species distributed among several highly divergent groups containing each a very small number of species such as Polypteriformes (bichirs), Acipenseriformes (sturgeons and paddlefish), Semoniotiformes (gars), Amiiformes (*Amia calva*), and on the other hand the very large and diverse group of Teleostei (teleosts) that contain all the other species (Eschmeyer & Fricke, 2012; Helfman, Collette, Facey, & Bowen, 2009) (Fig. 1). Teleosts represent thus roughly half of extant vertebrate species and during the ~300 million years of their evolutionary history (Near et al., 2013, 2012) they have diversified into different lineages showing a great diversity at all levels of organization, from the genome to the anatomical, physiological, and behavioral levels. Whereas the diversity of tetrapods has long been recognized, the diversity of teleosts, which is likely as large, has been much less explored. For these reasons, teleosts are outstanding models to study a multitude of questions about vertebrate evolution.

The reasons for the evolutionary success of teleosts are probably manifold but two main possible explanations have been proposed.

First, their genomes are highly plastic, as exemplified by a high variability in chromosome numbers and genome size (from the super compact 390 Mb pufferfish *Takifugu rubripes* to the 1.37 Gb zebrafish *Danio rerio*), or by their wide spectrum of sex determination (SD) systems. Second, whereas tetrapods retained the ancestral genome organization of gnathostomes that issued from two rounds of duplication (R1, R2) of the ancestral vertebrate genome, a whole genome duplication event (R3) occurred at the base of the teleost tree (Amores et al., 1998; Postlethwait et al., 1998) and may have provided valuable raw genetic material for diversification in the group (Roest Crollius & Weissenbach, 2005). Indeed, according to Ohno's hypothesis, a duplicated gene may acquire new function(s) and generate novelty, bypassing the need for cumulative mutations on already existing genes. Although duplicated genes probably rarely undergo such neofunctionalization but rather more often follow the pseudogenization/loss or the subfunctionalization paths, genome duplications appear to serve as a major seed for evolution (Holland, 2003; Ohno, 1970). Gene duplication might also be directly involved in fish diversification by creating "speciation genes" reducing hybrid fitness (reviewed in Volff, 2005). Nonetheless, several lines of evidence show that there is not such a simple correlation between genome duplication and increase of species richness and diversity. An impressive evolutionary radiation occurred in African cichlids that have an unoriginal R3 genome. On the contrary, Salmoniformes, that underwent

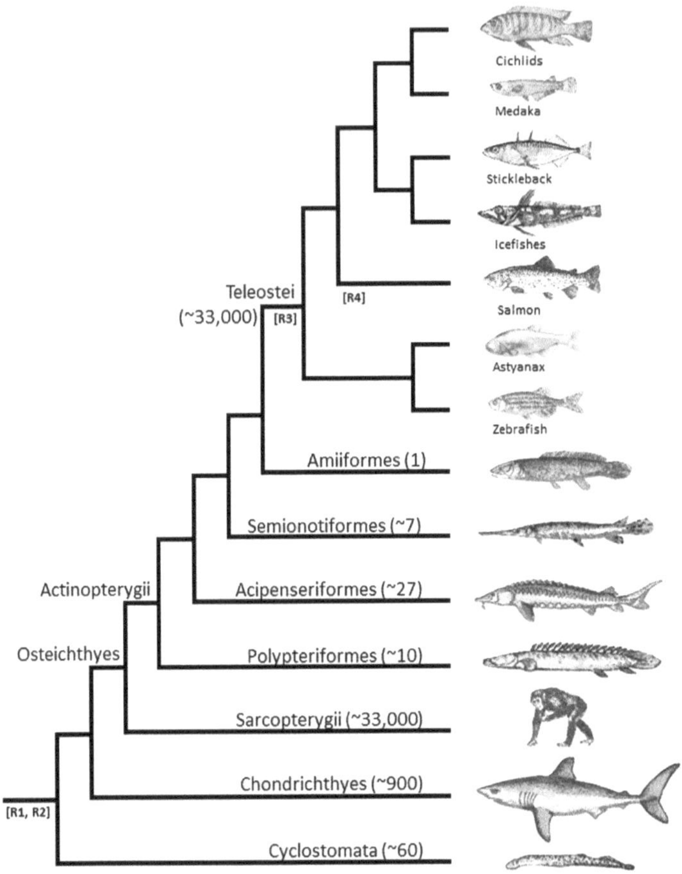


**Figure 1** ***Main teleost models for evolutionary genetics.*** Phylogenetic relationships are shown to the left. The number of species described in each clade is given between parentheses; R1, R2, R3, R4: whole genome duplications; Tree topology is based on *Near, T.J., Dornburg, A., Eytan, R.I., Keck, B.P., Smith, W.L., Kuhn, K.L., … Wainwright, P.C. (2013). Phylogeny and tempo of diversification in the superradiation of spiny-rayed fishes. Proceedings of the National Academy of Sciences of the United States of America, 110(31), 12738–12743; Near, T.J., Eytan, R.I., Dornburg, A., Kuhn, K.L., Moore, J.A., Davis, M.P., … Smith, W.L. (2012). Resolution of ray-finned fish phylogeny and timing of diversification. Proceedings of the National Academy of Sciences of the United States of America, 109(34), 13698–13703.* (See color plate)

one more whole genome duplication (R4), show neither a noticeable increased number of species nor a high species diversity (Mungpakdee et al., 2008; Pascual-Anaya, D'Aniello, Kuratani, & Garcia-Fernandez, 2013). Accordingly, it has also been shown that an ancient whole genome duplication occurred in horseshoe crabs despite the fact that these animals are very well known for their very slow morphological evolution and the very small number of extant species (Kenny et al., 2015).

Third, teleost genomes contain the highest diversity of transposable element (TE) superfamilies in vertebrates. A recent survey counted an average of 24 TE superfamilies present in the genomes of nine teleosts species studied, to be compared with only 12 in the human genome (Chalopin, Naville, Plard, Galiana, & Volff, 2015). While TEs can disrupt genes and lead to various types of chromosomal rearrangements that are highly deleterious, there is no doubt that TEs are also drivers of genome evolution, and evidence for recent and frequent activity has been reported for numerous families of fish TEs (reviewed in Volff, 2005).

Among teleosts, several species have been particularly well studied by ichthyologists, geneticists, or evolutionary and developmental biologists because they exhibit a combination of useful features that make them attractive as animal models. Besides the zebrafish *Danio rerio*, now the second most employed animal model in laboratories around the world, and the medaka *Oryzias latipes* that is used as an alternative to zebrafish in many labs (also representing the first two fish genomes to have been sequenced after the genome of the fugu (Aparicio et al., 2002)), many species have attracted interest. Below, as an introduction, we mention three examples to show how powerful fish are for unraveling the underlying genetic mechanisms of evolution. The most impressive results have been obtained with species showing great physiological and/or morpho-anatomical modifications associated with environmental shift. Another very important characteristic of these species is that they can be kept in the lab and they allow genetic, developmental, physiological, behavioral analyses at a much lower cost than with any tetrapod species. Finally, in the past years, the revolution of next-generation sequencing techniques has been exploited for the in-depth study of teleost genomes and transcriptomes. Sixteen teleosts genomes are now sequenced and will necessitate in-depth analyses to understand the genomic (and developmental) bases of their diversity. Teleosts have entered the postgenomic era, inspiring Braasch and colleagues to qualify them as a "new model army" swimming in a "genomic tsunami" (Braasch et al., 2015).

*Example 1*: The adaptive radiation of cichlids in East African lakes led to several hundreds of endemic species in each lake, which have diversified within different time frames: in the case of Lake Victoria, more than 500 species arose within a period of 15,000 to 100,000 years, while it took 10–12 million years in Lake Tanganyika (Brawand et al., 2014; Kuraku & Meyer, 2008). These fish show an astonishingly precise adaptation to particular food items and ecological niches. This is achieved through a very flexible oral skeleton and dentition and the evolution of novel arrangements of their pharyngeal jaws, a trait which seems to depend upon bone morphogenetic protein signaling (Albertson, Streelman, Kocher, & Yelick, 2005; Brawand et al., 2014). Sexual selection has also been suggested to serve as a major diversification force. Indeed, cichlids are richly and beautifully colored. Genes responsible for color pigments and patterns show high rates of amino acid replacement among cichlid species (Sugie, Terai, Ota, & Okada, 2004), and opsin genes involved in vision show high polymorphism in their coding sequences with a correlation between the frequency of particular wavelength-specific opsin alleles and male coloration (Terai et al., 2006). Quantitative trait loci (QTL) approaches have started to identify the genetic factors controlling opsin expression and how divergence of visual systems has contributed to the adaptive radiation of African cichlids (Schulte, O'Brien, Conte, O'Quin, & Carleton, 2014).

*Example 2*: Teleosts show an amazing diversity and rapid turnover of their SD systems and in genes involved in SD (Bachtrog et al., 2014; Kikuchi & Hamaguchi, 2013). This includes pure environmental SD, genetic SD, or a mixture of both (Herpin & Schartl, 2011a, 2011b; Ospina-Alvarez & Piferrer, 2008). The first fish SD gene to be discovered was *dmrt1bY* in the medaka *O. latipes* (Matsuda et al., 2002). However, this SD system is not shared even within the *Oryzias* genus as two other SD genes have been identified in two other *Oryzias* species. In *Oryzias luzonensis*, an allelic variation of the *gsdf* gene, a gene that acts downstream of *dmrt1bY* in medaka, determines male development (Myosho et al., 2012); and in *Oryzias dancena*, a Y specific *cis*-regulatory element upstream of the *sox3* gene drives male SD (Takehana et al., 2014). In the pufferfish *T. rubripes*, a single nucleotide allelic variation in the *amh-receptor 2* gene, again a downstream component of the SD cascade in other vertebrates, controls phenotypic sex (Kamiya et al., 2012). In the Patagonian pejerry *Odontesthes hatcheri*, a Y-linked duplicate of the *amh* gene drives male SD (Hattori et al., 2012), a situation that mirrors the evolutionary scenario of the medaka SD gene. In the rainbow trout *Oncorhynchus mykiss*, gene duplication and neofunctionalization led to the emergence of a new SD gene, *sdY* (Yano et al., 2012); surprisingly this duplicated gene is not derived from a previously known component of the SD cascade but from the *interferon regulatory factor 9*, an "immunity gene." In addition to this variety of SD mechanisms, transitions between male (XX/XY) and female heterogamety (ZZ/ZW) have been reported and are generally thought to reflect the important evolutionary instability of sex chromosomes in teleosts. Such rapid turnover in SD systems and genes are susceptible to promote speciation.

*Example 3*: Finally, teleosts display amazing adaptability to extreme environments, including extreme cold, dry, pressure, acid, or dark. Antarctic notothenoids icefish thrive at subzero temperatures, thanks to the production of an antifreeze glycoprotein encoded by a gene derived from a trypsinogen gene "created" through complex and multiple recombination/deletion/duplication events (Chen, DeVries, & Cheng, 1997a). So do their Arctic cod counterparts, although the Arctic antifreeze peptide originates from another ancestral sequence (Chen, DeVries, & Cheng, 1997b), highlighting a singular convergence. They also evolved a unique mechanism for cold adaptation of proteins, by de novo addition of domain allowing the 3D structure to compensate for the lower thermal kinetic energy available at low temperatures (Rizzello et al., 2013). On the other end of the climate spectrum, fish have even adapted to deserts and lack of water: the annual killifish of the genus *Nothobranchius* have a very short life span as they live in temporary savannah pools in Africa and the Neotropics. Their populations survive in a habitat that desiccates annually during the dry season, thanks to the evolution of peculiar life history traits, including developmental diapauses and ultrafast maturation and aging (recently reviewed in Cellerino, Valenzano, & Reichard, 2016). On the subterranean side, more than 150 fish species belonging to many orders have adapted to the absence of light and prosper in darkness. They have been found everywhere except in Antarctica and Europe so far and again they exemplify the phenomenon of independent and convergent evolution: like cave animals from all other phyla, they are depigmented and blind (or have strongly reduced eyes). Some emblematic cases include the following. The Ambliopsids of North America both intrigued and excited biologists since the discovery and description of *Amblyopsis spelaea* (a percopsiform), the first troglobitic fish ever described in the early 1840s (Niemiller & Poulson, 2010). The genus *Lucifuga* (ophidiiform) is one of the most diverse cavefish described thus far (Garcia-Machado et al., 2011; Romero, 2001). The Somalian cavefish *Phreatichthys andruzzii* and the Chinese cavefish of the genus *Sinocyclocheilus* (both cypriniforms) are used to compare developmental and physiological mechanisms of evolution with the zebrafish (Cavallari et al., 2011; Meng et al., 2013; Stemmer, Schuhmacher, Foulkes, Bertolucci, & Wittbrodt, 2015). Strikingly, some teleost cave species are real champions of adaptation since they are able to cope not only with the absence of light but also with additional challenges: *Cryptotora thamicola* (a cypriniform), the Thailand waterfall climbing loach or "cave angelfish," is also known for its large and modified pectoral fins which can grapple onto wet rocks; the cichlid *Lamprologus lethops* (a perciform) is the only blind and depigmented cichlid found in Congo River rapids and survives in extremely energetic flow regimes; the cave molly *Poecilia mexicana* (a cyprinodontiform) resists the high sulfur toxicity of the Mexican Cueva del Azufre (Plath et al., 2007); and for many aspects, deep sea fish can be considered as similar to cavefish as they live in the absence of light, together with high pressure, low temperature, and food scarcity (Priede & Froese, 2013; Warrant & Locket, 2004).

Facing the diversity of cavefishes and deep-dwellers one could wonder why *Astyanax mexicanus* (a characiform) has been chosen as the model species for the study of evolution and adaptation in dark environment. The reason for that is simple: this model allows approaches of evolutionary genetics and comparative biology at the microevolution scale because we have the good fortune that both the surface-dwelling and the cave-dwelling morphs of the species are available, that they can breed together, and that their progeny are viable and fertile. In the following text, we review the current knowledge in evolutionary genetics and genomics on *Astyanax*, a model that has allowed very significant advances in identifying mutations involved in cave phenotypes.

## 2. THE *ASTYANAX MEXICANUS* SURFACE FISH AND CAVEFISH MODEL SYSTEM

The most common form of *A. mexicanus,* or Sardinita Mexicana or Mexican Tetra, is the river-dwelling "surface fish," which spans a wide geographic distribution zone in Central America. Surface *Astyanax* tolerate a variety of habitats (wetlands, rivers, springs, canals, and backwaters). They have an affinity for bedrock, rocky substrates, and sand-bottomed pools. They are omnivores, feeding on zooplankton and small fishes, but also consume insects and filamentous algae.

On the other hand, 29 populations of *Astyanax* cavefish are currently known (Mitchell, Russell, & Elliott, 1977). They live in the total and permanent darkness of underground rivers and pools located in caves of the Sierra del Abra, the Sierra de Guatemala, and the Sierra de Colmena in North-East Mexico. At the adult stage, they are all depigmented and blind (or have severely reduced eyes). Except for temperature and some water parameters, not much is known about the exact ecological conditions in which they live. The cave environment is thought to be nutrient poor, predator free, and relatively stable, but each cave that hosts *Astyanax* should probably be considered as a specific case. Specifically, the conditions vary greatly depending, for example, on the presence or absence of bat colonies, or on the occurrence of seasonal flooding that can carry organic matter into the cave (Culver & Pipan, 2009; Mitchell et al., 1977).

Importantly, some aspects of cavefish evolutionary history are known. In particular, the polarity of the change is certain: the two extant forms of *A. mexicanus* derive from a common surface-like, eyed, ancestor. This is supported by three lines of evidence. First, developmental biology: during embryogenesis cavefish first develop eyes, which are secondarily lost by degeneration during larval and juvenile stages (Alunni et al., 2007; Jeffery, 2001; Jeffery & Martasian, 1998). Moreover, cavefish eyes can be rescued by embryonic transplantation of a surface fish lens (Yamamoto & Jeffery, 2000). This shows that cavefish have inherited an "eye genetic program" from their eyed ancestors. Second, genetics: crosses between cavefish originating from different caves can give rise to progeny with small eyes (Borowsky, 2008b; Wilkens & Strecker, 2003) (reviewed in Jeffery, 2009). This type of partial complementation also shows that a partially different set of genes is involved in the loss of eyes in different cave populations, and sets the basis for the evolutionary genetic approach reviewed in the next sections. Third, phylogeography: although the origins of cave *Astyanax* are complex (Gross, 2012) and the number and timing of cave colonization events is still a matter of intense investigation, available data converge to suggest an epigean ancestry of cave populations (Avise & Selander, 1972; Bradic, Beerli, Garcia-de Leon, Esquivel-Bobadilla, & Borowsky, 2012; Strecker, Faundez, & Wilkens, 2004).

Cave *Astyanax* and surface morphs differ in many respects. Not only have they lost their eyes and pigmentation during their evolution in the absence of light, but also they have evolved constructive traits at the morphological, physiological, and behavioral levels. The determinants of most of these phenotypic changes, listed below, have been addressed through genetic approaches and will be reviewed in the next sections.

Cavefish general morphology is clearly different from its conspecific surface morph. Besides an eyeless head and a white/pink color, cavefish have a large mouth and more teeth, their body shape and stoutness is slightly different, their craniofacial skeleton is modified, and curiously they possess one less rib (Dowling, Martasian, & Jeffery, 2002; Hinaux et al., 2011; Yamamoto, Espinasa, Stock, & Jeffery, 2003). They also have more taste buds (Varatharasan, Croll, & Franz-Odendaal, 2009; Yamamoto, Byerly,

Jackman, & Jeffery, 2009) and more neuromasts (Schemmel, 1967; Yoshizawa, Goricki, Soares, & Jeffery, 2010), which are the components of the lateral line, a specific sensory system of most aquatic vertebrates for detecting water movements and vibrations, and a larger olfactory epithelium (Bibliowicz et al., 2013). The latter changes are considered as evolutionary sensory compensations for the loss of vision in the dark.

Cavefish metabolism has also evolved. Their eggs are slightly larger and contain more yolk that surface fish eggs (Hüppop & Wilkens, 1991), as adults they possess more fat reserves than surface fish (Rose & Mitchell, 1982) and are more resistant to starvation (Hüppop, 1988). They have a lower standard metabolic rate than epigean *Astyanax* (Salin, Voituron, Mourin, & Hervant, 2010), and they have eliminated circadian rhythms in their metabolism, which confers an energy saving of about 30% (Moran, Softley, & Warrant, 2014).

Finally, cavefish exhibit motor, sensory, as well as complex behavioral changes. They show increased locomotor activity and they are bottom feeders, exploring the substrate at a precise 45 degrees angle, an efficient method to find food in the dark (Schemmel, 1980); they have evolved a "vibration attraction behavior" mediated by their enhanced lateral line system to locate food dropping onto the water surface (Yoshizawa et al., 2010; Yoshizawa, Yamamoto, O'Quin, & Jeffery, 2012); they have increased chemosensory capabilities (Bibliowicz et al., 2013); when put together in competition experiments in the dark with limited amounts of food, cavefish do well whereas surface fish starve (Espinasa, Bibliowicz, Jeffery, & Retaux, 2014; Hüppop, 1987); they have lost the aggressive behavior, which is a trademark of their surface conspecifics and they have gained sustained foraging behavior (Burchards, Dölle, & Parzefall, 1985; Elipot, Hinaux, Callebert, & Retaux, 2013; Espinasa, Yamamoto, & Jeffery, 2005; Parzefall, 1985); contrary to surface fish, they do not school (Kowalko, Rohner, Rompani, et al., 2013; Parzefall & Senkel, 1986); they sleep very little, supposedly to increase wakefulness and their chances to find food (Duboué & Borowsky, 2012; Duboué, Keene, & Borowsky, 2011). Far from supporting the long-held view that cavefish are degenerated species, this combination of adaptations more likely constitutes a so-called "cavefish behavioral syndrome." Thus, the many, subtle behavioral adaptations that have occurred in cavefish are likely to contribute to their survival and success during cave colonization. These behavioral characteristics all represent complex traits, probably due to several subtle changes in the wiring, size, and/or neurochemistry of their nervous system.

# 3. GENETICS IN THE PREMOLECULAR ERA

## 3.1 Several Morphs, One Species

Soon after the discovery of several populations of *A. mexicanus* cavefish, a series of breeding experiments on the reproductive isolation among cavefish populations and with river-dwelling fish were undertaken to clarify taxonomic and evolutionary genetics issues. Indeed, the first population to be discovered, found in La Cueva Chica, was assigned as a new species belonging to a new genus and accordingly named *Anoptichthys jordani* (Hubbs & Innes, 1936). The second population, found in La Cueva de los Sabinos, was assigned as a new species belonging to the same genus and named *Anoptichthys hubbsi*; then a third population found in La cueva de El Pachón was named *Anoptichthys antrobius*. However, *Anoptichthys* shares with the genus *Astyanax* all apparent characters other than those associated with blindness and subterranean life, and the different *Anoptichthys* species are very similar. It appeared that Chica cavefish and river fish could be crossed and the first generation all resembled the latter parent (Breder, 1943). A decade later, extensive analyses were performed using Chica, Sabinos, Pachón, and river fish (Sadoglu, 1957b). It was demonstrated that they could all be easily crossed and thus there was no reason for separating the cavefish generically from *Astyanax* and specifically from one another. In addition, the fertile F1 hybrids opened up a new avenue to explore the genetic determinism of troglomorphic traits.

## 3.2 Eyes and Pigmentation, and More

River/cave F1 hybrids were pigmented but they were not as dark as river fish, regardless of the cave population the cavefish parent came from. The size of eyes of the F1 hybrids was intermediate between the large eyes of the river fish and the very reduced and sunken eyes of the cavefish, but the mean diameter was greatest in the Chica/River F1 hybrids, more reduced in the Sabinos/River F1 hybrids, and most reduced in the Pachón/River F1 hybrids. These observations suggested that different genetic changes operated in the different caves. F2 fish obtained from crosses within each F1 hybrid population showed large variance in the phenotypic traits, from little to full pigmentation and from reduced and sunken to normal eyes. In addition, different combinations of pigmentation level and eye size were observed. These observations suggest that both traits have a polygenic determinism and they are controlled by two gene sets that are largely independent. Moreover, different morphological traits that characterize the eye shape, such as the size of the lens and the width of the pupillary opening, also appear to be controlled by two gene sets that are largely independent. Finally, the hybrids between Chica and Sabinos cavefish were as blind as their parents but some of them were much darker. This suggests that the genes involved in eye reduction are identical at least to a large extent in these two cave populations, but that different genes control pigmentation to allow complementation. More recently in 2008, Richard Borowsky (Borowsky, 2008b) crossed fish from other caves (Molino, Pachón, Japonés, Tinaja, Curva) and estimated the percentage of sighted individuals in the offspring. On average, 39% of Molino/Tinaja, 8% of Pachón/Tinaja, and 1% of Molino/Pachón F1 hybrids could see (the Molino/Pachón fry were albino, which might account for the low percentage). Offspring of Tinaja/surface crossed to Pachón/surface hybrids (ET/EP) and offspring of Molino/surface crossed to Pachón/surface hybrids (EM/EP) had high proportions of sighted individuals ($0.940 \pm 0.026$ and $0.950 \pm 0.022$, respectively); a significantly lower proportion of offspring from Curva/surface crossed to Tinaja/surface hybrids (EC/ET) could see ($0.409 \pm 0.074$). Based on complementation in hybrids, Borowsky's interpretation of the high proportions of sighted individuals in ET/EP and EM/EP crosses is that the Pachón population is genetically distinct from both the Molino and Tinaja populations, while noncomplementation and low proportion of sighted ET/EC individuals is due to a shared cave ancestry of Tinaja and Curva populations. This interpretation is consistent with geography: the Molino and Pachón caves as well as Pachón and Tinaja are distant, while Tinaja and Curva are close (Mitchell et al., 1977).

Eyed *Astyanax* surface fish have a bony sclera that is the tough outer covering of the eye and that provides structural support and helps the maintenance of intraocular pressure. The two scleral ossicles found in surface fish generally do not form in blind *Astyanax* cavefish populations. The independent origin of several groups of cavefish populations suggested that scleral ossification has been lost more than once. This hypothesis has been confirmed by a demonstration of the complementation of scleral ossification among the F1 hybrid offspring of two independently derived cavefish populations from the Pachón and Sabinos caves (O'Quin et al., 2015). Half of the F1 progeny examined (2/4) exhibited small, pigmented eyes with ossified sclera on both sides of their head. Although the sample size is small, this result demonstrates genetic complementation of scleral ossification and confirms experimentally the convergent loss of sclera ossification among these *Astyanax* cavefish populations (O'Quin et al., 2015).

The experimental crosses between river fish and cavefish also allowed the identification of two genes with a major effect on pigmentation. First, a recessive allele (albino) that is polymorphic in the Pachón cave was found to be responsible for albinism, ie, loss of black pigment in the eye and in all other parts of the body (Sadoglu, 1957a). A second recessive allele (brown) at an independent locus was later identified (Sadoglu & McKee, 1969). It seems to delay and reduce the amount of pigment in the eye and reduces the number of larval melanophores. At the adult stage, it has a uniform effect on the eye pigmentation, whereas body pigmentation was variable among fish. During the next decade, the studies of Perihan Sadoglu were refined and expanded to other characters, using the very same approach applied to a wider sample of cave populations (Wilkens, 1971). Cavefish, especially Pachón cavefish, have a larger number of taste buds than river fish, and this difference has a genetic basis (Schemmel, 1974). Cavefish and river fish have different behavior, the former being less aggressive and having lost the schooling behavior. Cavefish have also developed a specific feeding posture. They search for and pick up food while swimming at an angle of about 45° from the bottom. In contrast, the epigean form can only feed on the bottom while maintaining a steep angle of about 90°. These behavioral differences between epigean and hypogean forms are genetically based (Schemmel, 1980). At the end of the 1980s, Horst Wilkens published a comprehensive compilation of the genetic determinism of cavefish phenotypes performed until that time (Wilkens, 1988). Using the mean and the variance of a quantitative character in parental populations, F1, F2, and backcrosses, the minimal number of genes involved in the determinism of that character can be estimated (Lande, 1981). It was estimated that the variance of eye size involves about 6–7 genes, the variance of melanophore number involves 2–3 genes, the variance of feeding behavior involves 2 genes, and the variance of the number of taste buds involves 11–12 genes (Wilkens, 1988). It is worth noting that these estimations can be gross underestimates as they rely on several hypotheses that do not necessarily hold, such as a low variance of gene effects and no linkage between the loci. Nevertheless, it suggests that most troglomorphic traits involve changes at multiple loci.

## 3.3 Sleep Loss

The convergent evolution of sleep loss has been described in three cave populations, Pachón, Tinaja, and Molino (Duboué et al., 2011). The sleep phenotype in surface x Pachón F1 hybrids (or surface x Tinaja F1 hybrids) is indistinguishable from that of the pure Pachón (or pure Tinaja) cavefish. This strongly suggests that the global effect of the alleles in both cave populations is dominant to surface alleles. The hypothesis of a dominant allele at a single locus was rejected as in a backcross surface_Pachón_F1_hybrid x surface the expected 1 (surface): 1 (cave) phenotypic distribution with the modes at the extremes of the distribution was not observed. An alternative model of many loci with small additive effects was also rejected as it predicts a symmetrical unimodal distribution with the mode close to the mean, which was also not observed. It is thus likely that mutations at a small number of loci determine sleep loss. Moreover, eye and pupil sizes and the presence or absence of melanin were measured in surface x Pachón F2 hybrids that had been assayed for sleep. The relationships between sleep and the other three phenotypes were not significant, suggesting that genes underlying sleep loss are independent from those responsible for eye and pigment regression. All three cave populations have evolved a decreased sleep phenotype but it varies in subtle but significant ways, suggesting that the underlying genetic mechanisms might also vary.

## 3.4 Space-Dependent Growth

Most fish when raised in confined space do not grow as large as when they are in unconfined space. The effects of confined space on growth could be due to poor water quality, lower oxygen and food availability, higher level of excretion products, and limited exercise space. An alternative explanation is that fish may have evolved an intrinsic mechanism to regulate their growth according to the space available. In *A. mexicanus*, it has been shown that environmental size has a major impact on surface fish growth, but this impact is much reduced in cavefish (Gallo & Jeffery, 2012). Mating experiments between surface and Pachón cavefish have shown that space dependent growth (SDG) is inherited as a dominant trait, as hybrids show SDG like in surface fish. In contrast to F1 hybrids, F2 hybrids showed an intermediate SDG phenotype with significant SDG in terms of mass, but no significant SDG in terms of length. In addition, there was a great deal of variation and decoupling in the SDG phenotype among F2 hybrids. Some individuals showed SDG in terms of mass but not in terms of length, whereas others showed SDG in terms of length but not in terms of mass, indicating that SDG segregated in the F2 generation. The results also imply that SDG is a multigenic trait.

# 4. GENETIC DETERMINISM OF TROGLOMORPHIC TRAITS: INSIGHTS FROM A QUANTITATIVE TRAIT LOCI APPROACH

## 4.1 Linkage Maps for quantitative trait loci Analyses

The search for QTL is a straightforward approach to identify at the molecular level the genetic differences that are involved in phenotypic differences (Fig. 2). However, a high-resolution mapping of QTLs requires the use of a high-density linkage map of polymorphic markers (Fig. 3), which have only recently been obtained for *A. mexicanus* (Carlson, Onusko, & Gross, 2015). Nevertheless, the search for QTL started in the early 2000s and lead to very important progress in the identification of genes involved in cavefish evolution.

The first linkage map and QTL analyses were based on the progeny of an F1 hybrid female (Pachón x surface) mated with a Pachón male and used markers generated from random amplified polymorphic DNA fingerprinting (Borowsky & Wilkens, 2002). Three QTL involved in eye size, two QTL involved in melanophore number, and two QTL involved in condition factor (the residual of the log of weight after regression of log weight on log length, a surrogate of metabolic efficiency) were identified. As the authors expected to find about half of the QTL involved in the variance of these traits, considering that the map was about half complete, they concluded that these results were consistent with previous quantitative genetic estimates reported above. In addition, they found two pairs of linked QTL for constructive and regressive traits, ie, eye size/condition factor and melanophore number/condition factor. Such close linkage of regressive and constructive QTL could arise from pleiotropy. If a gene controlling eye size also influenced metabolism, selection on metabolism could also affect the eyes. Seemingly "neutral" changes on one trait could be indeed driven by natural selection on another correlated trait.

A higher-resolution map using more individuals and markers composed of polymorphic microsatellites was obtained soon after (Protas et al., 2006). Using this linkage map and a candidate gene approach, different loss-of-function mutations (ie, different exon deletions) were found in the *Oca2* (ocular and cutaneous albinism 2) gene that are responsible for albinism in Pachón and Molino individuals. Albino individuals from Japonés were crossed with Pachón albino and the offspring were albino too, suggesting that the same gene was involved. Nevertheless no loss-of-function mutation

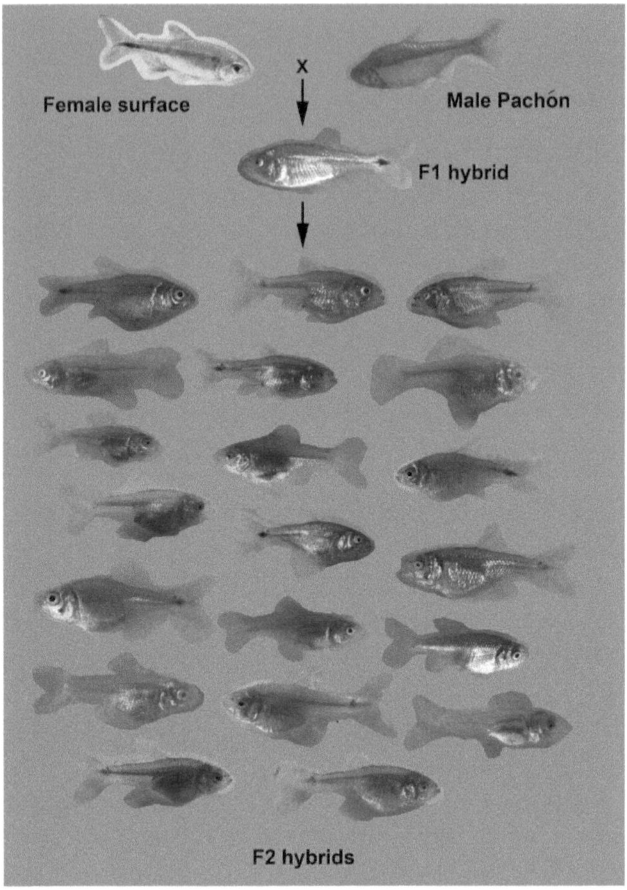


**Figure 2** ***Crosses for implementing quantitative trait loci (QTL) approach in* Astyanax.** Establishing crosses is the first step of the QTL approach. An example is shown here, with a Surface X Pachón cross giving rise to a first generation hybrid family (F1). These F1 fish are pigmented and have small eyes but are fully sighted. Subsequent crossing between two F1s (brother X sister cross) generates a second generation hybrid family (F2). Shown here are 20 randomly picked individuals from the F2 panel, photographed at 2 months of age. The segregation of traits is evident concerning pigmentation, eye and lens size. For example, a pigmented individual can have a small nearly degenerated eye, and a depigmented albino individual can have large eyes. (See color plate)

in *Oca2* was found in Japonés, suggesting another mutation in a regulatory sequence.

Using this linkage map, 48 other QTL were found: 8 affecting eye size, 6 affecting lens size, 18 affecting pigmentation, 7 affecting the length of the jaw bones, 6 affecting the number of maxillary teeth, and 3 affecting the

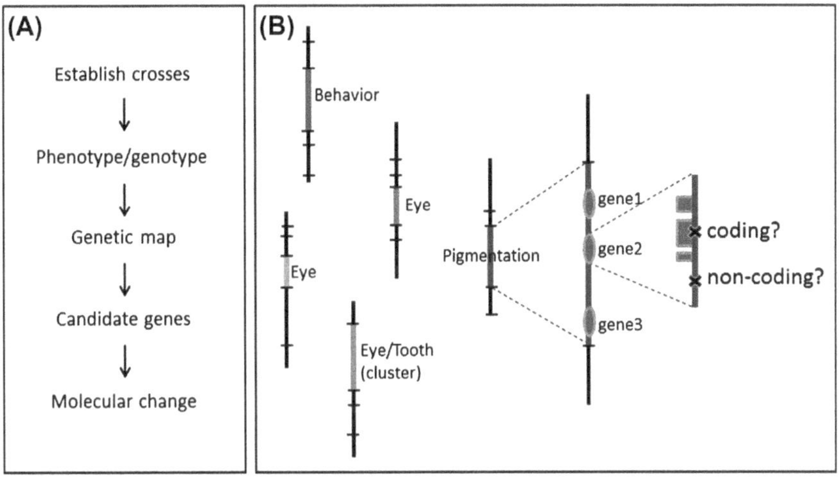


**Figure 3** ***Quantitative trait loci (QTL) approach in*** **Astyanax.** Scheme highlighting the steps of the quantitative genetics approach in this model system. *The figure is inspired from Peichel, C.L. (2005). Fishing for the secrets of vertebrate evolution in threespine sticklebacks.* Developmental Dynamics, 234*(4), 815–823., reporting the same type of QTL pipeline analysis for sticklebacks.* (See color plate)

number of taste buds (Protas, Conrad, Gross, Tabin, & Borowsky, 2007). Again, several sets of comapping QTL were identified. Interestingly, cave alleles at all 12 eye or lens QTL had relatively modest effects but all decreased eye or lens size, whereas cave alleles at QTL affecting melanophore number had positive ($n = 5$) as well as negative slopes ($n = 8$), and their substitution effects were much larger. The interpretation of the differences in the polarity of the effects of cave alleles between these characters was that regression of eyes resulted from natural selection, whereas decreases in numbers of melanophores resulted from genetic drift or indirect selection through gene pleiotropy. Indeed, if eye reduction were accomplished through genetic drift, it is unlikely that the pattern of effects would contrast so radically with that for melanophores. It was thus asserted that eyes in a cave could be actually disadvantageous. The argument against selection is that the cost of making an eye is probably trivial compared to the cost of its replacement tissue in the socket, or that the developmental cost is paid by cavefish anyway because eyes start developing and only degenerate after many cell cycles of tissue growth and replacement (Jeffery, 2005). However, it has been recently suggested that these arguments might be considering the wrong costs. The cost of vision was calculated to be 15% of resting metabolism in young fish and about 5% in older fish, demonstrating that the loss of the visual system in the cave could substantially lower the amount of energy expended on eye and brain structure such as the optic tectum, in particular for juvenile fish (Moran, Softley, & Warrant, 2015).

## 4.2 Quantitative Trait Loci Clusters

The same type of study as above was extended to assess 12 traits that differ between cave and surface fish (Protas et al., 2008). Clusters of QTL for different traits were observed. Whereas it cannot be excluded that these clusters of QTL could result from the effects of clusters of tightly linked genes each controlling a different trait, this observation suggested that several QTL clusters may reflect pleiotropic effects of alleles at loci involved in several phenotypic differences observed between cave and surface fish. These authors hypothesized that when the surface fish first entered the cave environment, numerous loci were potentially capable of giving rise to QTL clusters, but only a small number eventually sustained new mutations (or had "precave" rare allelic variants) with relevant sets of pleiotropic effects. These alleles were fixed by selection on the consensus of their strongest positive effects, creating the modern day associations. This scenario would account for the large number of clusters observed, their richly diverse phenotypic contents, as well as the sometimes counterintuitive polarity of substitution effects of some of their constituents.

Without the *A. mexicanus* genome sequence, it was impossible to identify the genes underlying the QTL, with the exception of *Oca2*. However, a comparative genomic analysis demonstrated that extensive synteny is conserved between *A. mexicanus* and *D. rerio* (Gross et al., 2008) allowing the search for candidate genes in the *D. rerio* genome. Thereby, one putative gene (*Hoxc8a*) for rib number and three putative genes (*Crystaa*, *Rom1*, *Shroom2*) for eye size reduction were identified.

## 4.3 Refining Quantitative Trait Loci Analyses for Morphological and Behavioral Traits

Applying the approach used to identify mutations in the *Oca2* gene involved in albinism, ie, a combination of QTL search and candidate gene sequencing, mutations in the *Mc1r* (melanocortin receptor 1) gene were identified as being responsible for the brown phenotype (Gross, Borowsky, & Tabin, 2009). A 2-bp deletion in Pachón cavefish was predicted to cause a frameshift leading to the introduction of a premature stop codon and to produce a truncated, most likely nonfunctional protein. Another mutation was found in Yerbaniz and Japonés individuals, C490T, causing a cysteine substitution at position R164, which is homologous to the R160W mutation known in human to cause loss of Mc1r receptor function. Complementation tests indicated that other cave populations, such as the Piedras and Curva caves, have brown phenotypes caused by mutations at the same locus as in Pachón and Yerbaniz. As no mutation was found in their Mc1r coding sequence, these populations likely carry regulatory mutations leading to a decrease or loss of Mc1r expression and activity.

With the goal of linking the regression of eyes to the evolution of adaptive traits, a search for QTL involved in the enhancement of vibration attraction behavior (VAB) and the greater number of superficial neuromasts (SN) in the eye orbit (EO) that mediate this behavior was undertaken (Yoshizawa et al., 2012). Using F2 progeny of a Pachón x surface F1 hybrid, the cave alleles at two loci were found to increase VAB in cavefish. These cavefish alleles at the same two loci or linked loci were found to increase the number of SN in the eye orbit. In addition, the cave alleles at the same two loci or closely linked loci were found to decrease the eye diameter. Three other QTL were only involved in eye size variation. The authors of this study concluded that natural selection for the enhancement of VAB and EO SN indirectly promotes eye regression in the Pachón cavefish population through an antagonistic relationship involving genetic linkage or pleiotropy among the genetic factors underlying these traits. This hypothesis was soon challenged (Borowsky, 2013; Yoshizawa, O'Quin, & Jeffery, 2013) because many QTL map to the same two very large chromosomic regions as many other characters, and therefore pleiotropy was not demonstrated. Moreover, even if pleiotropy is involved in eye reduction, it does not exclude the possibility that other evolutionary mechanisms such as direct selection for eye reduction as well as random fixation of alleles that reduce eye size could be involved, too.

In order to better understand the mechanism involved in eye degeneration, a QTL search was performed on retinal degeneration (O'Quin, Yoshizawa, Doshi, & Jeffery, 2013). Four QTL were identified, each one associated with the thickness of the ganglion, inner nuclear, outer plexiform, and outer nuclear layers of the retina. For all but one QTL, cavefish alleles resulted in a clear reduction in the thickness of the corresponding layer. Comparative mapping of genetic markers within each QTL revealed that each QTL corresponds to an approximately 35-Mb region of the zebrafish genome. Within each region, several candidate genes were identified that are associated with the function of each affected retinal layer but none have yet been validated.

A QTL search was also performed on loss of scleral ossicles (O'Quin et al., 2015). A strongly skewed distribution of scleral ossicles found among surface fish x Pachón cavefish F2 hybrids suggests that scleral ossification is a threshold trait with a complex genetic basis. Quantitative genetic mapping identified a single QTL for scleral ossification on *Astyanax* linkage group 1 and the threshold for this trait is likely determined by at least three genetic factors, which may also control the severity and onset of lens degeneration in cavefish (O'Quin et al., 2015).

Several QTL involved in feeding posture were identified (Kowalko, Rohner, Linden, et al., 2013). Two QTL were found using F2 fish from a surface x Pachón F1 cross. Only one QTL was found using F2 fish from a surface x Tinaja F1 cross. Both Pachón cave alleles reduced the feeding angle as expected, whereas the Tinaja allele increased the feeding angle. In both F2 fish, only a small fraction of the variance was explained by the QTL, ~13 and ~11%, respectively. The QTL were on different linkage groups suggesting that the feeding posture evolved, at least partially, through genetic changes at different loci.

Looking for craniofacial changes that could have a genetic determinism in cavefish (Gross, Krutzler, & Carlson, 2014), QTL were found for 7 craniofacial phenotypes among 33 that were scored. Four of these were present on both the left and right sides of the head, including (1) total bony area of the second suborbital bone (SO2)(3 QTL), (2) total bony area of the third suborbital bone (SO3)(2 QTL), (3) total bony area of the fourth suborbital bone (SO4)(3 QTL), and (4) total bony area of the fifth suborbital bone (SO5)(1 QTL). Asymmetry in the genetic architecture was observed for three traits when comparing these traits on the right and left sides of the cranium, including (1) SO1+2 fusion (2 QTL), (2) SO3 fragmentation (2 QTL), and (3) SO4+5 fusion (1 QTL). Of these craniofacial phenotypes that yielded QTL, only four linked markers did not map to regions associated with previously identified troglomorphic phenotypes. This may indicate that certain regions of the cavefish genome, populated with multiple genes, have accumulated derived alleles during cave adaptation. Alternatively, the craniofacial QTL identified, along with previously discovered QTL, may be mediated by the same alleles of pleiotropic genes or may be closely linked to the causative genes controlling other cave-associated traits (Protas et al., 2008).

In summary, about 10 years of QTL analyses have provided a wealth of candidate genes that may underlie cavefish troglomorphic traits. They are summarized in Table 1. Fig. 4 is an attempt to summarize and compile those

**Table 1** QTL mapped after crosses between cavefish and surface fish of *Astyanax mexicanus*

| Year of publication | 2002 | 2006 | 2007 | 2008 | 2009 | 2012 | 2013 | 2013 | 2014 | 2014 | 2015 |
|---|---|---|---|---|---|---|---|---|---|---|---|
| References | 1 | 2 | 3 | 4 | 5 | 6 | 7 | 8 | 5 | 10 | 11 |
| Crosses | Pa × SF | Mo × SF Pa × SF | Pa × SF | Pa × SF | Mo × SF Pa × SF | Pa × SF | Pa × SF | Pa × SF | Ti × SF | Pa × SF | Pa × SF |
| Eye size | 3 | | 8 | 8 | | 5 | | 5 | 1 | 2 | 2 |
| Eye area | | | | | | | | | | | 5 |
| Lens size | | | 6 | | | | | | 1 | | 1 |
| Pigmentation | 2 | | 18 | 4 | | | | | | | |
| Albinism | 1 | 1 | | | | 1 | | | | 1 | |
| Condition factor | 2 | | | 9 | | | | | | | |
| Jaw bone(s) length | | | 7 | | | | | | | | |
| Maxillary tooth number | | | 6 | | | | | | | | |
| Taste bud number | | | 6 | | | | | 2 | | | |
| Weight loss | | | | 1 | | | | | | | |
| Tooth number | | | | 5 | | | | | | | |
| Peduncle depth | | | | 3 | | | | | | | |
| Fin placement | | | | 9 | | | | | | | |
| Anal fin rays | | | | 2 | | | | | | | |
| SO3 width | | | | 6 | | | | | | | |
| Rib number | | | | 6 | | | | | | | |
| Length | | | | 5 | | | | | | | |
| Chemical sense | | | | 4 | | | | | | | |

(*Continued*)

**Table 1** QTL mapped after crosses between cavefish and surface fish of *Astyanax mexicanus*—cont'd

| **Year of publication** | **2002** | **2006** | **2007** | **2008** | **2009** | **2012** | **2013** | **2013** | **2014** | **2014** | **2015** |
|---|---|---|---|---|---|---|---|---|---|---|---|
| **References** | **1** | **2** | **3** | **4** | **5** | **6** | **7** | **8** | **5** | **10** | **11** |
| **Crosses** | **Pa × SF** | **Mo × SF**<br>**Pa × SF** | **Pa × SF** | **Pa × SF** | **Mo × SF**<br>**Pa × SF** | **Pa × SF** | **Pa × SF** | **Pa × SF** | **Ti × SF** | **Pa × SF** | **Pa × SF** |
| Brown melanophore | | | | | 1 | | | | | | |
| Vibration attraction | | | | | | 2 | | | | | |
| Superficial neuromasts | | | | | | 2 | | | | | |
| Retinal thickness | | | | | | | 4 | | | | |
| Feeding angle | | | | | | | | 2 | | | |
| Jaw angle | | | | | | | | 1 | | | |
| Head depth | | | | | | | | 1 | | | |
| Body depth | | | | | | | | 1 | | | |
| Schooling behavior | | | | | | | | | 2 | | |
| Dark preference | | | | | | | | | 1 | | |
| Craniofacial bones | | | | | | | | | | 12 | |
| Sex | | | | | | | | | | 1 | |
| Standard length | | | | | | | | | | 1 | |
| Scleral ossification | | | | | | | | | | | 1 |

Origin of the fish used in the crosses: SF = Surface fish; Pa = Pachón cave; Mo = Molino cave; Ti = Tinaja cave.
1 - Borowsky and Wilkens (2002), 2 - Protas et al. (2006), 3 - Protas et al. (2007), 4 - Protas et al. (2008), 5 - Gross et al. (2009), 6 - Yoshizawa et al. (2012), 7 - O'Quin et al. (2013), 8 - Kowalko, Rohner, Linden, et al. (2013), 9 - Kowalko, Rohner, Rompani, et al. (2013), 10 - Gross et al. (2014), 11 - O'Quin et al. (2015).

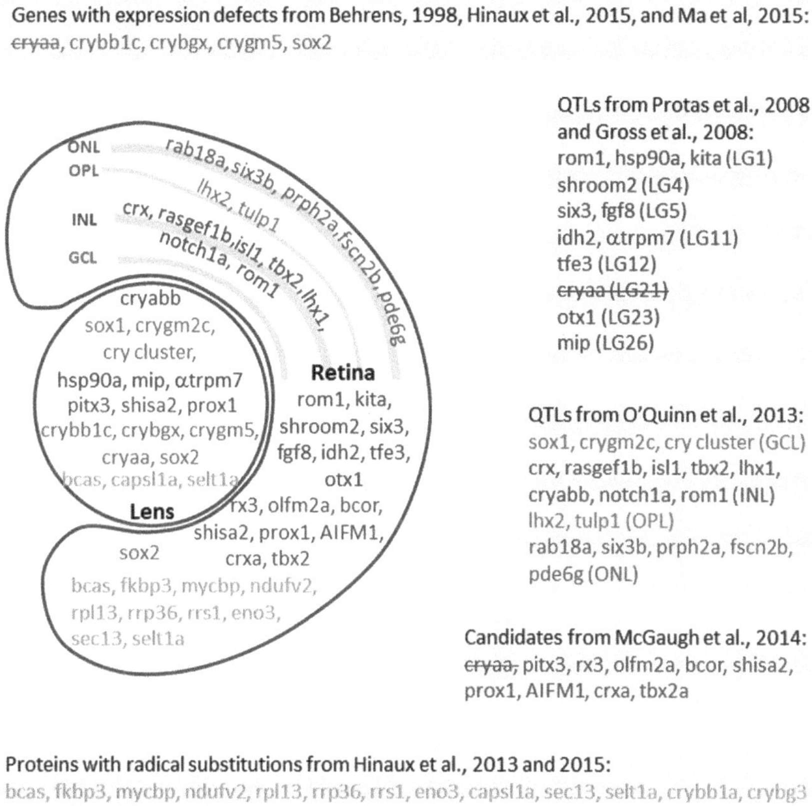


**Figure 4** ***Genetics of eye loss in cavefish.*** Candidate genes potentially involved in eye loss in cavefish are reported on a schematic eye, including a retina and a lens. The original papers from which data were taken are reported in gray boxes, and the candidate genes from these papers are color coded. Only genes that are expressed in the eye, according to the ZFIN zebrafish expression database, are reported. On the schematic eye, genes are indicated according to their expression patterns, ie, in the retina, in the lens, or both. Of note, cryaa (in red, orange, and brown) is crossed out in all cases because there is evidence that *cryaa* is not the causal gene in the identified quantitative trait loci (QTL) interval, although absence of *cryaa* expression in the cavefish lens is indeed deleterious to the lens (see text). (See color plate)

candidate genes found in "eye QTL" intervals and that are actually expressed in the eye. Clearly, researchers are left with a wealth of interesting, potentially biologically relevant candidates. The discovery of the exact molecular changes and their functional validation now awaits, and should speed up with the help of genomic information and the genetic tools now available (see next paragraphs).

## 5. GENETIC DETERMINISM OF TROGLOMORPHIC TRAITS: A CANDIDATE GENE APPROACH

A candidate gene approach, without preliminary QTL mapping, has been successfully applied several times. This approach is based on prior knowledge of gene function in other species suggesting that coding sequence differences in these genes, in their *cis*-regulatory sequences or in their *trans*-regulatory factors, could be involved in cavefish versus surface fish phenotypic differences.

### 5.1 Usual Suspects: Pax6 and Hedgehogs

Several lines of evidence suggest that Pax6 is a prime candidate for mediating cavefish eye degeneration (Gehring & Ikeo, 1999; Strickler, Yamamoto, & Jeffery, 2001). The *Pax6* gene encodes a paired-class homeodomain transcription factor expressed during early eye development in a pattern suggesting a critical role in lens and retina specification and patterning in vertebrates (Puschel, Gruss, & Westerfield, 1992). Ectopic expression of *Pax6* results in the formation of ectopic eyes in *Drosophila* (Halder, Callaerts, & Gehring, 1995) and *Xenopus* (Chow, Altmann, Lang, & Hemmati-Brivanlou, 1999), and the morphology of the micropthalmic cavefish eye resembles the Small eye phenotype in mouse, which is caused by a mutation in the *Pax6* gene (Hill et al., 1991). In early surface fish embryos, two bilateral *Pax6* expression domains are present in the anterior neural plate, which extends across the midline and fuse to form the forebrain and optic primordia. In cavefish embryos, these *Pax6* domains are diminished in size and remain separated, resulting in an anterior gap in *Pax6* expression and presumably the formation of smaller optic primordia (Behrens, Langecker, Wilkens, & Schmale, 1997; Strickler et al., 2001). Moreover, similar anterior gaps in Pax6 expression are observed in independently derived cavefish populations, suggesting that they may be important in eye degeneration. However, and somehow disappointingly, the *Pax6* gene itself is not found in regions encompassing QTLs for eye size or retinal layers.

Since it is well known that hedgehog proteins diffusing from the prechordal plate regulate the size of the optic primordia by suppressing *Pax6* expression in the neural plate (Ekker et al., 1995), the diminutive *Pax6* domains and optic primordia in cavefish embryos may be a consequence of a higher level of hedgehog signaling emanating from the prechordal plate. Hedgehog genes are thus obvious candidates for acting upstream of *Pax6*. As expected, the two hedgehog paralogs analyzed, *sonic hedgehog* (*shh*) *and tiggy-winkle hedgehog* (*twhh*), have an expression domain that is expanded along the anterior embryonic midline. The expansion of hh signaling results in hyperactivation of downstream genes, lens apoptosis, and arrested eye growth and development. These features can be mimicked in surface fish by *twhh* and/or *shh* overexpression, supporting the role of hh signaling in the evolution of cavefish eye regression (Yamamoto et al., 2009; Yamamoto, Stock, & Jeffery, 2004). Similarly, expanded *hh* genes expression was observed in independently derived cavefish populations, suggesting a key role in eye degeneration. Shh signaling also has an important role in oral and taste bud development and hyperactive Shh signaling increases oral and taste bud amplification in cavefish at the expense of eyes. Therefore, selection for constructive oral—pharyngeal traits may be responsible for eye loss during cavefish evolution via pleiotropic function of the Shh signaling pathway. However, once again neither *twhh* nor *shh* genes are found within eye QTL regions.

## 5.2 Lens Defects: Alpha-A Crystallin and Sox2

Cavefish eyes are lost through apoptosis of the lens, which in turn promotes the degeneration of other optic tissues. As *alpha-A-crystallin* (*cryaa*) expression is missing in the cavefish lens (Behrens, Wilkens, & Schmale, 1998; Hinaux et al., 2015; Ma, Parkhurst, & Jeffery, 2014; Strickler, Byerly, & Jeffery, 2007) and the *cryaa* gene is located in a genomic region (QTL) responsible for eye loss (Gross et al., 2008; McGaugh et al., 2014), *cryaa* has been proposed as a candidate for regulating cavefish eye degeneration. The involvement of *cryaa* in eye degeneration was confirmed by knocking down its expression in surface fish, which led to apoptosis of the lens (Ma et al., 2014) and by reexpressing it in transgenic Pachón CF, which protected lens cells from apoptosis (Hinaux et al., 2015).

The underlying reason for absence of *cryaa* expression in cavefish was investigated by comparing genomic *cryaa* sequences in surface fish and cavefish. However, no obvious changes in *cis*-regulatory sequences were identified (Ma et al., 2014). Furthermore, the cavefish *cryaa* allele is expressed in surface fish x cavefish F1 hybrids, indicating that changes in upstream genes are most likely responsible for *cryaa* downregulation (Ma et al., 2014), and that *cryaa* is unlikely to be the gene associated with an *Astyanax* eye QTL.

In other species, Sox2 is one of the transcription factors that regulate lens crystallin genes during eye development. Comparison of *Sox2* expression patterns during surface fish and cavefish development showed that *Sox2* is not expressed in the cavefish lens. The upstream regulatory function of

Sox2 was demonstrated by knockdown in surface fish, which abolished *cryaa* expression and induced lens apoptosis (Ma et al., 2014).

### 5.3 The Quiet Fish: Monoamine Oxidase and the Cavefish Behavioral Syndrome

Regarding the cavefish "behavioral syndrome," researchers have investigated their aminergic neuromodulatory transmission systems. A "hyperaminergic" phenotype in cavefish brains was found, including high levels and neurotransmission indexes for serotonin, dopamine and noradrenaline, and low monoamine oxidase (MAO) activity. Although MAO expression is unchanged in cavefish, a pro106leu mutation was identified in the MAO coding sequence. This mutation is responsible for low MAO activity and high serotonin neurotransmission index in cavefish brains. The same mutated allele was found in several natural cavefish populations, some having evolved independently. The occurrence of the same allele in several cave populations suggests that low MAO activity is advantageous for cave life and could provide the genetic basis for several aspects of the cavefish "behavioral syndrome," including loss of social hierarchical order and aggressiveness, loss of schooling behavior, and persistent food searching behavior (Elipot, Hinaux, et al., 2014).

### 5.4 MC4R and Metabolism

Finally, concerning the physiological changes observed in cavefish regarding energy homeostasis and feeding behavior, a candidate gene-based screen led to the identification of coding mutations at conserved residues of the melanocortin 4 receptor gene (*MC4R*) of Tinaja cavefish that could contribute to the insatiable appetite found in some populations of cavefish. One of these mutations has been shown to be linked to obesity in humans. The allele results in both reduced maximal response and reduced basal activity of the receptor in vitro. It also contributes to elevated appetite, growth, and starvation resistance in vivo. This derived allele appears to be fixed in cave populations in which the overeating phenotype is present. The presence of the same allele in multiple caves could be due to selection from standing genetic variation present in surface populations (Aspiras, Rohner, Martineau, Borowsky, & Tabin, 2015).

## 6. GENOMICS AND EVOLUTIONARY MOLECULAR GENETICS

The karyotypic variability of the genus *Astyanax* is relatively high, with diploid numbers ranging from 36 to 50. Both surface and cave

*A. mexicanus* (cave origin was not reported in the publications) have been karyotyped and have a diploid number of $2n = 50$ chromosomes for both sexes (Kavalco & De Almeida-Toledo, 2007; Kirby, Thompson, & Hubbs, 1977). The karyotypic formula of cavefish is 8M + 18SM + 12ST + 12A (Kavalco & De Almeida-Toledo, 2007). Thus, each of the 25 linkage groups analyzed in most QTL studies corresponds to one chromosome (Carlson et al., 2015).

## 6.1 Cavefish Transcriptomics

With the advent of next-generation sequencing technologies, efforts have initially concentrated on the comparative analyses of surface fish and cavefish transcriptomes, during development and in adults (Gross, Furterer, Carlson, & Stahl, 2013; Hinaux et al., 2015, 2013). These papers have concentrated on the Pachón cavefish.

Hinaux and colleagues sequenced the transcriptomes of pooled embryos and larvae, which corresponds to crucial stages for the understanding of eye development and degeneration in cavefish. Their aim was (1) to generate cDNA libraries from biologically relevant developmental stages for surface fish and cavefish, as a clone resource and (2) to generate transcriptome data for analysis of the genetic basis of cavefish developmental evolution through polymorphism analysis (Hinaux et al., 2013). The level of polymorphism was about two times higher in surface fish than in cavefish, confirming previous findings based on microsatellites (Bradic et al., 2012). A simple explanation is that cavefish have a smaller effective population size than surface fish, and possibly underwent population bottlenecks due to environmental variations. A significant number of these polymorphic sites were shared between the two morphs (207 in the data set), which indicate either a recent origin of the Pachón cave population, or migrations and mixing of the two populations. Cavefish and surface fish have also fixed different alleles (940 population-specific, fixed, alleles in the data set). Divergent polymorphism, as expected, was very rare.

Relatively surprisingly, no indels and no stop codons were found in cavefish "eye genes" (Hinaux et al., 2015, 2013) (see also Yokoyama, Meany, Wilkens, & Yokoyama, 1995). Rather, the only stop codon detected in a cavefish sequence affected a member of the *srd5a* gene family, encoding steroid 5α-reductases involved in steroid hormone metabolism (Hinaux et al., 2013). In humans, these enzymes influence sperm quality, and polymorphisms in SRD5A are associated with prostate cancer. The consequence of its loss of function in cavefish remains unknown. In addition, a GO term enrichment analysis of proteins carrying radical mutations in cavefish showed an overrepresentation of proteins involved in carbohydrate metabolism. This may be related to differences that have been reported between the two morphs concerning energy storage and metabolism (Moran et al., 2014; Salin et al., 2010). Finally and interestingly, among those proteins showing radical substitutions in cavefish, about a third corresponded to "eye genes" (expressed in the developing eye, based on their expression patterns in the ZFIN zebrafish in situ hybridization database). This study therefore provided a list of 11 genes representing good candidates for playing a role in cavefish eye degeneration (Hinaux et al., 2013).

Gross and colleagues sequenced the transcriptomes of two cave and two surface adult *Astyanax* (Gross et al., 2013). Their data set identified sequence polymorphisms between the two morphotypes, yielding numerous markers for genetic studies, some of which were validated and used through a QTL reanalysis of albinism in a Surface x Pachón F2 pedigree. They also adapted an RNA-seq-based method to identify "exclusive" genes that are expressed in cavefish but absent from surface fish, and vice versa. As expected, they observed that numerous vision-related genes fail to be expressed in adult eyeless cavefish. More interestingly, they also found an overrepresentation of metabolism-related genes among those genes that are expressed exclusively in Pachón cavefish but not detected in surface fish. They discussed this finding as providing genetic support for the notion that the impoverished cave environment selects for metabolic efficiency in cave-dwelling organisms (Gross et al., 2013).

These pioneering transcriptomic studies therefore converged on a global picture where "eye genes" and "metabolic genes" are particularly subjected to changes in cavefish, both in terms of molecular evolution (Hinaux et al., 2015, 2013) and in terms of expression levels (Gross et al., 2013). That visual genes are affected can shed light on the evolutionary forces at play during cavefish evolution, but was anticipated. Conversely, the insights on metabolic processes have opened new windows on the genetic mechanisms of cavefish evolution and adaptation, which are currently the focus of new research directions in the *Astyanax* community.

## 6.2 Cavefish Genomics

The "cavefish draft genome paper" reporting the first de novo assembly of the genome of a Pachón female individual was published in October 2014 (McGaugh et al., 2014). The cavefish genome is about 1 Gb in length, and a total of 23,042 protein-coding genes were predicted. One-third of the cavefish genome is composed of TEs, a content that is comparable to most published fish genomes with the exception of zebrafish (52.2%) and *Fugu* (2.8%).

The genome was examined for genes within the QTL region for eye size from Pachón cavefish x surface fish crosses from previous studies (detailed above). A total of 2408 genes were associated with these genomic regions. Using transcriptomics expression data from Pachón and surface fish and integrated pathway analysis to refine this list, specific candidate genes under eye-related QTL were highlighted, and considered as particularly promising: *cryaa*, *pitx3*, *rx3*, *olfm2a* and *olfml2a*, *BCoR*. Some candidate genes within the QTL with potentially pleiotropic effects, ie, for which multiple troglomorphic phenotypes colocalize with eye size, were also proposed. For *shisa2*, validation through anatomical analysis detected a lack of *shisa2* expression in the cavefish lens, which suggests that changes in the regulatory region of this gene may contribute to the loss of eyes in cavefish. Other candidates included *prox1*, *AIFM1*, *crxa*, and *Tbx2a*. It should be noted that coding sequence changes in all these candidate genes were very modest or absent, and never corresponded to frameshifts, deletions, or stop codons.

Analyses of putative gene losses by comparison with zebrafish and eight other teleost genomes available in Ensembl also provided a very preliminary list, in the absence of full genome coverage, of 550 genes. Notably, these genes were not enriched for the 305 gene ontology accessions related to eye development or function. This is in sharp contrast to the case of another subterranean animal, the naked mole rat *Heterocephalus glaber*, for which 19 visual perception genes are inactivated or missing in the genome, either through complete loss detected using synteny information, a frameshift, or a premature stop codon (Kim et al., 2011).

The availability of the Pachón cavefish genome has revolutionized the study of the genetic basis of troglomorphism in *Astyanax* (McGaugh et al., 2014), and has helped the research community enter the genomic era. Population genomics studies, as well the search for signs of genome-wide selection, or epigenetic marks, will now be feasible. However, the possibilities in terms of comparative genomics are still limited, due to the lack of a corresponding surface fish genome, as well as the genomes of other cavefish populations.

In particular, the surface fish genome may soon unravel the genetic bases of a potentially ongoing speciation process between surface and cave types of *Astyanax*. This is suggested for example by the 100% female F1 progeny obtained after some crosses of Pachón females with either surface fish or

Molino males (Borowsky, 2008a; Elipot, Legendre, Pere, Sohm, & Retaux, 2014). Indeed, Borowsky and colleagues recently discussed that with standing genetic variation providing numerous alleles combined with strong ecological pressure, adaptation to cave life may have been relatively rapid. In turn, these factors would provide the conditions for the evolution of numerous correlated (pleiotropic) effects throughout the genome, including those promoting reproductive isolation. Indeed, they observed numerous departures from Mendelian expectations for allelic frequencies in hybrids between surface and cave *Astyanax* (Borowsky & Cohen, 2013). Both transmission ratio distortions of blocks of linked loci and epistatic interactions between pairs of unlinked loci were observed, which is likely to reduce hybrid fitness. Transmission ratio distortion occurs mainly between the diploid stage and that of fully mature sperm, as shown by the differences in allelic content of DNA from somatic tissue versus that from mature sperm; however, the underlying mechanism is unknown. The network of epistatic interactions between loci included seven genes thought to be involved in cavefish evolution: *igfbp5* and *tfe3*, which comap with cave trait QTL (Protas et al., 2008), *oca2* and *mc1r*, which control melanin phenotypes in cave populations (Gross et al., 2009; Protas et al., 2006) and *hsp90a*, *fgf8*, and *shh*, which have variant patterns of expression in cave embryos that influence the development of cave phenotypes (Hooven, Yamamoto, & Jeffery, 2004; Pottin, Hinaux, & Retaux, 2011; Yamamoto et al., 2004). The authors' working hypothesis was that the biases against heterozygote combinations reflect functional incompatibilities of alleles from the two populations, and that these incompatibilities have deleterious effects on sperm as well as zygote function. Thus, interacting loci may be good candidates for "speciation genes," with interacting cave-specific alleles fixed during adaptation to cave life.

## 7. FUNCTIONAL GENETICS AND MOLECULAR TOOLS

As an "emergent" model organism, *A. mexicanus* cannot compare with established fish models such as zebrafish or medaka in terms of the development of genetic tools including catalogs of transgenic lines, driver lines, or mutants. Indeed, in the case of *Astyanax*, the goal of researchers is to understand the evolutionary mechanisms underlying the phenotype of cavefish, which are considered as (complex) natural mutants. Hence, the small *Astyanax* community is united by an absolute interest to generate specific tools and reagents to help understand the genetic basis of cave-specific traits, but large scale programs have not yet been discussed.

## 7.1 Resources

In terms of molecular resources available, the situation has progressively and significantly improved during the past 10 years. First, a bacterial artificial chromosome (BAC) genomic library for the surface form of *A. mexicanus* has been made available (Di Palma, Kidd, Borowsky, & Kocher, 2007). It provides a 3.6X coverage of the surface fish genome, obtained from pooled specimens from Arroyo Sarco near the village of Encino, Tamaulipas, Mexico, with an average insert size of 104 kb. It can be used for various applications, including sequencing of BACs of interest as well as BAC transgenesis. In addition, cDNA libraries have been constructed from pooled embryos and larvae, both for surface fish and cavefish (Hinaux et al., 2013). The mean insert size is 900 bp, and sequences correspond to Pachón cavefish and Texas surface fish originally collected in San Salomon Spring, Balmorhea State Park. They have been used for comparative transcriptomics and also constitute an invaluable resource for gene expression studies, bypassing the need for tedious PCR cloning of cDNA probes. Then, the available Pachón cavefish genome (reviewed in the previous section) now provides the necessary "omics" tools to the *Astyanax* community (McGaugh et al., 2014).

## 7.2 Functional Genetics, Transgenesis, Genome-Editing

Functional genetic analysis started about 10 years ago in the *Astyanax* system, with classical gene expression perturbation experiments during development. As the *Astyanax* eggs are transparent, develop rapidly at about the same pace as zebrafish eggs during the first 20 h of development, show synchronous development between surface and cave embryos, and are relatively easy to obtain in numbers either through natural spawning or by in vitro fertilization (Hinaux et al., 2011), most methods used in other fish model species such as zebrafish and medaka can be applied.

A seminal work for cavefish evo-devo used *shh* and *twhh* mRNA injections to investigate the impact of increased early midline hedgehog signaling on eye development and degeneration in cavefish (Yamamoto et al., 2004). *Hedgehog*-injected surface fish embryos mimicked (generally on one side only) the cavefish phenotype in terms of neural tube patterning, lens defects, and eye morphogenesis. Transient and conditional transgenesis was further used to demonstrate the pleiotropic functions of embryonic *shh* expression and thereby to link jaw and taste bud amplification with eye loss during cavefish evolution (Yamamoto et al., 2009). This was achieved through injection of a linearized *hsp70:shh:GFP* DNA expression construct at the two to four cell stage, followed by a 1 h heat shock at 36°C performed at various stages of development to induce *shh* overexpression (itself monitored by following GFP expression). Hence, these functional experiments were the first to suggest that selection for constructive oral-pharyngeal traits may be responsible for eye loss during cavefish evolution, via the pleiotropic roles of the hedgehog signaling pathway.

In parallel, efforts have concentrated on understanding the functional genetic basis of the autonomous apoptotic lens defect in cavefish, itself previously demonstrated in classical embryological experiments through transplantation procedures (Yamamoto & Jeffery, 2000). Using standard tools borrowed from the fish developmental biology community, a morpholino gene knockdown approach was applied to demonstrate the role of the transcription factor Sox2 in the lack of expression of the *cryaa* gene in the cavefish lens, and further, the role of the lack of *cryaa* expression in lens apoptosis (Ma et al., 2014). These data, coupled to a nicely designed "*cis*-trans F1 hybrid" genetic test, also prompted the discussion that although the candidate gene *cryaa* is located in a QTL interval identified for controlling eye size (see above), it is unlikely to harbor a mutation responsible for cavefish eye degeneration (Ma et al., 2014). Instead, one or more of the other genes linked to this QTL may be involved. However, *cryaa*, which happens to be an antiapoptotic small heat-shock chaperone, is probably the end point of a defective developmental differentiation cascade involving misregulation of genes such as *Pax6* and *Sox2* in the lens (Ma et al., 2014; Strickler et al., 2001). Indeed, rescue experiments through Tol2-mediated transient transgenesis in Pachón cavefish have demonstrated that mosaic *cryaa* reexpression in lens cells protects these cells from apoptosis, in a cell-autonomous manner (Hinaux et al., 2015). Further studies are now required, in particular to establish stable transgenic lines with rescued *cryaa* expression, in the hope to obtain genetically engineered cavefish with eyes. Although it is doubtful that such animals will actually see, since too many of their "eye genes" including several crystallins carry nonsynonymous, potentially deleterious mutations (Hinaux et al., 2015, 2013), they would nevertheless constitute excellent tools to decipher eye-dependent from eye-independent traits that have evolved in the cavefish head and brain. The transgenesis methods and requirements to generate such *Astyanax* cavefish or surface fish stable lines have been established (Elipot, Legendre, et al., 2014). Both I-SceI meganuclease-mediated (Thermes et al., 2002) and Tol2 transposase-mediated (Kawakami, 2005) transgenesis have proved efficient: in proof of principle experiments using *blue lens* or *red muscle* generic control transgenesis vectors, both methods generate 37–46% mosaic transgenic embryos in FO-injected individuals, and the male F0 founders show transgene transmission rates to their F1 progeny comprised between 15% and 24% (Elipot, Legendre, et al., 2014).

Finally, genome-editing methods are also now being applied to *Astyanax*. Transcription activator like effector nucleases (TALENs) have recently been designed to target two "pigmentation genes" which contain coding changes in cavefish relative to surface fish and which map to QTLs for pigmentation: *oca2* and *mc1r*. TALEN-induced mutation in *oca2* in surface fish results in mosaic loss of melanin pigmentation visible as albino patches in FO founder fish, suggesting biallelic gene mutations in FOs (Ma, Jeffery, Essner, & Kowalko, 2015). In addition, these patches behave similarly to cavefish pigment cells in that they can produce melanin upon treatment with L-DOPA. Thus, the TALEN technology, as well as the more recent and even more promising CRISPR-Cas9 technology (Hwang et al., 2013), not yet published in *Astyanax*, have great potential for studying the role of candidate genes responsible for the evolution of cavefish traits.

Although a cavefish will probably never be comparable to a zebrafish in terms of functional genomics amenability, all modern available tools for manipulation of genome and gene expression are applicable to *Astyanax*, making the researcher's life every day more exciting, and hopefully attracting new groups to work on the *Astyanax* model.

## 8. CONCLUSION

We have shown that during the last past 10 years, very significant advances have been made in identifying proximal mechanisms (Mayr, 1961), which are the mutations that have changed the physiological, developmental, and behavior traits of cavefish. New molecular tools available today will allow us to identify such mutations at an ever increasing rate.

However it is much more tricky, not to say impossible, to disentangle distal mechanisms (Mayr, 1961), which are the de novo or standing variation origin of these mutations, and the impact of recombination, epistasis, genetic drift, selection, and migration in driving their evolution. Accordingly, these evolutionary mechanisms have fueled discussions on their relative importance

(Jeffery, 2010; Retaux & Casane, 2013; Wilkens, 2010). Very often, as evolutionary biology tries to reconstruct the history of life, many different scenarios can explain the same observation and they can only be ranked according to their likelihoods. When the same evolutionary process occurs several times independently, this can help to evaluate the effect of different mechanisms such as stochastic mechanisms (eg, genetic drift) and deterministic mechanisms (eg, selection). It can also help the identification of the impact of de novo mutations versus standing genetic variation. Such cases of repeated evolution are rare; however, cavefish evolved independently in different caves, and each cave corresponds to a similar environmental shift (indeed one of the main attractions of studying *Astyanax* cavefish). As a result of their evolution in the cave environment, *Astyanax* populations converged towards several traits, such as loss of eyes and pigmentation, improved nonvisual senses such as enhanced olfactory and neuromast sensitivity and other physiological and behavioral changes related to food scarcity. Not every cavefish population exhibits every trait, but convergence is very often observed. Increased sensitivity to vibration or soluble chemicals and a more efficient metabolism are very likely constructive changes and are thus probably the result of natural selection. Regressive changes, such as eye and pigmentation losses could have involved both selection and genetic drift. Selection could have been direct or indirect, ie, due to the pleiotropic effects of selection for a constructive trait. Analysis of the polarities of mutation effects of QTL related to eye size and numbers of melanophores led to the hypotheses that eye regression was driven by selection whereas melanophore loss was due to genetic drift (Protas et al., 2007). Moreover, selection and genetic drift are not exclusive mechanisms. Whereas selection may not be involved in the evolution of a given polymorphism, genetic drift is always involved in allele frequency changes because real populations have always a finite size. As cave population sizes are often quite small, genetic drift is expected to be a major component of cavefish genome evolution. The issue is thus to evaluate in which case selection could have prevailed over genetic drift.

Genome scans for selection could be applied to *Astyanax*. For example, large $F_{ST}$, a measure of population differentiation, is commonly used as evidence for outlier polymorphisms, that is, polymorphic sites where different alleles are fixed—or at least occur with very different frequencies—in different populations. In the case of putative adaptive mutations in CF, a convincing outlier caused by selection in CF would be one with a high frequency of the derived allele in CF and low frequency in SF. In addition, around this polymorphic site, the level of polymorphism in SF should be on par with the median for the whole SF genome, but the level of polymorphism in CF should be lower than for the whole CF genome. This last observation would indicate a selective sweep in CF, ie, the fixation of one or a few derived alleles through selection and the fixation of other neutral alleles at closely linked loci through genetic hitchhiking. The efficiency of this method to detect genomic regions under selection depends on the age of the fixation of the selected allele. If the allele reached fixation a long time ago, then the molecular cues left by the selective sweep would probably be erased. On the other hand, the observation of a selective sweep can be taken as good evidence that selection occurred recently. Such analysis of selective sweep has not yet been reported in *Astyanax*.

Another important and still unanswered question is whether the derived alleles involved in the evolution of cave *Astyanax* originated from de novo mutations, or if they were already present in the standing genetic variation of surface fish. Mutation de novo refers to a change that follows the event of surface fish being swept into the caves. The change in the environment comes first, and subsequently, the isolated population acquires new mutations, which, if helpful for survival, will eventually be selected for and fixed within the population. In contrast, standing genetic variation refers to mutations that were already present at low frequencies within the ancestral population, and were selected only after the environmental change. In the case of *A. mexicanus*, this would mean the beneficial mutations were present in the population of surface fish that were trapped inside the cave, and the fish carrying such mutations had a selective advantage in this new environment. We would expect to find the same mutations involved in the same phenotypic change in different caves if these mutations were already present in surface fish, whereas it is very unlikely that the very same mutation would appear and be fixed several times independently.

Finally, another question that remains to be solved, and which is closely related to the two previous ones, is when cave colonization occurred, and how long it took for different groups of surface fish to become adapted to the cave environment. Rough estimations of the divergence times between cave and surface fish populations have been performed, but we do not know very much about when cavefish populations were established and how long the process of adaptation takes. Selection pressure should have been very strong at the beginning, as surface fish do not survive for long in the cave environment and there was room for phenotypic improvement. Under strong selective pressure, evolution can be very fast. It is reasonable to assume that cavefish evolution has been a very fast process that occurred very recently, and that standing genetic variation was particularly important. To answer these questions, we have to use different population genomic approaches and the wealth of new genomic data that are now available. We also need improved and adapted analytical tools to tackle such difficult but central questions.

## ACKNOWLEDGMENTS

This work was supported by a collaborative ANR grant [Blindtest]. We thank members of the Rétaux and Casane groups for their enthusiasm on *Astyanax* studies. Thanks to Patrick Laurenti for art in Fig. 1 and to Victor Simon for help in photographing fishes in Fig. 2.

## REFERENCES

Albertson, R. C., Streelman, J. T., Kocher, T. D., & Yelick, P. C. (2005). Integration and evolution of the cichlid mandible: the molecular basis of alternate feeding strategies. *Proceedings of the National Academy of Sciences of the United States of America, 102*(45), 16287–16292.

Alunni, A., Menuet, A., Candal, E., Penigault, J. B., Jeffery, W. R., & Retaux, S. (2007). Developmental mechanisms for retinal degeneration in the blind cavefish *Astyanax mexicanus*. *Journal of Comparitive Neurology, 505*(2), 221–233.

Amores, A., Force, A., Yan, Y. L., Joly, L., Amemiya, C., Fritz, A., … Postlethwait, J. H. (1998). Zebrafish hox clusters and vertebrate genome evolution. *Science, 282*(5394), 1711–1714.

Aparicio, S., Chapman, J., Stupka, E., Putnam, N., Chia, J., Dehal, P., … Brenner, S. (2002). Whole-genome shotgun assembly and analysis of the genome of *Fugu rubripes*. *Science, 297*(5585), 1301–1310.

Aspiras, A. C., Rohner, N., Martineau, B., Borowsky, R. L., & Tabin, C. J. (2015). Melanocortin 4 receptor mutations contribute to the adaptation of cavefish to nutrient-poor conditions. *Proceedings of the National Academy of Sciences of the United States of America, 112*(31), 9668–9673.

Avise, J. C., & Selander, R. K. (1972). Evolutionary genetics of cave-dwelling fishes of genus *Astyanax*. *Evolution, 26*(1), 1–19.

Bachtrog, D., Mank, J. E., Peichel, C. L., Kirkpatrick, M., Otto, S. P., Ashman, T.-L., … Tree Sex, C. (2014). Sex determination: why so many ways of doing it? *PLoS Biology, 12*(7).

Behrens, M., Langecker, T. G., Wilkens, H., & Schmale, H. (1997). Comparative analysis of Pax-6 sequence and expression in the eye development of the blind cave fish *Astyanax fasciatus* and its epigean conspecific. *Molecular Biology and Evolution, 14*(3), 299–308.

Behrens, M., Wilkens, H., & Schmale, H. (1998). Cloning of the alphaA-crystallin genes of a blind cave form and the epigean form of *Astyanax fasciatus*: a comparative analysis of structure, expression and evolutionary conservation. *Gene, 216*(2), 319–326.

Bibliowicz, J., Alie, A., Espinasa, L., Yoshizawa, M., Blin, M., Hinaux, H., … Retaux, S. (2013). Differences in chemosensory response between eyed and eyeless *Astyanax mexicanus* of the Rio Subterraneo cave. *Evodevo, 4*(1), 25.

Borowsky, R. (2008a). *Astyanax mexicanus*, the blind mexican cave fish: a model for studies in development and morphology. *CSH Protocols, 2008*. pdb emo107.

Borowsky, R. (2008b). Restoring sight in blind cavefish. *Current Biology, 18*(1), R23–R24.

Borowsky, R. (2013). Eye regression in blind *Astyanax* cavefish may facilitate the evolution of an adaptive behavior and its sensory receptors. *BMC Biology, 11*(1), 81.

Borowsky, R., & Cohen, D. (2013). Genomic consequences of ecological speciation in *Astyanax* cavefish. *PLoS One, 8*(11), e79903.

Borowsky, R., & Wilkens, H. (2002). Mapping a cave fish genome: polygenic systems and regressive evolution. *The Journal of Heredity, 93*(1), 19–21.

Braasch, I., Peterson, S. M., Desvignes, T., McCluskey, B. M., Batzel, P., & Postlethwait, J. H. (2015). A new model army: emerging fish models to study the genomics of vertebrate Evo-Devo. *Journal of Experimental Zoology Part B-Molecular and Developmental Evolution, 324*(4), 316–341.

Bradic, M., Beerli, P., Garcia-de Leon, F. J., Esquivel-Bobadilla, S., & Borowsky, R. L. (2012). Gene flow and population structure in the Mexican blind cavefish complex (*Astyanax mexicanus*). *BMC Evolutionary Biology, 12*, 9.

Brawand, D., Wagner, C. E., Li, Y. I., Malinsky, M., Keller, I., Fan, S. H., … Di Palma, F. (2014). The genomic substrate for adaptive radiation in African cichlid fish. *Nature, 513*(7518), 375–381.

Breder, C. M. J. (1943). Problems in the behavior and evolution of a species of blind cave fish. *Transactions of the New York Academy of Sciences, 5*, 168–176.

Burchards, H., Dölle, A., & Parzefall, J. (1985). Aggressive behaviour of an epigean population of *Astyanax mexicanus* (Characidae, Pisces) and some observations of three subterranean populations. *Behavioural Processes, 11*(3), 225–235.

Carlson, B. M., Onusko, S. W., & Gross, J. B. (2015). A high-density linkage map for *Astyanax mexicanus* using genotyping-by-sequencing technology. *G3: Genes Genomes Genetics, 5*(2), 241–251.

Cavallari, N., Frigato, E., Vallone, D., Froehlich, N., Fernando Lopez-Olmeda, J., Foa, A., … Foulkes, N. S. (2011). A blind circadian clock in cavefish reveals that opsins mediate peripheral clock photoreception. *PLoS Biology, 9*(9), e1001142.

Cellerino, A., Valenzano, D. R., & Reichard, M. (2016 May). From the bush to the bench: the annual *Nothobranchius* fishes as a new model system in biology. *Biol Rev Camb Philos Soc, 91*(2), 511–533. http://dx.doi.org/10.1111/brv.12183.

Chalopin, D., Naville, M., Plard, F., Galiana, D., & Volff, J.-N. (2015). Comparative analysis of transposable elements highlights mobilome diversity and evolution in vertebrates. *Genome Biology and Evolution*, 7(2), 567–580.

Chen, L., DeVries, A. L., & Cheng, C.-H. C. (1997a). Evolution of antifreeze glycoprotein gene from a trypsinogen gene in Antarctic notothenioid fish. *Proceedings of the National Academy of Sciences of the United States of America, 94*(8), 3811–3816.

Chen, L. B., DeVries, A. L., & Cheng, C.-H. C. (1997b). Convergent evolution of antifreeze glycoproteins in Antarctic notothenioid fish and Arctic cod. *Proceedings of the National Academy of Sciences of the United States of America, 94*(8), 3817–3822.

Chow, R. L., Altmann, C. R., Lang, R. A., & Hemmati-Brivanlou, A. (1999). Pax6 induces ectopic eyes in a vertebrate. *Development, 126*(19), 4213–4222.

Culver, D. C., & Pipan, T. (2009). *The biology of caves and other subterranean habitats*. Oxford: Oxford University Press.

Di Palma, F., Kidd, C., Borowsky, R., & Kocher, T. D. (2007). Construction of bacterial artificial chromosome libraries for the Lake Malawi Cichlid (*Metriaclima zebra*), and the blind cavefish (*Astyanax mexicanus*). *Zebrafish, 4*(1), 41–47.

Dowling, T. E., Martasian, D. P., & Jeffery, W. R. (2002). Evidence for multiple genetic forms with similar eyeless phenotypes in the blind cavefish, *Astyanax mexicanus*. *Molecular Biology and Evolution, 19*(4), 446–455.

Duboué, E. R., & Borowsky, R. L. (2012). Altered rest-activity patterns evolve via circadian independent mechanisms in cave adapted balitorid loaches. *PLoS One*, 7(2), e30868.

Duboué, E. R., Keene, A. C., & Borowsky, R. L. (2011). Evolutionary convergence on sleep loss in cavefish populations. *Current Biology: CB, 21*(8), 671–676.

Ekker, S. C., Ungar, A. R., Greenstein, P., Vonkessler, D. P., Porter, J. A., Moon, R. T., & Beachy, P. A. (1995). Patterning activities of vertebrate hedgehog proteins in the developing eye and brain. *Current Biology, 5*(8), 944–955.
Elipot, Y., Hinaux, H., Callebert, J., Launay, J.-M., Blin, M., & Rétaux, S. (2014). A mutation in the enzyme monoamine oxidase explains part of the Astyanax cavefish behavioural syndrome. *Nature Communications, 5*, 3647.
Elipot, Y., Hinaux, H., Callebert, J., & Retaux, S. (2013). Evolutionary shift from fighting to foraging in blind cavefish through changes in the serotonin network. *Current Biology, 23*(1), 1–10.
Elipot, Y., Legendre, L., Pere, S., Sohm, F., & Retaux, S. (2014b). *Astyanax* transgenesis and husbandry: how cavefish enters the laboratory. *Zebrafish, 11*(4), 291–299.
Eschmeyer, W. N., & Fricke, R. (2012). *Catalog of fishes*. San Francisco: California Academy of Sciences.
Espinasa, L., Bibliowicz, J., Jeffery, W., & Retaux, S. (2014). Enhanced prey capture skills in *Astyanax* cavefish larvae are independent from eye loss. *Evodevo, 5*(1), 35.
Espinasa, L., Yamamoto, Y., & Jeffery, W. R. (2005). Non-optical releasers for aggressive behavior in blind and blinded *Astyanax* (Teleostei, Characidae). *Behavioural Processes, 70*(2), 144–148.
Foulkes, N. S., Whitmore, D., Vallone, D., & Bertolucci, C. (2016). Studying the evolution of the vertebrate circadian clock: the power of fish as comparative models. In T. Friedmann, J. Dunlap, & S. F. Goodwin (Eds.). *Genetics, Genomics and Fish Phenomics* (Vol. 95).
Gallo, N. D., & Jeffery, W. R. (2012). Evolution of space dependent growth in the teleost *Astyanax mexicanus. PLoS One, 7*(8), e41443.
Garcia-Machado, E., Hernandez, D., Garcia-Debras, A., Chevalier-Monteagudo, P., Metcalfe, C., Bernatchez, L., & Casane, D. (2011). Molecular phylogeny and phylogeography of the Cuban cave-fishes of the genus *Lucifuga*: evidence for cryptic allopatric diversity. *Molecular Phylogenetics and Evolution, 61*(2), 470–483.
Gehring, W. J., & Ikeo, K. (1999). Pax 6-mastering eye morphogenesis and eye evolution. *Trends in Genetics, 15*(9), 371–377.
Gross, J. B. (2012). The complex origin of *Astyanax* cavefish. *BMC Evolutionary Biology, 12*, 105.
Gross, J. B., Borowsky, R., & Tabin, C. J. (2009). A novel role for Mc1r in the parallel evolution of depigmentation in independent populations of the cavefish *Astyanax mexicanus. PLoS Genetics, 5*(1), e1000326.
Gross, J. B., Furterer, A., Carlson, B. M., & Stahl, B. A. (2013). An integrated transcriptome-wide analysis of cave and surface dwelling *Astyanax mexicanus. PLoS One, 8*(2), e55659.
Gross, J. B., Krutzler, A. J., & Carlson, B. M. (2014). Complex craniofacial changes in blind cave-dwelling fish are mediated by genetically symmetric and asymmetric loci. *Genetics, 196*(4), 1303–1319.
Gross, J. B., Protas, M., Conrad, M., Scheid, P. E., Vidal, O., Jeffery, W. R., … Tabin, C. J. (2008). Synteny and candidate gene prediction using an anchored linkage map of *Astyanax mexicanus. Proceedings of the National Academy of Sciences of the United States of America, 105*(51), 20106–20111.
Halder, G., Callaerts, P., & Gehring, W. J. (1995). Induction of ectopic eyes by targeted expression of the eyeless gene in *Drosophila. Science, 267*(5205), 1788–1792.
Hattori, R. S., Murai, Y., Oura, M., Masuda, S., Majhi, S. K., Sakamoto, T., … Struessmann, C. A. (2012). A Y-linked anti-Mullerian hormone duplication takes over a critical role in sex determination. *Proceedings of the National Academy of Sciences of the United States of America, 109*(8), 2955–2959.
Helfman, G., Collette, B. B., Facey, D. E., & Bowen, B. W. (2009). *The diversity of fishes: Biology, evolution, and ecology*. Wiley-Blackwell.

Herpin, A., & Schartl, M. (2011a). Dmrt1 genes at the crossroads: a widespread and central class of sexual development factors in fish. *FEBS Journal, 278*(7), 1010–1019.
Herpin, A., & Schartl, M. (2011b). Sex determination: switch and suppress. *Current Biology, 21*(17), R656–R659.
Hill, R. E., Favor, J., Hogan, B. L. M., Ton, C. C. T., Saunders, G. F., Hanson, I. M., … Vanheyningen, V. (1991). Mouse small eye results from mutations in a paired-like homeobox-containing gene. *Nature, 354*(6354), 522–525.
Hinaux, H., Blin, M., Fumey, J., Legendre, L., Heuze, A., Casane, D., & Retaux, S. (2015). Lens defects in *Astyanax mexicanus* cavefish: evolution of crystallins and a role for alphaA-crystallin. *Developmental Neurobiology, 75*(5), 505–521.
Hinaux, H., Pottin, K., Chalhoub, H., Pere, S., Elipot, Y., Legendre, L., & Retaux, S. (2011). A developmental staging table for *Astyanax mexicanus* surface fish and Pachon cavefish. *Zebrafish, 8*(4), 155–165.
Hinaux, H., Poulain, J., Da Silva, C., Noirot, C., Jeffery, W. R., Casane, D., & Retaux, S. (2013). De novo sequencing of *Astyanax mexicanus* surface fish and Pachon cavefish transcriptomes reveals enrichment of mutations in cavefish putative eye genes. *PLoS One, 8*(1), e53553.
Holland, P. W. H. (2003). More genes in vertebrates? *Journal of Structural and Functional Genomics, 3*(1–4), 75–84.
Hooven, T. A., Yamamoto, Y., & Jeffery, W. R. (2004). Blind cavefish and heat shock protein chaperones: a novel role for hsp90alpha in lens apoptosis. *The International Journal of Developmental Biology, 48*(8–9), 731–738.
Hubbs, C. L., & Innes, W. T. (1936). The first known blind fish of the family Characidae: a new genus from Mexico. *Occasional Papers of the Museum of Zoology, University of Michigan, 342*, 1–7.
Hüppop, K. (1987). Food finding ability in cave fish *(Astyanax fasciatus)*. *International Journal of Speleology, 18*, 59–66.
Hüppop, K. (1988). *Phänomene und Bedeutung der Energieersparnis bei dem Höhlenfisch* Astyanax fasciatus *(Characidae, Pisces)* (Thesis). Universität Hamburg.
Hüppop, K., & Wilkens, H. (1991). Bigger eggs in subterranean *Astyanax fasciatus* (Characidae, Pisces) – their significance and genetics. *Zeitschrift für Zoologische Systematik und Evolutionsforschung, 29*(4), 280–288.
Hwang, W. Y., Fu, Y., Reyon, D., Maeder, M. L., Tsai, S. Q., Sander, J. D., … Joung, J. K. (2013). Efficient genome editing in zebrafish using a CRISPR-Cas system. *Nature Biotechnology, 31*(3), 227–229.
Jeffery, W. R. (2001). Cavefish as a model system in evolutionary developmental biology. *Developmental Biology, 231*(1), 1–12.
Jeffery, W. R. (2005). Adaptive evolution of eye degeneration in the Mexican blind cavefish. *The Journal of Heredity, 96*(3), 185–196.
Jeffery, W. R. (2009). Regressive evolution in *Astyanax* cavefish. *Annual Review of Genetics, 43*, 25–47.
Jeffery, W. R. (2010). Pleiotropy and eye degeneration in cavefish. *Heredity (Edinburg), 105*(5), 495–496.
Jeffery, W. R., & Martasian, D. P. (1998). Evolution of eye regression in the cavefish *Astyanax*: apoptosis and the Pax-6 gene. *American Zoologist, 38*, 685–696.
Kamiya, T., Kai, W., Tasumi, S., Oka, A., Matsunaga, T., Mizuno, N., … Kikuchi, K. (2012). A trans-species Missense SNP in Amhr2 is associated with sex determination in the tiger pufferfish, *Takifugu rubripes* (fugu). *PLoS Genetics, 8*(7).
Kavalco, K. F., & De Almeida-Toledo, L. F. (2007). Molecular cytogenetics of blind mexican tetra and comments on the karyotypic characteristics of genus *Astyanax* (Teleostei, Characidae). *Zebrafish, 4*(2), 103–111.

Kawakami, K. (2005). Transposon tools and methods in zebrafish. *Developmental Dynamics, 234*(2), 244–254.
Keene, A. C., Yoshizawa, M., & McGaugh, S. E. (2015). *Biology and evolution of the Mexican cavefish*. Academic Press.
Kenny, N. J., Chan, K. W., Nong, W., Qu, Z., Maeso, I., Yip, H. Y., … Hui, J. H. L. (2016). Ancestral whole-genome duplication in the marine chelicerate horseshoe crabs. *Heredity, 116*, 190–199. http://dx.doi.org/10.1038/hdy.2015.89.
Kikuchi, K., & Hamaguchi, S. (2013). Novel sex-determining genes in fish and sex chromosome evolution. *Developmental Dynamics, 242*(4), 339–353.
Kim, E. B., Fang, X., Fushan, A. A., Huang, Z., Lobanov, A. V., Han, L., … Gladyshev, V. N. (2011). Genome sequencing reveals insights into physiology and longevity of the naked mole rat. *Nature, 479*(7372), 223–227.
Kirby, R. F., Thompson, K. W., & Hubbs, C. (1977). Karyotypic similarities between Mexican and blind tetras. *Copeia*, (3), 578–580.
Kowalko, J. E., Rohner, N., Linden, T. A., Rompani, S. B., Warren, W. C., Borowsky, R., … Yoshizawa, M. (2013). Convergence in feeding posture occurs through different genetic loci in independently evolved cave populations of *Astyanax mexicanus*. *Proceedings of the National Academy of Sciences of the United States of America, 110*(42), 16933–16938.
Kowalko, J. E., Rohner, N., Rompani, S. B., Peterson, B. K., Linden, T. A., Yoshizawa, M., … Tobin, C. J. (2013). Loss of schooling behavior in cavefish through sight-dependent and sight-independent mechanisms. *Current Biology, 23*(19), 1874–1883.
Kuraku, S., & Meyer, A. (2008). Genomic analysis of cichlid fish natural mutants. *Current Opinion in Genetics & Development, 18*(6), 551–558.
Lande, R. (1981). The minimum number of genes contributing to quantitative variation between and within populations. *Genetics, 99*(3–4), 541–553.
Lecointre, G., & Le Guyader, H. (2006). *The tree of life: A phylogenetic classification*. Belknap Press of Harvard University Press.
Ma, L., Jeffery, W. R., Essner, J. J., & Kowalko, J. E. (2015). Genome editing using TALENs in blind mexican cavefish, *Astyanax mexicanus*. *PLoS One, 10*(3), e0119370.
Ma, L., Parkhurst, A., & Jeffery, W. (2014). The role of a lens survival pathway including sox2 and alphaA-crystallin in the evolution of cavefish eye degeneration. *Evodevo, 5*(1), 28.
Matsuda, M., Nagahama, Y., Shinomiya, A., Sato, T., Matsuda, C., Kobayashi, T., … Sakaizumi, M. (2002). DMY is a Y-specific DM-domain gene required for male development in the medaka fish. *Nature, 417*(6888), 559–563.
Mayr, E. (1961). Cause and effect in biology. *Science, 134*(3489), 1501–1506.
McGaugh, S. E., Gross, J. B., Aken, B., Blin, M., Borowsky, R., Chalopin, D., … Warren, W. C. (2014). The cavefish genome reveals candidate genes for eye loss. *Nature Communications, 5*, 5307.
Meng, F., Braasch, I., Phillips, J. B., Lin, X., Titus, T., Zhang, C., & Postlethwait, J. H. (2013). Evolution of the eye transcriptome under constant darkness in *Sinocyclocheilus* cavefish. *Molecular Biology and Evolution, 30*(7), 1527–1543.
Mitchell, R. W., Russell, W. H., & Elliott, W. R. (1977). Mexican eyeless characin fishes, genus *Astyanax*: environment, distribution and evolution. *Special Publications Museum Texas Tech University, 12*, 1–89.
Moran, D., Softley, R., & Warrant, E. J. (2014). Eyeless mexican cavefish Save energy by eliminating the circadian rhythm in metabolism. *PLoS One, 9*(9), e107877.
Moran, D., Softley, R., & Warrant, E. J. (2015). The energetic cost of vision and the evolution of eyeless Mexican cavefish. *Science Advances, 1*(8), e1500363.
Mungpakdee, S., Seo, H. C., Angotzi, A. R., Dong, X., Akalin, A., & Chourrout, D. (2008). Differential evolution of the 13 Atlantic salmon Hox clusters. *Molecular Biology and Evolution, 25*(7), 1333–1343.

Myosho, T., Otake, H., Masuyama, H., Matsuda, M., Kuroki, Y., Fujiyama, A., … Sakaizumi, M. (2012). Tracing the emergence of a novel sex-determining gene in medaka, *Oryzias luzonensis*. *Genetics, 191*(1), 163–170.

Near, T. J., Dornburg, A., Eytan, R. I., Keck, B. P., Smith, W. L., Kuhn, K. L., … Wainwright, P. C. (2013). Phylogeny and tempo of diversification in the superradiation of spiny-rayed fishes. *Proceedings of the National Academy of Sciences of the United States of America, 110*(31), 12738–12743.

Near, T. J., Eytan, R. I., Dornburg, A., Kuhn, K. L., Moore, J. A., Davis, M. P., … Smith, W. L. (2012). Resolution of ray-finned fish phylogeny and timing of diversification. *Proceedings of the National Academy of Sciences of the United States of America, 109*(34), 13698–13703.

Niemiller, M. L., & Poulson, T. L. (2010). Subterranean fishes of North America: Amblyopsidae. In E. Trajano, M. E. Bichuette, & B. G. Kapoor (Eds.), *The biology of subterranean fishes* (pp. 1–112). Science.

O'Quin, K. E., Doshi, P., Lyon, A., Hoenemeyer, E., Yoshizawa, M., & Jeffery, W. R. (2015). Complex evolutionary and genetic patterns characterize the loss of scleral ossification in the blind cavefish *Astyanax mexicanus*. *PLoS One, 10*(12), e0142208.

Ohno, S. (1970). *Evolution by gene duplication*. Berlin: Springer-Verlag.

O'Quin, K. E., Yoshizawa, M., Doshi, P., & Jeffery, W. R. (2013). Quantitative genetic analysis of retinal degeneration in the blind cavefish *Astyanax mexicanus*. *PLoS One, 8*(2), e57281.

Ospina-Alvarez, N., & Piferrer, F. (2008). Temperature-dependent sex determination in fish revisited: prevalence, a single sex ratio response pattern, and possible effects of climate change. *PLoS One, 3*(7).

Parzefall, J. (1985). On the heredity of behavior patterns in cave animals and their epigean relatives. *National Speleological Society Bulletin, 47*, 128–135.

Parzefall, J., & Senkel, S. (1986). Schooling behavior in cavernicolous fish and their epigean conspecifics. *Congres International Espeleology, 2*, 107–109.

Pascual-Anaya, J., D'Aniello, S., Kuratani, S., & Garcia-Fernandez, J. (2013). Evolution of Hox gene clusters in deuterostomes. *BMC Developmental Biology, 13*, 26.

Peichel, C. L. (2005). Fishing for the secrets of vertebrate evolution in threespine sticklebacks. *Developmental Dynamics, 234*(4), 815–823.

Plath, M., Hauswaldt, J. S., Moll, K., Tobler, M., Garcia De Leon, F. J., Schlupp, I., & Tiedemann, R. (2007). Local adaptation and pronounced genetic differentiation in an extremophile fish, *Poecilia mexicana*, inhabiting a Mexican cave with toxic hydrogen sulphide. *Molecular Ecology, 16*(5), 967–976.

Postlethwait, J. H., Yan, Y. L., Gates, M. A., Horne, S., Amores, A., Brownlie, A., … Talbot, W. S. (1998). Vertebrate genome evolution and the zebrafish gene map. *Nature Genetics, 18*(4), 345–349.

Pottin, K., Hinaux, H., & Retaux, S. (2011). Restoring eye size in *Astyanax mexicanus* blind cavefish embryos through modulation of the Shh and Fgf8 forebrain organising centres. *Development, 138*(12), 2467–2476.

Priede, I. G., & Froese, R. (2013). Colonization of the deep sea by fishes. *Journal of Fish Biology, 83*(6), 1528–1550.

Protas, M., Conrad, M., Gross, J. B., Tabin, C., & Borowsky, R. (2007). Regressive evolution in the Mexican cave tetra, *Astyanax mexicanus*. *Current Biology, 17*(5), 452–454.

Protas, M., Tabansky, I., Conrad, M., Gross, J. B., Vidal, O., Tabin, C. J., & Borowsky, R. (2008). Multi-trait evolution in a cave fish, *Astyanax mexicanus*. *Evolution & Development, 10*(2), 196–209.

Protas, M. E., Hersey, C., Kochanek, D., Zhou, Y., Wilkens, H., Jeffery, W. R., … Tabin, C. J. (2006). Genetic analysis of cavefish reveals molecular convergence in the evolution of albinism. *Nature Genetics, 38*(1), 107–111.

Puschel, A. W., Gruss, P., & Westerfield, M. (1992). Sequence and expression pattern of pax-6 are highly conserved between zebrafish and mice. *Development, 114*(3), 643–651.
Retaux, S., & Casane, D. (2013). Evolution of eye development in the darkness of caves: adaptation, drift, or both? *Evodevo, 4*(1), 26.
Rizzello, A., Romano, A., Kottra, G., Acierno, R., Storelli, C., Verri, T., … Maffia, M. (2013). Protein cold adaptation strategy via a unique seven-amino acid domain in the icefish (*Chionodraco hamatus*) PEPT1 transporter. *Proceedings of the National Academy of Sciences of the United States of America, 110*(17), 7068–7073.
Roest Crollius, H., & Weissenbach, J. (2005). Fish genomics and biology. *Genome Research, 15*(12), 1675–1682.
Romero, A. (2001). *The biology of hypogean fishes*. Kluwer Acaemic Publishers.
Rose, F. L., & Mitchell, R. W. (1982). Comparative lipid values of epigean and cave-adapted *astyanax*. *Southwestern Naturalist, 27*(3), 357–358.
Sadoglu, P., & McKee, A. (1969). A second gene that affects eye and body color in Mexican blind cave fish. *Journal of Heredity, 60*(1), 10–14.
Sadoglu, P. (1957a). A mendelian gene for albinism in natural cave fish. *Experientia, 13*(10), 394.
Sadoglu, P. (1957b). Mendelian inheritance in the hybrids between the Mexican blind fishes and their overground ancestor. *Verhandlungen der Deutschen Zoologischen Gesellschaft*, 432–439.
Salin, K., Voituron, Y., Mourin, J., & Hervant, F. (2010). Cave colonization without fasting capacities: an example with the fish *Astyanax fasciatus mexicanus*. *Comparative Biochemistry and Physiology Part A Molecular & Integrative Physiology, 156*(4), 451–457.
Schemmel, C. (1967). Vergleichende Untersuchungen an den Hautsinnesorganen ober- und unterirdisch lebender *Astyanax*-Formen. *Zeitschrift für Morphologie der Tiere, 61*(2), 255–316.
Schemmel, C. (1974). Genetische Untersuchungen zur Evolution des Geschmacksapparates bei cavernicolen Fischen. *Journal of Zoological Systematics Evolutionary Research, 12*, 169–215.
Schemmel, C. (1980). Studies on the genetics of feeding behavior in the cave fish *Astyanax mexicanus* f. *Anoptichthys*. An example of apparent monofactorial inheritance by polygenes. *Zeitschrift Fur Tierpsychologie — Journal of Comparative Ethology, 53*(1), 9–22.
Schulte, J. E., O'Brien, C. S., Conte, M. A., O'Quin, K. E., & Carleton, K. L. (2014). Interspecific variation in Rx1 expression controls opsin expression and causes visual system diversity in African cichlid fishes. *Molecular Biology and Evolution, 31*(9), 2297–2308.
Stemmer, M., Schuhmacher, L.-N., Foulkes, N. S., Bertolucci, C., & Wittbrodt, J. (2015). Cavefish eye loss in response to an early block in retinal differentiation progression. *Development, 142*(4), 743–752.
Strecker, U., Faundez, V. H., & Wilkens, H. (2004). Phylogeography of surface and cave *Astyanax* (Teleostei) from Central and North America based on cytochrome b sequence data. *Molecular Phylogenetics and Evolution, 33*(2), 469–481.
Strickler, A. G., Byerly, M. S., & Jeffery, W. R. (2007). Lens gene expression analysis reveals downregulation of the anti-apoptotic chaperone alphaA-crystallin during cavefish eye degeneration. *Development Genes and Evolution, 217*(11–12), 771–782.
Strickler, A. G., Yamamoto, Y., & Jeffery, W. R. (2001). Early and late changes in Pax6 expression accompany eye degeneration during cavefish development. *Development Genes and Evolution, 211*(3), 138–144.
Sugie, A., Terai, Y., Ota, R., & Okada, N. (2004). The evolution of genes for pigmentation in African cichlid fishes. *Gene, 343*(2), 337–346.
Takehana, Y., Matsuda, M., Myosho, T., Suster, M. L., Kawakami, K., Shin, T. I., … Naruse, K. (2014). Co-option of Sox3 as the male-determining factor on the Y chromosome in the fish *Oryzias dancena*. *Nature Communications, 5*, 4157.

Terai, Y., Seehausen, O., Sasaki, T., Takahashi, K., Mizoiri, S., Sugawara, T., ... Okada, N. (2006). Divergent selection on opsins drives incipient speciation in Lake Victoria cichlids. *PLoS Biology, 4*(12), 2244–2251.

Thermes, V., Grabher, C., Ristoratore, F., Bourrat, F., Choulika, A., Wittbrodt, J., & Joly, J. S. (2002). I-SceI meganuclease mediates highly efficient transgenesis in fish. *Mechanisms of Development, 118*(1–2), 91–98.

Varatharasan, N., Croll, R. P., & Franz-Odendaal, T. (2009). Taste bud development and patterning in sighted and blind morphs of *Astyanax mexicanus. Developmental Dynamics, 238*(12), 3056–3064.

Volff, J. N. (2005). Genome evolution and biodiversity in teleost fish. *Heredity (Edinburg), 94*(3), 280–294.

Warrant, E. J., & Locket, N. A. (2004). Vision in the deep sea. *Biological Reviews, 79*(3), 671–712.

Wilkens, H. (1971). Genetic interpretation of regressive evolutionary processes: studies of hybrid eyes of two *Astyanax* cave populations (Characidae, Pisces). *Evolution, 25*(3), 530–544.

Wilkens, H. (1988). Evolution and genetics of epigean and cave *Astyanax fasciatus* (Characidae, Pisces): support for the neutral mutation theory. *Evolutionary Biology, 23*, 271–367.

Wilkens, H. (2010). Genes, modules and the evolution of cave fish. *Heredity (Edinburg), 105*(5), 413–422.

Wilkens, H., & Strecker, U. (2003). Convergent evolution of the cavefish *Astyanax* (Characidae, Teleostei): genetic evidence from reduced eye-size and pigmentation. *Biological Journal of the Linnean Society, 80*(4), 545–554.

Yamamoto, Y., Byerly, M. S., Jackman, W. R., & Jeffery, W. R. (2009). Pleiotropic functions of embryonic sonic hedgehog expression link jaw and taste bud amplification with eye loss during cavefish evolution. *Developmental Biology, 330*(1), 200–211.

Yamamoto, Y., Espinasa, L., Stock, D. W., & Jeffery, W. R. (2003). Development and evolution of craniofacial patterning is mediated by eye-dependent and -independent processes in the cavefish *Astyanax. Evolution & Development, 5*(5), 435–446.

Yamamoto, Y., & Jeffery, W. R. (2000). Central role for the lens in cave fish eye degeneration. *Science, 289*(5479), 631–633.

Yamamoto, Y., Stock, D. W., & Jeffery, W. R. (2004). Hedgehog signalling controls eye degeneration in blind cavefish. *Nature, 431*(7010), 844–847.

Yano, A., Guyomard, R., Nicol, B., Jouanno, E., Quillet, E., Klopp, C., ... Guiguen, Y. (2012). An immune-related gene evolved into the master sex-determining gene in rainbow trout, *Oncorhynchus mykiss. Current Biology, 22*(15), 1423–1428.

Yokoyama, S., Meany, A., Wilkens, H., & Yokoyama, R. (1995). Initial mutational steps toward loss of opsin gene function in cavefish. *Molecular Biology and Evolution, 12*(4), 527–532.

Yoshizawa, M., Goricki, S., Soares, D., & Jeffery, W. R. (2010). Evolution of a behavioral shift mediated by superficial neuromasts helps cavefish find food in darkness. *Current Biology, 20*(18), 1631–1636.

Yoshizawa, M., Yamamoto, Y., O'Quin, K. E., & Jeffery, W. R. (2012). Evolution of an adaptive behavior and its sensory receptors promotes eye regression in blind cavefish. *BMC Biology, 10*, 108.

Yoshizawa, M., O'Quin, K. E., & Jeffery, W. R. (2013). Evolution of an adaptive behavior and its sensory receptors promotes eye regression in blind cavefish: response to Borowsky (2013). *BMC Biology, 11*(1), 82.

CHAPTER SIX

# Transcriptional Regulation During Zygotic Genome Activation in Zebrafish and Other Anamniote Embryos

**J. Wragg and F. Müller**[1]
University of Birmingham, Birmingham, United Kingdom
[1]Corresponding author: E-mail: f.mueller@bham.ac.uk

## Contents



## Abstract

Embryo development commences with the fusion of two terminally differentiated haploid gametes into the totipotent fertilized egg, which through a series of major cellular and molecular transitions generate a pluripotent cell mass. The activation of the zygotic genome occurs during the so-called maternal to zygotic transition and

*Advances in Genetics*, Volume 95
ISSN 0065-2660
http://dx.doi.org/10.1016/bs.adgen.2016.05.001

prepares the embryo for zygotic takeover from maternal factors, in the control of the development of cellular lineages during differentiation. Recent advances in next generation sequencing technologies have allowed the dissection of the genomic and epigenomic processes mediating this transition. These processes include reorganization of the chromatin structure to a transcriptionally permissive state, changes in composition and function of structural and regulatory DNA-binding proteins, and changeover of the transcriptome as it is overhauled from that deposited by the mother in the oocyte to a zygotically transcribed complement. Zygotic genome activation in zebrafish occurs 10 cell cycles after fertilization and provides an ideal experimental platform for elucidating the temporal sequence and dynamics of establishment of a transcriptionally active chromatin state and helps in identifying the determinants of transcription activation at polymerase II transcribed gene promoters. The relatively large number of pluripotent cells generated by the fast cell divisions before zygotic transcription provides sufficient biomass for next generation sequencing technology approaches to establish the temporal dynamics of events and suggest causative relationship between them. However, genomic and genetic technologies need to be improved further to capture the earliest events in development, where cell number is a limiting factor. These technologies need to be complemented with precise, inducible genetic interference studies using the latest genome editing tools to reveal the function of candidate determinants and to confirm the predictions made by classic embryological tools and genome-wide assays. In this review we summarize recent advances in the characterization of epigenetic regulation, transcription control, and gene promoter function during zygotic genome activation and how they fit with old models for the mechanisms of the maternal to zygotic transition. This review will focus on the zebrafish embryo but draw comparisons with other vertebrate model systems and refer to invertebrate models where informative.

## 1. ZYGOTIC GENOME ACTIVATION

Embryogenesis starts with a single fertilized egg cell, with cytoplasm and cytoplasmic factors entirely derived from the maternal ovum. The embryo is initially transcriptionally quiescent with cellular behavior and division almost entirely controlled by maternal factors. In most animals, this maternal control directs a series of rapid synchronous mitoses, lacking gap phases, with the overall volume of the embryo remaining fairly constant as it forms a blastula. In zebrafish (*Danio rerio*) and clawed frog (*Xenopus* species), the activation of the zygotic genome coincides with a sequence of events commonly termed the mid-blastula transition (MBT) (Foe & Alberts, 1983; Gerhart, 1980; Kane & Kimmel, 1993; Newport & Kirschner, 1982a; Signoret & Lefresne, 1975) (Fig. 1A—C). MBT is characterized by the loss of synchrony in cell divisions, elongation of the cell cycle, and activation of the zygotic genome. The timing of MBT varies considerably among species, occurring after 10–12 cell divisions in anamniotes. At the MBT the rate of cell divisions follow a pattern typical of somatic cells, coinciding with the switching of the embryo from maternal to zygotic control; the maternal-to-zygotic transition (Fig. 1D). This process is mediated by two major molecular activities: firstly maternal clearance, which is the regulated and stepwise degradation of maternally inherited messenger RNAs (reviewed in Barckmann & Simonelig, 2013; Lee, Bonneau, & Giraldez, 2014; and Tadros & Lipshitz, 2009); secondly, global transcriptional activation of embryonic genes, also termed zygotic genome activation (ZGA). These processes collectively contribute to the first major overhaul of the transcriptome in the developing embryo. The initiation of zygotic transcription is a vital process for subsequent embryo development. When this process is inhibited in zebrafish and *Xenopus* embryos they continue with cell divisions, but fail to undergo gastrulation (Kane et al., 1996; Newport & Kirschner, 1982a). The molecular mechanisms that govern the MBT and the coincidental activation of the zygotic genome remain mostly unknown despite the first model of this process being proposed by Newport and Kirschner over 30 years ago. This model still remains the preeminent theory explaining this process and proposes that the MBT is initiated by the achievement of a threshold nucleoplasm to cytoplasm ratio. In the early embryo the genome is thought to be held in a transcriptionally repressive chromatin state by maternally deposited factors (Newport & Kirschner, 1982a) (Fig. 1E). This is alleviated by rapid cell divisions, leading to the serial dilution of these factors and resulting in the opening of the genome for the binding of transcriptional machinery to gene regulatory elements.

## 2. TWO PHASES OF ZYGOTIC GENOME ACTIVATION ARE REGULATED BY DISTINCT MECHANISMS

### 2.1 Mid-Blastula Transition and the Major Phase of ZGA

Important progress has been made with regards to (1) the characterization the epigenome changes before and during MBT, (2) the identification of maternally deposited transcription factors active at the ZGA, and (3) the identification of candidate molecules, which fit into the nucleocytoplasmic ratio model and when manipulated, affect the MBT. Two sets of candidate molecules have been identified and experimentally verified. Firstly, replication factors, which regulate the elongation of the cell cycle (Collart, Allen, Bradshaw, Smith, & Zegerman, 2013) and secondly, maternally deposited

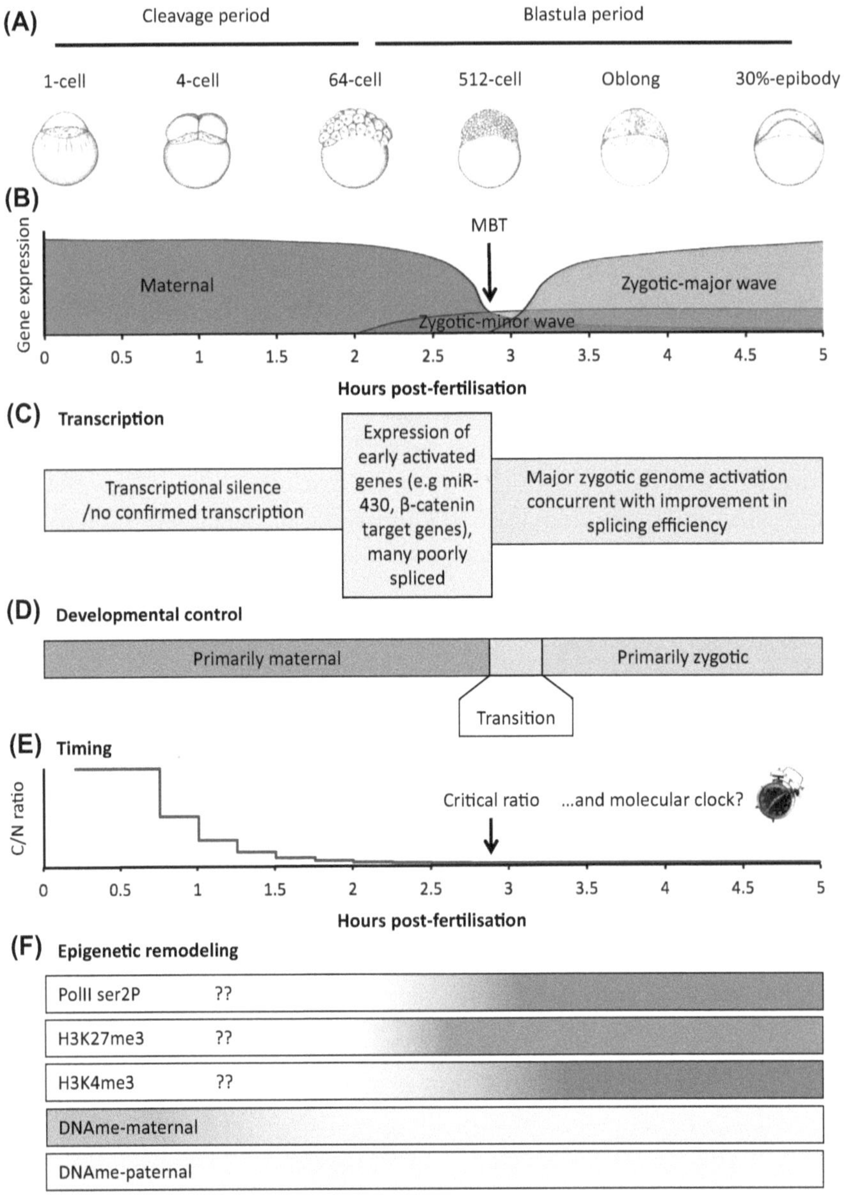


**Figure 1** Outline of the main events occurring during the process of zygotic genome activation. (A), Timeline of zygotic development, highlighting the major macroscopic changes occurring in the embryo during the cleavage and blastula periods. *(Adapted from (Kimmel et al., (1995), used with permission.)* (B), Gene expression level of maternal (*red*) and zygotic (*blue*) derived transcripts from 0 to 5 hours post fertilization.

free histones, acting as cytoplasmic repressors of transcription at ZGA (Amodeo, Jukam, Straight, & Skotheim, 2015).

Besides a specific nucleus-to-cytoplasmic ratio regulating the timing of MBT, additional theories have been proposed to explain the activation of certain genes during this process. These include a kind of a molecular clock, which initiates gene activation at a set time point postfertilization, and involve the sequential modification of DNA during replication to a state permissible to regulated transcription (Hara, Tydeman, & Kirschner, 1980; Sawai, 1979). The molecular clock hypothesis is supported by the observation that the degradation of maternal cyclins, factors vital for transcription-free cell division (Groisman, Jung, Sarkissian, Cao, & Richter, 2002), depend on the time elapsed since fertilization, rather than the nucleus-to-cytoplasmic ratio (Howe, Howell, Hunt, & Newport, 1995; Howe & Newport, 1996; Stack & Newport, 1997) (Fig. 1E). In further support of this model, the majority of zygotic transcription in *Drosophila* was found to be dependent on absolute time postfertilization or developmental stage, with only a small group of genes dependent on the nucleus-to-cytoplasmic ratio (Lu, Li, Elemento, Tavazoie, & Wieschaus, 2009). Another model proposes that the embryo is prevented from initiating transcription before MBT due to the binding of the DNA replication machinery during the period of rapid cell cycling, early in development (Edgar & Schubiger, 1986; Kimelman, Kirschner, & Scherson, 1987). Harvey et al. (2013) used cycloheximide to block cell cycle progression in zebrafish before normal ZGA and found that certain sets of zygotic genes were activated at around the same time point as control embryos, but at an earlier cycle number. This result suggests that separate mechanisms are at play in regulating the timing of activation of different genes or parts of the genome (Harvey et al., 2013). It is therefore likely that a combination of the above proposed mechanisms is at play in regulating the timing of ZGA.

Mid-blastula transition (MBT) is marked. (C), Description of transitional events from 0 to 5 hours post fertilization. (D), Overview of developmental control from 0 to 5 hours post fertilization. (E), Plot of the cytoplasmic to nuclear (C/N) ratio within the embryo from 0 to 5 hours post fertilization. *(Plotted using data from (Kimmel et al., (1995).)* (F), Illustration of epigenetic remodeling events from 0 to 5 hours hours post fertilization (*darker color* indicates a greater level of DNA/chromatin modification, ??, status of modification unknown). (See color plate)

## 2.2 A Minor, First Wave of Transcription Precedes Global Genome Activation

One of the most important findings emanating from the use of genomic approaches in recent years is the confirmation of an old observation that the major wave of genome activation at the MBT is preceded by low level, yet distinct, early activation of transcription, well before the major wave of ZGA (Blythe, Cha, Tadjuidje, Heasman, & Klein, 2010; Heyn et al., 2014; Kimelman et al., 1987; Leung, Soll, Arnold, Kemler, & Driever, 2003; Yang, Tan, Darken, Wilson, & Klein, 2002). This raises questions about (1) the causal relationship between epigenetic remodeling, maternal cytoplasmic and nuclear factors, defining MBT, (2) the distinct waves of transcription regulation, acting in the genome, and (3) what the function and relationship of the two waves of genome activation are in preparing the embryo for its subsequent differentiation processes.

To better understand the relationship between the two waves of ZGA, it is necessary to determine the timing and order of genomic activation and transcription. This is complicated, however, by the facts that (1) maternally contributed RNAs greatly outnumber zygotic transcripts, accounting for 60–70% of mRNA molecules even at the peak of zygotic expression in zebrafish (Lee et al., 2013) and (2) the posttranscriptional regulation, including polyadenylation and clearance of maternal transcript copies of constitutively expressed genes, occurs in parallel with their de novo transcription, affecting de novo transcriptional analysis (Aanes et al., 2013; Ferg et al., 2007; Giraldez et al., 2006; Mathavan et al., 2005). The identification of de novo transcripts either by transcription block using pharmacological tools, or by incorporation of labeled nucleotides followed by transcript sequencing, identified differences in the level and integrity of mRNAs of genes, which were therefore concluded to be the result of zygotic transcription (Aanes et al., 2013; Hamatani, Carter, Sharov, & Ko, 2004; Heyn et al., 2014; Lee et al., 2013; Ulitsky et al., 2012). Allelic differences and differential promoter utilization between the mother and zygotically active embryo also allowed the distinction of maternal and zygotic components of otherwise identical transcripts (Haberle et al., 2014; Harvey et al., 2013). These analyses revealed extensive variability in the timing of activation of genes and parts of the genome during ZGA. Heyn et al. (2014) identified 592 genes transcribed three cell cycles before the main ZGA in zebrafish by zygotic transcript labeling, followed by Illumina sequencing. Consistent with the very short cell cycle at this stage of embryo development, the majority of transcripts were found to be short, and often unspliced. Transcription mostly shuts down at mitosis; therefore it has been proposed that if an RNA product is not completely transcribed and processed before the commencement of the cell cycle, it will be aborted (McHale et al., 2011; Shermoen & O'Farrell, 1991; Tadros & Lipshitz, 2009). If this theory is true, it would suggest that early embryos may only be able to generate short genes with simple exon–intron architecture. However, although the majority of early expressed transcripts do fall into this category, sequencing by Mathavan et al. (2005) in zebrafish, identified some longer genes with many introns to be among the earliest transcribed genes. Indeed, growing evidence points to transcription continuing and spanning across several cell cycles in other model systems (Helmrich, Ballarino, & Tora, 2011), suggesting that early embryos could also use a mechanism, by which transcriptional abortion can be overridden, allowing the formation of longer, more complex transcripts.

The first phase of zygotic transcription, at the mid-one cell stage in mice, is uncoupled from splicing and 3′ processing, leading to the formation of nonfunctional transcripts (Abe et al., 2015; Graf et al., 2014). The apparent inefficiency of the spliceosome could (beside cell cycle time constraints mentioned previously) lead to the enrichment for short, intron-poor transcripts (Heyn et al., 2014) similar to that found in *Drosophila* (De Renzis, Elemento, Tavazoie, & Wieschaus, 2007). Splicing activity is affected by chromatin structure around the exon–intron boundary (reviewed in Bentley, 2014 and de Almeida & Carmo-Fonseca, 2014). The lack of efficient splicing could therefore highlight an early embryo-specific chromatin signature, nonpermissive for splicing. After MBT, concurrent with the major wave of zygotic genome activation, efficient splicing of newly activated genes becomes vital for the progression of epiboly and gastrulation (Aanes et al., 2013; Konig, Matter, Bader, Thiele, & Muller, 2007).

## 2.3 The Identity and Function of Early Zygotic Transcripts

There are a few hints about the potential function of the earliest transcripts, as well as the reasoning behind the timing of this transcriptional activity. The genes encoding the microRNA, miR-430 are among the earliest activated, with miR-430 transcripts already detectable at the 64-cell stage (Heyn et al., 2014). miR-430 plays a crucial role in the deadenylation and subsequent degradation of a large proportion of maternally deposited mRNAs in the embryo, before, or during the ZGA (Giraldez et al., 2006). Moreover, target genes of β-catenin signaling, required for the induction of the mesoderm, are among the first zygotic genes to be activated, presumably to initiate the process of mesoderm induction soon after the main wave of ZGA (Blythe et al., 2010; Leung et al., 2003). Despite these examples, however, the overall functional significance of most pre-MBT transcription remains elusive. There is considerable interspecies variation in the content of this first (minor) wave of genome activation, possibly due to differences in the timing and developmental stage at which ZGA occurs. In rapidly dividing *Drosophila*, where early zygotic transcripts appear after six cell divisions, they primarily encode developmental patterning and cell cycle factors, vital for proper cell division synchrony (Karr, Ali, Drees, & Kornberg, 1985). A similar pattern of expression is observed in *Xenopus* (Paranjpe, Jacobi, van Heeringen, & Veenstra, 2013; Yang et al., 2002). In zebrafish, where the first wave of zygotic transcription occurs after seven cell divisions, both housekeeping genes and embryo patterning genes, including transcription factors and chromatin modifiers, are expressed at this early stage (Aanes et al., 2011; Lee et al., 2013). Beyond this very broad pattern of early-activated gene selection, there appears to be surprisingly little cross-species conservation of which genes are activated early in development (Heyn et al., 2014). As discussed, the majority of early transcribed zygotic genes are short, a characteristic associated with recently evolved genes (Grzybowska, 2012; Neme & Tautz, 2013). In support of this, early expressed zygotic genes in zebrafish and *Drosophila* were found to be significantly younger than those expressed maternally or during later stages of embryo development (Heyn et al., 2014). Strikingly, comparison of early expressed zygotic genes across species identified that the profile of gene expression in the zebrafish is considerably closer to that used by *Drosophila* than its fellow vertebrate, mouse. This suggests that the cellular dynamics of embryogenesis maybe a more influential determinant of gene expression profiles than the evolutionary proximity of species.

The variability of genes selected for earliest activation in various animal models could have several explanations: (1) it may suggest a robustness in early development, founded on widespread genetic redundancy consistent with the early phylotypic period of the developmental hourglass model (Kalinka et al., 2010), (2) it may indicate plasticity in the roles played by genes of the first wave of ZGA in morphologically divergent blastulas, or (3) it may simply suggest that many or most of the earliest expressed genes are not critically required for the developmental program at this early stage but represent transcriptional noise as a result of a shared chromatin opening event required for a small number of key zygotic gene activities, but affecting many. Nevertheless, it is crucial to understand the mechanism that allows their early activation, as it will reflect on how genes are primed for transcription before fertilization in the gametes or after fertilization via chromatin reorganization. Evidence for both potential mechanisms exist and will be discussed further below. Taken together, the first wave of zygotic transcription, activated by maternally deposited factors, likely augments the maternal contribution with zygotically derived transcripts, which may contribute to signaling the further activation of the genome and the major wave of ZGA (Harvey et al., 2013; Lee et al., 2014).

## 3. PROMOTER REGULATION MECHANISMS DURING ZYGOTIC GENOME ACTIVATION

The core promoter, a stretch of DNA surrounding the transcription start site (TSS) of all genes, is a key site of transcriptional regulation. It constitutes the minimal requirement to initiate transcription, integrating regulatory inputs and recruiting general transcription factors (D'Alessio, Wright, & Tjian, 2009; Kadonaga, 2012). The core promoter as a regulatory element has been extensively investigated as a key topic, vital for the understanding of the processes governing transcriptional regulation throughout an organism's life cycle, but also in diseases such as cancer. It is during early embryo development, however, where the core promoter regulation plays its most central role and where it is most extensively studied.

The key questions concerning how ZGA is established at the promoter level are much the same as those asked about transcription regulation in other biological contexts. What defines the promoters where transcription will start? What are the sequence determinants and proteins that mark and define the sites of the genome where transcription will be activated? Why are these sites transcriptionally inactive before the ZGA, despite the transcriptional competence of the early embryo? What are the causal relationships acting between epigenetic signals on the chromatin and transcription factors inherited by the newly divided cells, in defining the temporal sequence of events leading to the two phases of gene expression activation during ZGA? Are there epigenetic signals instructive for the transcriptional regulatory program during ZGA and what is their identity and how does the transcription machinery overcome the repressive effects of early chromatin conformations? Before these questions can be answered, detailed annotation and characterization of the epigenomes, including transcription factor binding profiles, is required. In this regard, the last few years have witnessed substantial progress particularly in zebrafish embryo models.

## 3.1 Changes to Chromatin State During ZGA

Transcription is initiated by the assembly of RNA polymerase II, general transcription factors and associated factors, into a preinitiation complex at the transcription start site within the core promoter of a gene. For the activation of zygotic gene transcription to occur, the core promoter must be made permissive for RNA polymerase II binding and activity. It has long been established that the early embryo is transcriptionally competent, in so far as the machinery required for transcription activation is mostly available from the very beginning of embryo development. This was proven by experiments in which exogenous plasmids were injected into *Xenopus* embryos, which are initially transcriptionally active, then get repressed by chromatin assembly (Almouzni & Wolffe, 1995; Harvey et al., 2013; Lund & Dahlberg, 1992; Prioleau, Huet, Sentenac, & Méchali, 1994; Wiekowski, Miranda, & DePamphilis, 1991). This competitive chromatin repression can only be circumvented by prebinding the plasmid with one of the key initiation complex nucleating factors, the TATA-binding protein (TBP) (Prioleau et al., 1994). While TBP mRNA is abundantly present as mRNA from the oocyte, its translation is rate limiting in *Xenopus* (Veenstra, Destree, & Wolffe, 1999) and delayed until just before the main wave of the ZGA in zebrafish (Muller, Lakatos, Dantonel, Strahle, & Tora, 2001). These results suggest that while transcriptional competence is present, it is inherently limited by the availability of general transcription factors such as TBP, which are necessary for the activation for a large number of genes at the ZGA (Ferg et al., 2007; Jacobi et al., 2007).

## 3.2 Nucleosome Occupancy and Its Sequence Determinants

Transcribed genes are characterized by their promoter region, with reduced nucleosome occupancy at what is called the nucleosome-depleted region (NDR) facilitating transcription factor binding at the promoter. The common model for ZGA in anamniotes suggests that chromatin assembly represses basal and activated transcription before the main wave of ZGA at MBT, by placement of nucleosomes over promoter regions, preventing the binding of transcription machinery (Newport & Kirschner, 1982b). Nucleosome positioning at promoters blocks transcription, unless transcriptional activators can displace nucleosomes in a dynamic process, which is influenced by the DNA sequence, the occupancy of transcription factors, histone variants and modifications and chromatin remodeling complexes (Hughes & Rando, 2014; Jiang & Pugh, 2009). The mechanism by which promoters become nucleosome depleted before and during the ZGA, are unclear. Accumulating evidence suggest that nucleosomes are not evenly positioned around transcription start sites before the ZGA and show a nonrandom and asymmetric distribution (Haberle et al., 2014; Zhang et al., 2014). At the major wave of ZGA, the +1 nucleosome is already positioned downstream of the future transcription start site, followed by phased nucleosomes (Zhang et al., 2014). Whether the repositioned +1 nucleosome is preceded by an NDR cannot be fully determined from the existing nucleosome mapping data. Nevertheless, this prepositioning of nucleosomes is consistent with the presence of an NDR or co-occupation of the promoter region by general transcription factors, or possibly by the presence of paused polymerase II (Adelman & Lis, 2012), which is not engaged in elongation. Importantly, at the dome stage the +1 nucleosome becomes more tightly positioned and is followed by a phased distribution of subsequent nucleosomes, either suggesting that transcription factor/Pol II binding pushes the +1 position into a specific space or that epigenetic regulatory mechanisms involving chromatin remodelers, driven by underlying sequence information and associated epigenetic marks, "prepare" the NDR for transcription factor binding, facilitating gene activation. The phased nucleosome positioning, downstream of the transcription start site, has been shown to be driven by a sequence profile characterized by (1) GC−AT sequence enrichment boundary, (2) intranucleosomal 10 bp oscillation of A/T with G/C dinucleotides, and (3) a larger order, internucleosomal GC/AT enrichment pattern (Haberle et al., 2014; Sasaki et al., 2009), previously described in human (Valouev et al., 2011), indicating that sequence determinants contribute to the nucleosomal phasing pattern. These phased nucleosomes have also been found to be enriched for the transcription activating H3K4me3 posttranslational modification (Haberle et al., 2014; Zhang et al., 2014). None of the nucleosome positioning studies in embryos, however, shed light on the predicted formation of the NDR before ZGA, except for identifying the pervasive existence of a distinct stretch of GC-rich sequence upstream to the +1 nucleosome and encompassing the NDR (Haberle et al., 2014). These findings are in sharp contrast with yeast, where an AT-rich sequence stretch was found to be favored by NDRs (Struhl & Segal, 2013).

The absence of nucleosomes upstream of the +1 nucleosome may alternatively indicate the presence of prebound components of the transcriptional machinery, in which RNA polymerase II is, by some mechanism, stalled before its engagement in transcript elongation. This phenomenon of polymerase II pausing in a position proximal to the core promoter leads to the formation of short, unstable 20–50 bp RNA products and is well described as a key regulator in *Drosophila* early development (Chen et al., 2013). In *Drosophila*, stalled transcriptional machinery has been found to be preferentially located on developmentally regulated genes, associated with the presence of specific, positionally constrained promoter elements, such as the Downstream Promoter Element (DPE) (Chen et al., 2013). PolII pausing has been proposed to prime the genes for synchronized, robust, timely, and topologically constrained activation of genes, important for developmental pattering in the embryo (Lagha et al., 2013) (reviewed in Gaertner & Zeitlinger, 2014 and Levine, 2011). However, the phenomenon of polymerase II pausing has yet to be extensively demonstrated in zebrafish or *Xenopus*. Lindeman et al. (2011) did observe promoter restricted PolII engagement on genes at the 256-cell stage in zebrafish, without evidence for transcription measured by RNA-seq; however, this study did not directly address whether bound PolII is engaged in elongating transcription.

### 3.3 Histone Variants as Placeholders for Regulatory Regions

Two key questions remain, however: (1) how are nucleosomes repositioned and (2) how are promoter regions recognized for establishment of NDRs? One mechanism to achieve this could be the deployment of variant histones, which lead to destabilization of nucleosomes at promoters and other regulatory elements. NDRs are often flanked by the variant histone, H2A.Z, in eukaryotic cells (Mavrich et al., 2008). Nucleosomes at promoters containing the double variants H2A.Z and H3.3 appear unstable and mark active genes in human (Jin et al., 2009). This pattern of histone deposition is suggested to facilitate access of regulatory complexes to regulatory regions in mammalian embryonic stem cells (Hu et al., 2013). Therefore variant histone-containing unstable nucleosomes could serve as a nucleation point for transcription factor binding and establish the NDRs necessary to allow the transcription machinery to build at promoters. So far, no data exists for the presence of these variant histones at promoters before the ZGA in vertebrates; nevertheless, H2A.Z is pervasive at the +1 nucleosomes in zebrafish promoters during the major wave of zygotic genome activation (Haberle et al., 2014; Nepal et al., 2013).

A potentially important transition of one key chromatin component is observed in a range of organisms, including sea urchins and *Xenopus* embryos; the switch from the expression of oocyte-specific forms of linker histone H1, to zygotic (somatic) forms (Dimitrov, Almouzni, Dasso, & Wolffe, 1993; Poccia, 1986). This replacement may contribute to nucleosome repositioning or to making the nucleosomes more mobile, by remodelers (Dimitrov et al., 1993; Poccia, 1986). In mice, the expression of maternally encoded forms of H1 is high immediately postfertilization but declines markedly from the four-cell to eight-cell embryo (Tanaka, Hennebold, Macfarlane, & Adashi, 2001). In contrast, somatic forms of H1 appear at the one-cell (Adenot et al., 2000), two-cell (Stein & Schultz, 2000), and four-cell (Clarke, Oblin, & Bustin, 1992) stages. The synthesis of zygotic forms of H2A and H3 is also upregulated between the one- and two-cell stages (Wiekowski et al., 1991) while histone H4 is still maternally encoded. In addition, the transplantation of post-ZGA, somatic cell nuclei into *Xenopus* oocytes results in the active replacement of somatic variants of H1, with oocyte-specific linker histones B4 and HMG1 (Dimitrov & Wolffe, 1996), mediated by the molecular chaperone, nucleoplasmin. Moreover, somatic linker histones are also selectively deposited in the frog and lead to lineage-restricted regulation of genes (Steinbach, Wolffe, & Rupp, 1997). The functional relevance of the replacement of histone variants is as yet unclear; however, they have been suggested to play a role in epigenetic memory in *Xenopus* (Ng & Gurdon, 2008).

## 3.4 Demethylated Regions Appear on Promoters Before Genome Activation

Methylation of cytosine at CpG dinucleotides in promoter regions is generally associated with transcriptional repression, through methyl CpG binding proteins (Hendrich & Bird, 1998). Such CpGs are enriched in most mammalian promoters and are called CpG islands, or CGIs (reviewed in Deaton & Bird, 2011). In fish and frogs CGIs are not detectable by algorithms designed for mammals; however, a similar enrichment for CpGs does exist in their promoters, despite the largely different GC content and CpG density and suggests a similar role to that in mammals (Jiang et al., 2013; Long et al., 2013; Potok, Nix, Parnell, & Cairns, 2013). Importantly, these CpG-enriched promoter regions coincide with nonmethylated islands

(NMIs) (Long et al., 2013). Nonmethylated CpG enrichments in promoters are bound by CXXC finger protein 1 (Cfp1) and related proteins and allow for the recruitment of Set domain proteins, such as Histone-lysine N-methyltransferases, which generate the methylation marks on Histone three tails at the K4 position (Thomson et al., 2010). Nonmethylated CpG-enriched regions have been observed within developmental promoters in zebrafish (Andersen, Reiner, Aanes, Alestrom, & Collas, 2012; Potok et al., 2013) and *Xenopus* (Bogdanovic et al., 2011). Developmentally active promoters appear mostly demethylated before genome activation in fish, which suggest a potential role for selective demethylation in preparing these promoters for the ZGA later. NMIs are a result of a remarkable overhaul of the DNA methylation state of gene promoters, observed in gametes and then in the early embryo in fish (Jiang et al., 2013; Potok et al., 2013; Wu, Zhang, & Cairns, 2011). Both the oocyte and the sperm carry distinct methylation profiles (hyper and hypo respectively). The oocyte methylation pattern appears to be dramatically overhauled by the 16-cell stage leading to a profile that mimics that of the sperm, by the MBT (Fig. 1F). The majority of developmentally active promoters appear demethylated before the ZGA, suggesting that NMIs contribute to the establishment of transcriptional competence. This is supported by experimental manipulation of the factors involved in DNA methylation. Gene activation at ZGA is affected by changing the methylation state of promoters in *Xenopus*, through the manipulation of methyltransferases (eg, DNMT1) and the methyl CpG binding protein, Kaiso (Ruzov et al., 2004; Stancheva, El-Maarri, Walter, Niveleau, & Meehan, 2002). How promoters are recognized by the factors that determine DNA methylation state, is unclear; however, it is unlikely that CpG density alone is the signal. Instead of enrichment for binding sites, coinciding with CpG enrichment, together may establish NMIs at promoters (Feldmann et al., 2013; Lienert et al., 2011) (reviewed in Schubeler, 2015).

## 3.5 Posttranslational Histone Modifications at Promoters

Zygotic genome activation is marked by extensive changes in posttranslational histone modifications, particularly at the promoter. The core histone H4, diacetylated in the *Xenopus* oocyte, becomes deacetylated upon assembly into the chromatin after fertilization (Dimitrov et al., 1993). This change may help establish and maintain a transcriptionally repressed state in the early embryo. In the lead up to genome activation, however, a marked increase in H4 acetylation has been observed at the nuclear periphery (Stein, Worrad,

Belyaev, Turner, & Schultz, 1997; Thompson, Legouy, Christians, & Renard, 1995; Worrad & Schultz, 1997), in addition to increased acetylation of H2A and H3, priming the genome for transcription initiation. Additionally, several posttranslational modifications (PTMs) of Histone 3 (eg, H3K4me3, H3K9ac, H3K27me3) occur on promoters around transcription start sites (reviewed in Vastenhouw & Schier, 2012). Temporal analysis suggests that they are potentially predictive for developmental gene expression during early embryo development. Genome-wide analysis by chromatin immunoprecipitation (ChIP—CHIP and ChIP—seq) from zebrafish and *Xenopus* demonstrated that histone PTMs on promoters get dramatically enriched just before and during the major wave of ZGA (Akkers et al., 2009; Vastenhouw et al., 2010). However, these promoter-associated histone PTMs appear to premark promoters at lower levels, well before gene activation at the main phase of ZGA (Blythe et al., 2010; Lindeman et al., 2011) and their presence is independent of transcription (Haberle et al., 2014; Hontelez et al., 2015). Many developmentally regulated genes, which are activated much later after the initiation of the ZGA, are premarked well before their activation (Haberle et al., 2014; Vastenhouw & Schier, 2012). These findings suggest that premarking by histone PTMs are potentially instructive for transcriptional activation at and after MBT. A low-level presence before gene activation could provide a nucleating signal for gene activation, possibly through TAF3 binding to promoters, which has been shown to promote H3K4me3 deposition (Vermeulen et al., 2007). Additionally, a second major wave of H3K4me3 deposition could be a consequence of active transcription and represents a marker of gene activation at MBT (Rybtsova et al., 2007).

A characteristic feature of developmentally regulated genes at ZGA is the distinct deposition of the H3K27me3 mark, which, in contrast to H3K4me3, is associated with inactivation of genes during embryo development, via the activity of the Polycomb proteins in the PRC1 and 2 complexes (reviewed in van Kruijsbergen, Hontelez, & Veenstra, 2015). This mark, similar to H3K4me3, is mostly deposited at the MBT in zebrafish and *Xenopus* (Akkers et al., 2009; Vastenhouw & Schier, 2012) but is also present at a smaller subset of genes before ZGA in fish (Lindeman et al., 2011), suggesting that it may contribute to the control of gene expression at the ZGA.

Sequential ChIP has revealed that many promoter regions of inactive genes are positive for both trimethylation marks in fish (Vastenhouw & Schier, 2012), suggesting a similar epigenetic profile to that previously described in embryonic stem cells, commonly termed, bivalency (Azuara et al., 2006; Bernstein et al., 2006; Mikkelsen et al., 2007; Pan et al., 2007; Zhao et al., 2007). These bivalent chromatin marks could therefore be priming the genes for activation (H3K4me3), while keeping them repressed (H3K27me3), as has been demonstrated in embryonic stem cells. However, bivalency is not a prevalent feature in frogs, where the two opposing marks are present mostly on a nonoverlapping set of gene promoters and topologically separated in different dorsoventral poles of the developing embryo (Akkers et al., 2009). Moreover, the H3K27me3 mark was shown to appear at a later stage than the H3K4me3. These apparent differences between zebrafish and *Xenopus* raises the question as to whether epigenetic regulation of transcription in *Xenopus* development follows fundamentally different paths or whether the cells of the *Xenopus* blastula stage embryo are epigenetically less plastic than those in equivalent stage zebrafish and the ES cells of mice? The deposition of the bivalency-associated histone modification marks has been linked with the function of stem cell factors such as Oct4 in mouse (Ang et al., 2011) and is a property of the promoters of development regulator genes in fish embryos (Lindeman et al., 2011), while a similar set of gene promoters are enriched for H3K27me3 after the ZGA in frog (van Heeringen et al., 2011; Hontelez et al., 2015). Whether the stem cell factors influence the function of histone modifiers and thus contribute to histone PTM profiles or whether an opposite causative relationship exists, whereby bivalent marked promoters attract stem cell factors, remains to be addressed, potentially by genetic interference experiments and subsequent genome-wide epigenetic analyses.

## 3.6 Maternal Control of ZGA by Pioneering Transcription Factors

The key outstanding question regarding the process of zygotic genome activation is the identity and function of the factors involved in the active remodeling of the chromatin around promoters and *cis*-regulatory elements, activating transcription at ZGA. Maternally supplied general transcription factors, such as TBP have previously been proposed to be among these factors (Prioleau et al., 1994; Veenstra et al., 1999). These factors, however, are unlikely to be sufficient to improve accessibility and to provide the specificity necessary to activate the correct set of genes for controlled reprogramming of the pluripotent embryo cell mass, ahead of activation of differentiation pathways.

The search for a specific genome activator was aided by the discovery in *Drosophila* that many early transcribed gene promoters share a *cis*-regulatory heptamer motif (CAGGTAG). Liang et al. (2008) identified the zinc-finger protein Zelda (zinc-finger early *Drosophila* activator), which specifically binds to these sites and activates transcription. In support of its role as an early genome activator, Zelda is detectable in the nucleus of the *Drosophila* embryo within 1 h of fertilization (Nien et al., 2011) and activates its target genes by remodeling the chromatin to make the gene promoter more accessible for transcription (Foo et al., 2014; Harrison, Li, Kaplan, Botchan, & Eisen, 2011; Sun et al., 2015; Xu et al., 2014; Yanez-Cuna, Dinh, Kvon, Shlyueva, & Stark, 2012). Zelda appears to fit the criteria for being a pioneer transcription factor (reviewed in Zaret & Carroll, 2011). Pioneer transcription factors have been proposed to be able to access regulatory elements before other transcription factors and either passively promote regulatory element function by prebinding or contribute to accessibility by actively removing nucleosomes. No Zelda homologs have been found in vertebrates. In mouse embryonic stem cells, Oct4 was shown to be able to establish NDRs (You et al., 2011). Additionally, its zebrafish homolog, Pou5f1 (Leichsenring, Maes, Mossner, Driever, & Onichtchouk, 2013), was shown to be the first to occupy binding sites in regulatory elements of early expressed genes at the MBT where the stem cell factors SoxB1 and Nanog were also shown to bind and together play a major role in activating a set of genes at the major wave of the ZGA (Lee et al., 2013). These stem cell factors are maternally deposited as mRNA and have been shown by genome-wide ribosome profiling to be translated early in blastula formation (Lee et al., 2013). These findings demonstrate that early transcribed genes could, at least in part, be activated by targeted binding of maternal proteins, which make promoter regions accessible for transcription. Besides the translation of maternal mRNA encoding pioneering and general transcription factors, other mechanisms of maternal control may also contribute to the timely availability of ZGA regulators. Additional regulatory events may include the controlled subcellular localization of transcription factors and chromatin modifiers. For example the chromatin-remodeling enzyme ISWI has been implicated in the removal of TBP from transplanted somatic nuclei in egg extracts (Kikyo, Wade, Guschin, Ge, & Wolffe, 2000), which suggests that chromatin remodeling factors may contribute to compartmentalization of factors. This could potentially explain the observed maintenance of an oocytelike inactive chromatin state before the ZGA.

## 3.7 Paternal Contribution to the ZGA by Chromatin-Associated Epigenetic Mechanisms

Several lines of investigation have highlighted the male germ line and its potential contribution to the regulation of the ZGA in vertebrates. In mammalian spermatogenesis, histones packaging the DNA are largely replaced with arginine-rich protamines (Ward & Coffey, 1991; Wykes & Krawetz, 2003), facilitating sperm head condensation (Balhorn, 2007). However, a small percentage of the paternal chromatin is retained (Brykczynska et al., 2010; Hammoud et al., 2009). This retained chromatin has been found to be primarily located around developmentally regulated genes containing DNA hypomethylation, marking them with activating (H3K4me2), as well as bivalent marks (H3K4me3, H3K27me3). The deposition of retained histone marks appears to correlate strongly with a gene activation state in the early embryo (Hammoud et al., 2009), suggesting a potential instructive role for these chromatin marks. After fertilization, the DNA in the paternal pronucleus must be repackaged with maternally derived histones before fusion with the maternal genome (Nonchev & Tsanev, 1990). During this repackaging the paternal DNA becomes uncondensed and hyperaccessible for transcriptional initiation (Cho, Sakai, Nagata, & Aoki, 2002). It is thought that during the mid to late S-phase of the first cell division, the transcriptionally competent regions of the paternal pronucleus permit a low level of endogenous transcription, contributing to the minor wave of ZGA (Aoki, Hara, & Schultz, 2003; Bouniol, Nguyen, & Debey, 1995; Ram & Schultz, 1993; Wiekowski et al., 1991; Xue et al., 2013). It is conceivable that the targeted retention of paternal histones could play a role in specifying the activated genes, although this has yet to be conclusively demonstrated. In contrast, the paternal chromatin packaging in zebrafish is retained through spermatogenesis (Wu et al., 2011). Potok et al. (2013) discovered that the DNA methylation of the embryonic genome at ZGA is highly correlated to that in the sperm, which suggests that (1) the maternal genome is structurally reorganized to conform to the paternal program, between fertilization and ZGA, as described previously and (2) it is this paternal program that provides the template for DNA methylation-associated gene regulation at the ZGA. Remarkably, there appears to be a large overlap in the histone PTMs observed in the sperm and in early blastula embryos before ZGA, which is consistent with the suggestion that the histone PTMs are either inherited or selectively redeployed, from the sperm to the embryo (Lindeman et al., 2011). In line with the suggestion that premarking by histone PTMs is instructive for developmental gene regulation, findings in mice have revealed transgenerational effects of disruption of histone methylation in sperm (Siklenka et al., 2015).

## 4. TRANSITIONS IN THE UTILIZATION OF TRANSCRIPTION MACHINERY AT ZGA

The overhaul of the transcriptome from maternal to zygotic origin is paralleled by notable changes in key components of the transcription machinery, both at the level of the regulatory sequences in the DNA and the transcription factors that bind to them (reviewed in Muller, Zaucker, & Tora, 2010). This transition in the utilization of transcription machinery may be a crucial regulatory component of the ZGA for several possible reasons. It may firstly be required to adapt the machinery to the requirements of early embryos in generating pluripotency from terminally differentiated gametes. Secondly, pluripotent and then multipotent cells of differentiating lineages need to respond to complex spatially distinct signals, in a temporally dynamic fashion, which may require adaptation of the transcription machinery to respond to rapidly changing and topologically varying signals, in a tightly controlled manner. The responsiveness of transcription machinery to signals depends on *cis*-regulatory elements, residing in both the core promoter and in distal regulatory elements, located away from the core promoter of PolII transcribed genes. Here we briefly overview core promoter features that may play distinct roles in regulating developmental transcription during and after the ZGA.

The binding of RNA polymerase II to the core promoter is regulated by positionally constrained sequence determinants, guiding the binding location of RNA polymerase II and thus the site of transcriptional initiation (reviewed in Müller, Demény, & Tora, 2007; Ohler & Wassarman, 2009; Sandelin et al., 2007; and Smale & Kadonaga, 2003). The developmental dynamics of core promoter functional elements have been best studied in *Drosophila* and several excellent reviews discuss these sequence properties (Juven-Gershon & Kadonaga, 2010). In *Drosophila,* core promoter motifs (eg, motif 10 element (MTE) or DPE) have been associated with developmental functions, such as regulation by the factor, Dorsal (Zehavi, Kuznetsov, Ovadia-Shochat, & Juven-Gershon, 2014), or promoter pausing in the generation of a spatially coordinated response of Decapentaplegic (Dpp) target genes in mesoderm induction (Gaertner & Zeitlinger, 2014;

Lagha et al., 2013). In vertebrates most of these sequences are either lacking or are too infrequent to allow statistically significant association with a developmental transcriptional event (Lenhard, Sandelin, & Carninci, 2012). Nevertheless, a small number of positionally constrained sequence motifs and dinucleotide profiles remain identifiable in vertebrate embryos, such as the TATA box, or the TCT initiator sequence, which may carry distinct regulatory roles during development. The latter is selectively enriched in a subset of translation-associated gene promoters during development (Nepal et al., 2013).

## 4.1 Transcription Initiation Profiles and Corresponding Promoter Classification

Evidence for differential behavior of promoters, with distinct core promoter features, comes from high-resolution analysis of transcription initiation sites by CAGE (Carninci et al., 2006; Sandelin et al., 2007). Promoters in which a single nucleotide is used for the majority of transcriptional initiation events are referred to as sharp promoters and usually contain a canonical TATA box sequence 30 bp upstream of the transcription start site, providing dominant positional information to the transcription initiation machinery, recruited by TBP (Ponjavic et al., 2006). With such a dominant initiation signal, genes regulated by these promoters tend to be expressed at a high level when the gene is activated and are associated with tissue-specific factors, important for terminal differentiation of a cell (Akalin et al., 2009; Plessy et al., 2012). Broad promoters (with multiple TSSs, over a region greater than 100 bp) often appear to have less-defined transcription factor binding sites, lack a TATA box, but carry a far higher frequency of CpG dinucleotides than sharp promoters (Deaton & Bird, 2011; Lenhard et al., 2012). This characteristic is thought to permit dynamic reorganization of the chromatin within the promoter, into multiple conformations and thus promote one TSS over another (Rach et al., 2011). TSS selection is therefore thought to be influenced more by nucleosome positioning, than the site of transcription factor binding (Rach et al., 2011). Genes regulated by broad promoters have been associated with ubiquitously expressed or developmentally regulated genes (Carninci et al., 2006; Yamashita, Suzuki, Sugano, & Nakai, 2005), the latter characterized by the presence of multiple distal regulatory elements, such as enhancers (Lenhard et al., 2012) and longer CpG enrichment, often spanning across the whole gene and associated with H3K27me occupancy and polycomb regulation (van Heeringen et al., 2014). Altogether, the simplistic promoter classification, based on TSS utilization, appears to be a useful model, which explains the transition of promoter usage at the ZGA.

CAGE analysis of the dynamics of promoter usage in zebrafish development has revealed a characteristic shift in the utilization of sequence determinants of TSS usage, in developmentally active promoters in the oocyte, reflecting differential reading of promoter information in the mother and zygotic transcription (Haberle et al., 2014; Nepal et al., 2013). The maternal TSS selection was found to be dependent on an A/T-rich (W-box) element. This is replaced by a zygotic TSS selection, which is associated with a GC/AT enrichment boundary and a nucleosome-wide phasing of GC/AT dinucleotide enrichment, which corresponds to the positioning of the downstream, H3K4me3-marked nucleosome. Similarly, a transition from TATA-type promoters to TATA-less appears to be a feature of the *Drosophila* and *Xenopus* promoterome (van Heeringen et al., 2011; Ni et al., 2010). These observations suggest that the maternal transcription machinery recognizes the same promoter, but in a different manner to the embryo. The likely explanation for this observation may stem from the apparent diversity of general transcription factors and their distinct composition in the oocyte, in comparison to the zygotically active embryo (reviewed in Muller & Tora, 2014 and Müller et al., 2007).

## 4.2 Maternal to Zygotic Transition in Transcription Factor Usage

Zebrafish and *Xenopus* studies have revealed the existence and activity of three members of the TBP family of general transcription factors, in the early anamniote embryo. Two of them, TLF/TRF2 and TBP2/TRF3, are paternally and maternally enriched in the gametes (Jallow, Jacobi, Weeks, Dawid, & Veenstra, 2004) and have been found to function and potentially replace TBP in regulating transcription in developmental contexts, where TBP was found to be redundant (Ferg et al., 2007; Muller et al., 2001; Veenstra, Weeks, & Wolffe, 2000). Indeed TBP2 has a similar DNA-binding capacity to TBP (Bartfai et al., 2004) and was shown to replace TBP in the frog oocyte (Akhtar & Veenstra, 2009), similar to mouse (Gazdag et al., 2009). TLF does not bind the TATA element, however, and has been suggested to act through the DPE element in Drosophila (Kedmi et al., 2014). The specificity of the vertebrate paralog TLF to promoters in the anamniote embryo remains to be discovered. The transcription machinery in the oocyte and during spermatogenesis differs in more than one way from that in the embryo (reviewed in Levine, Cattoglio, & Tjian, 2014). Several general transcription factors are replaced in the gametes, with homologs of somatic variants. For example, the TFIIA homolog, ALF and the TFIID complex components, and the TAFs are replaced by oocyte-specific homologs (Han, Xie, Hammes, & DeJong, 2003; Xiao, Kim, & DeJong, 2006). Similarly, several TAFs have male germ cell-specific paralogs in mammals (Pointud et al., 2003). Whether such alternative transcription factors drive a differential mechanism of promoter recognition or the variant proteins facilitate the recruitment of PolII to different positions on promoters due to steric reasons and thus explain the differences in the TSS profiles of promoter usage between maternal and zygotic transcripts remains to be established.

## 5. CONCLUSIONS AND PERSPECTIVES

The delayed nature of zygotic genome activation in the zebrafish and frog embryo provides an ideal experimental platform for elucidating the temporal dynamics of establishment of a transcriptionally active chromatin state and addressing the determinants of transcription activation at polymerase II transcribed gene promoters. The relatively large number of cells generated by the rapid cell divisions before ZGA allows for the analysis of epigenetic profiles, which precede transcriptional activity. Important advances have been made thanks to the continuing development of genome-wide assays that interrogate the epigenome and transcriptome with everincreasing resolution. Despite these advances, progress in understanding the mechanisms involved in ZGA has been relatively slow, due to the remaining major technical limitations of epigenomic analysis.

One of these problems arises from working with very small cell numbers at the earliest stages of embryo development (eg, in the morula, fertilized eggs, and oocytes). At these stages there is very little genetic material to analyze and existing genetic and epigenetic technologies are not sensitive enough to adequately explore the chromatin features and transcription regulation-associated nuclear processes occurring in these samples. Due to this limitation, very little is known about what is occurring in, for example, the zebrafish embryo before the 64-cell stage. The advent of single cell analysis tools, such as single cell chip analysis (Rotem et al., 2015; Xue et al., 2013) and small cell number CHIP protocols (Schmidl, Rendeiro, Sheffield, & Bock, 2015), with far greater sensitivity for small amounts of genetic material, will facilitate the future application of chromatin immunoprecipitation sequencing for very early embryos. These methods are most efficient with histones, but the detection of transcription factor binding in minute sample remains a major challenge and will need to be addressed before the parental contribution to transcription regulation in the embryo can be defined.

Characterization of the epigenomic events is not sufficient, however, for a complete understanding of the molecular mechanisms underlying zygotic genome activation. The application of temporally and spatially regulated genetic interference and manipulation methods will be needed for the dissection of the function of parent-derived transcription factors, chromatin remodelers, cell cycle, and DNA replication regulators and a variety of additional epigenetic modifiers. However, sufficiently simple genetic tools for generating efficient maternal loss of function in key genes are not available in anamniotes. The root of the problem is the same factor that makes the fish and frog embryos so resilient to gene interference. The maternally derived mRNA and proteins are associated with long lasting functions that persist well into embryo and larval development and can survive up to 5 days in zebrafish, spanning the whole term of larval development. Zygotic loss of function mutants are compensated by maternal gene products, leading to a large number of loss of function mutants of genes, which although lethal in yeast, show comparatively mild defects in zebrafish (Amsterdam et al., 2004). However, the generation of maternal loss of function or paternal loss of function is not a straightforward task in anamniote embryos. Several methods have been developed and demonstrated for parental gene interference: (1) morpholino antisense oligomers can be injected into oocytes to remove gene function in them before in vitro maturation and fertilization (Nair, Lindeman, & Pelegri, 2013) or injected in the embryo before the start of zygotic transcription (Lee et al., 2013); (2) the generation of homozygous zygotic mutants by one of the reverse genetic tools, such as TILLING (Wienholds & Plasterk, 2004), TALEN (Bedell et al., 2012), and most recently CrispR/Cas9 (Hwang et al., 2013), followed by a traditional, several generation breeding program, can lead to homozygous viable adults, which provide mutant gametes (Kumari et al., 2013); and (3) when the zygotic mutant is not viable, zygotic mutant donors can be used for primordial germ cell transplantation into somatically wild type recipients, which form gametic mutant gonad-containing adults for the generation of maternal effect mutants (Ciruna et al., 2002). This approach has led to far more dramatic phenotypes in early embryo development (Leichsenring et al., 2013). While it has not yet been demonstrated, the CRISPR/Cas9 system could theoretically be used in the future to selectively disrupt gene function during gametogenesis before fertilization by expressing the components of the system in a gamete-specific manner from transgenes.

A whole new range of genome targeting-based manipulation possibilities are now made possible by virtue of the plasticity of the Cas9 protein and the ability to add new functions to it via various protein fusions, involving, for example, transcriptional activators and chromatin modifiers (Konermann et al., 2013, 2015). Thus, besides function disruption, the CRISPR/Cas9 technique can now be used to target epigenetic modifiers to specific sites in the genome, allowing a far greater level of interrogation of the causal relationship between epigenetic programs and gene function than was previously possible. In this approach the endonuclease component of the CRISPR/Cas9 complex is replaced with an effector protein, such as a chromatin remodeler. This complex is then guided to the site of interest using appropriate target site-specific guide RNAs (gRNAs). Additionally, noncoding and protein-binding RNAs can be tethered to the targeted site through fusion to gRNA, affecting local protein–DNA interactions (Shechner, Hacisuleyman, Younger, & Rinn, 2015). However, these techniques are still at an early stage of development and have not yet been used in early embryos. Nevertheless, over the coming years, these novel approaches promise to revolutionize our understanding of the parallel processes acting together which define and prepare the sites of transcription at ZGA.

## ACKNOWLEDGMENTS

We would like to thank Ms. Ledean Cooper and Dr. Joseph Grice for their help and consultation on the planning of the early transcript. We would also like to acknowledge the Wellcome Trust for their funding of the chapter authors.

## REFERENCES

Aanes, H., Ostrup, O., Andersen, I. S., Moen, L. F., Mathavan, S., Collas, P., & Alestrom, P. (2013). Differential transcript isoform usage pre- and post-zygotic genome activation in zebrafish. *BMC Genomics, 14*, 331.

Aanes, H., Winata, C. L., Lin, C. H., Chen, J. P., Srinivasan, K. G., Lee, S. G., … Mathavan, S. (2011). Zebrafish mRNA sequencing deciphers novelties in transcriptome dynamics during maternal to zygotic transition. *Genome Research, 21*(8), 1328–1338.

Abe, K., Yamamoto, R., Franke, V., Cao, M., Suzuki, Y., Suzuki, M. G., … Aoki, F. (2015). The first murine zygotic transcription is promiscuous and uncoupled from splicing and 3′ processing. *EMBO Journal, 34*(11), 1523–1537.

Adelman, K., & Lis, J. T. (2012). Promoter-proximal pausing of RNA polymerase II: emerging roles in metazoans. *Nature Reviews Genetics, 13*(10), 720–731.

Adenot, P. G., Campion, E., Legouy, E., Allis, C. D., Dimitrov, S., Renard, J., & Thompson, E. M. (2000). Somatic linker histone H1 is present throughout mouse embryogenesis and is not replaced by variant H1 degrees. *Journal of Cell Science, 113*(Pt 16), 2897–2907.

Akalin, A., Fredman, D., Arner, E., Dong, X., Bryne, J. C., Suzuki, H., … Lenhard, B. (2009). Transcriptional features of genomic regulatory blocks. *Genome Biology, 10*(4), R38.

Akhtar, W., & Veenstra, G. J. (2009). TBP2 is a substitute for TBP in *Xenopus* oocyte transcription. *BMC Biology*, 7, 45.

Akkers, R. C., van Heeringen, S. J., Jacobi, U. G., Janssen-Megens, E. M., Francoijs, K. J., Stunnenberg, H. G., & Veenstra, G. J. (2009). A hierarchy of H3K4me3 and H3K27me3 acquisition in spatial gene regulation in *Xenopus* embryos. *Developmental Cell, 17*(3), 425–434.

de Almeida, S. F., & Carmo-Fonseca, M. (2014). Reciprocal regulatory links between cotranscriptional splicing and chromatin. *Seminars in Cell & Developmental Biology, 32*, 2–10.

Almouzni, G., & Wolffe, A. P. (1995). Constraints on transcriptional activator function contribute to transcriptional quiescence during early *Xenopus* embryogenesis. *EMBO Journal, 14*(8), 1752–1765.

Amodeo, A. A., Jukam, D., Straight, A. F., & Skotheim, J. M. (2015). Histone titration against the genome sets the DNA-to-cytoplasm threshold for the *Xenopus* midblastula transition. *Proceedings of the National Academy of Sciences of the Unites States of America, 112*(10), E1086–E1095.

Amsterdam, A., Nissen, R. M., Sun, Z., Swindell, E. C., Farrington, S., & Hopkins, N. (2004). Identification of 315 genes essential for early zebrafish development. *Proceedings of the National Academy of Sciences of the United States of America, 101*(35), 12792–12797.

Andersen, I. S., Reiner, A. H., Aanes, H., Alestrom, P., & Collas, P. (2012). Developmental features of DNA methylation during activation of the embryonic zebrafish genome. *Genome Biology, 13*(7), R65.

Ang, Y. S., Tsai, S. Y., Lee, D. F., Monk, J., Su, J., Ratnakumar, K., … Lemischka, I. R. (2011). Wdr5 mediates self-renewal and reprogramming via the embryonic stem cell core transcriptional network. *Cell, 145*(2), 183–197.

Aoki, F., Hara, K. T., & Schultz, R. M. (2003). Acquisition of transcriptional competence in the 1-cell mouse embryo: requirement for recruitment of maternal mRNAs. *Molecular Reproduction and Development, 64*(3), 270–274.

Azuara, V., Perry, P., Sauer, S., Spivakov, M., Jorgensen, H. F., John, R. M., … Fisher, A. G. (2006). Chromatin signatures of pluripotent cell lines. *Nature Cell Biology, 8*(5), 532–538.

Balhorn, R. (2007). The protamine family of sperm nuclear proteins. *Genome Biology, 8*(9), 227.

Barckmann, B., & Simonelig, M. (2013). Control of maternal mRNA stability in germ cells and early embryos. *Biochimica et Biophysica Acta, 1829*(6–7), 714–724.

Bartfai, R., Balduf, C., Hilton, T., Rathmann, Y., Hadzhiev, Y., Tora, L., … Muller, F. (2004). TBP2, a vertebrate-specific member of the TBP family, is required in embryonic development of zebrafish. *Current Biology, 14*(7), 593–598.

Bedell, V. M., Wang, Y., Campbell, J. M., Poshusta, T. L., Starker, C. G., Krug, R. G., 2nd, … Ekker, S. C. (2012). In vivo genome editing using a high-efficiency TALEN system. *Nature, 491*(7422), 114–118.

Bentley, D. L. (2014). Coupling mRNA processing with transcription in time and space. *Nature Reviews Genetics, 15*(3), 163–175.

Bernstein, B. E., Mikkelsen, T. S., Xie, X., Kamal, M., Huebert, D. J., Cuff, J., … Lander, E. S. (2006). A bivalent chromatin structure marks key developmental genes in embryonic stem cells. *Cell, 125*(2), 315–326.

Blythe, S. A., Cha, S. W., Tadjuidje, E., Heasman, J., & Klein, P. S. (2010). beta-Catenin primes organizer gene expression by recruiting a histone H3 arginine 8 methyltransferase, Prmt2. *Developmental Cell, 19*(2), 220–231.

Bogdanovic, O., Long, S. W., van Heeringen, S. J., Brinkman, A. B., Gomez-Skarmeta, J. L., Stunnenberg, H. G., … Veenstra, G. J. (2011). Temporal uncoupling of the DNA methylome and transcriptional repression during embryogenesis. *Genome Research, 21*(8), 1313–1327.

Bouniol, C., Nguyen, E., & Debey, P. (1995). Endogenous transcription occurs at the 1-cell stage in the mouse embryo. *Experimental Cell Research, 218*(1), 57–62.

Brykczynska, U., Hisano, M., Erkek, S., Ramos, L., Oakeley, E. J., Roloff, T. C., … Peters, A. H. (2010). Repressive and active histone methylation mark distinct promoters in human and mouse spermatozoa. *Nature Structural & Molecular Biology, 17*(6), 679–687.

Carninci, P., Sandelin, A., Lenhard, B., Katayama, S., Shimokawa, K., Ponjavic, J., … Hayashizaki, Y. (2006). Genome-wide analysis of mammalian promoter architecture and evolution. *Nature Genetics, 38*(6), 626–635.

Chen, K., Johnston, J., Shao, W., Meier, S., Staber, C., & Zeitlinger, J. (2013). A global change in RNA polymerase II pausing during the *Drosophila* midblastula transition. *eLife, 2*, e00861.

Cho, T., Sakai, S., Nagata, M., & Aoki, F. (2002). Involvement of chromatin structure in the regulation of mouse zygotic gene activation. *Animal Science Journal, 73*(2), 113–122.

Ciruna, B., Weidinger, G., Knaut, H., Thisse, B., Thisse, C., Raz, E., & Schier, A. F. (2002). Production of maternal-zygotic mutant zebrafish by germ-line replacement. *Proceedings of the National Academy of Sciences of the United States of America, 99*(23), 14919–14924.

Clarke, H. J., Oblin, C., & Bustin, M. (1992). Developmental regulation of chromatin composition during mouse embryogenesis: somatic histone H1 is first detectable at the 4-cell stage. *Development, 115*(3), 791–799.

Collart, C., Allen, G. E., Bradshaw, C. R., Smith, J. C., & Zegerman, P. (2013). Titration of four replication factors is essential for the *Xenopus laevis* midblastula transition. *Science, 341*(6148), 893–896.

D'Alessio, J. A., Wright, K. J., & Tjian, R. (2009). Shifting players and paradigms in cell-specific transcription. *Molecular Cell, 36*(6), 924–931.

De Renzis, S., Elemento, O., Tavazoie, S., & Wieschaus, E. F. (2007). Unmasking activation of the zygotic genome using chromosomal deletions in the *Drosophila* embryo. *PLoS Biology, 5*(5), e117.

Deaton, A. M., & Bird, A. (2011). CpG islands and the regulation of transcription. *Genes & Development, 25*(10), 1010–1022.

Dimitrov, S., Almouzni, G., Dasso, M., & Wolffe, A. P. (1993). Chromatin transitions during early *Xenopus* embryogenesis: changes in histone H4 acetylation and in linker histone type. *Developmental Biology, 160*(1), 214–227.

Dimitrov, S., & Wolffe, A. P. (1996). Remodeling somatic nuclei in *Xenopus laevis* egg extracts: molecular mechanisms for the selective release of histones H1 and H1(0) from chromatin and the acquisition of transcriptional competence. *EMBO Journal, 15*(21), 5897–5906.

Edgar, B. A., & Schubiger, G. (1986). Parameters controlling transcriptional activation during early *Drosophila* development. *Cell, 44*(6), 871–877.

Feldmann, A., Ivanek, R., Murr, R., Gaidatzis, D., Burger, L., & Schubeler, D. (2013). Transcription factor occupancy can mediate active turnover of DNA methylation at regulatory regions. *PLoS Genetics, 9*(12), e1003994.

Ferg, M., Sanges, R., Gehrig, J., Kiss, J., Bauer, M., Lovas, A., … Muller, F. (2007). The TATA-binding protein regulates maternal mRNA degradation and differential zygotic transcription in zebrafish. *EMBO Journal, 26*(17), 3945–3956.

Foe, V. E., & Alberts, B. M. (1983). Studies of nuclear and cytoplasmic behaviour during the five mitotic cycles that precede gastrulation in *Drosophila* embryogenesis. *Journal of Cell Science, 61*, 31–70.

Foo, S. M., Sun, Y., Lim, B., Ziukaite, R., O'Brien, K., Nien, C. Y., … Rushlow, C. A. (2014). Zelda potentiates morphogen activity by increasing chromatin accessibility. *Current Biology, 24*(12), 1341–1346.

Gaertner, B., & Zeitlinger, J. (2014). RNA polymerase II pausing during development. *Development, 141*(6), 1179–1183.

Gazdag, E., Santenard, A., Ziegler-Birling, C., Altobelli, G., Poch, O., Tora, L., & Torres-Padilla, M. E. (2009). TBP2 is essential for germ cell development by regulating transcription and chromatin condensation in the oocyte. *Genes & Development, 23*(18), 2210–2223.

Gerhart, J. C. (1980). Mechanisms regulating pattern formation in the amphibian egg and early embryo. *Biological Regulation and Development*, 133–316 (Chapter 4).

Giraldez, A. J., Mishima, Y., Rihel, J., Grocock, R. J., Van Dongen, S., Inoue, K., … Schier, A. F. (2006). Zebrafish MiR-430 promotes deadenylation and clearance of maternal mRNAs. *Science, 312*(5770), 75–79.

Graf, A., Krebs, S., Zakhartchenko, V., Schwalb, B., Blum, H., & Wolf, E. (2014). Fine mapping of genome activation in bovine embryos by RNA sequencing. *Proceedings of the National Academy of Sciences of the United States of America, 111*(11), 4139–4144.

Groisman, I., Jung, M. Y., Sarkissian, M., Cao, Q., & Richter, J. D. (2002). Translational control of the embryonic cell cycle. *Cell, 109*(4), 473–483.

Grzybowska, E. A. (2012). Human intronless genes: functional groups, associated diseases, evolution, and mRNA processing in absence of splicing. *Biochemical and Biophysical Research Communications, 424*(1), 1–6.

Haberle, V., Li, N., Hadzhiev, Y., Plessy, C., Previti, C., Nepal, C., … Lenhard, B. (2014). Two independent transcription initiation codes overlap on vertebrate core promoters. *Nature, 507*(7492), 381–385.

Hamatani, T., Carter, M. G., Sharov, A. A., & Ko, M. S. H. (2004). Dynamics of global gene expression changes during mouse preimplantation development. *Developmental Cell, 6*(1), 117–131.

Hammoud, S. S., Nix, D. A., Zhang, H., Purwar, J., Carrell, D. T., & Cairns, B. R. (2009). Distinctive chromatin in human sperm packages genes for embryo development. *Nature, 460*(7254), 473–478.

Han, S., Xie, W., Hammes, S. R., & DeJong, J. (2003). Expression of the germ cell-specific transcription factor ALF in *Xenopus* oocytes compensates for translational inactivation of the somatic factor TFIIA. *Journal of Biological Chemistry, 278*(46), 45586–45593.

Hara, K., Tydeman, P., & Kirschner, M. (1980). A cytoplasmic clock with the same period as the division cycle in *Xenopus* eggs. *Proceedings of the National Academy of Sciences of the United States of America,* 77(1), 462–466.

Harrison, M. M., Li, X.-Y., Kaplan, T., Botchan, M. R., & Eisen, M. B. (2011). Zelda binding in the early *Drosophila melanogaster* embryo marks regions subsequently activated at the maternal-to-zygotic transition. *PLoS Genetics,* 7(10), e1002266.

Harvey, S. A., Sealy, I., Kettleborough, R., Fenyes, F., White, R., Stemple, D., & Smith, J. C. (2013). Identification of the zebrafish maternal and paternal transcriptomes. *Development (Cambridge, England), 140*(13), 2703–2710.

van Heeringen, S. J., Akhtar, W., Jacobi, U. G., Akkers, R. C., Suzuki, Y., & Veenstra, G. J. (2011). Nucleotide composition-linked divergence of vertebrate core promoter architecture. *Genome Research, 21*(3), 410–421.

van Heeringen, S. J., Akkers, R. C., van Kruijsbergen, I., Arif, M. A., Hanssen, L. L., Sharifi, N., & Veenstra, G. J. (2014). Principles of nucleation of H3K27 methylation during embryonic development. *Genome Research, 24*(3), 401–410.

Helmrich, A., Ballarino, M., & Tora, L. (2011). Collisions between replication and transcription complexes cause common fragile site instability at the longest human genes. *Molecular Cell, 44*(6), 966–977.
Hendrich, B., & Bird, A. (1998). Identification and characterization of a family of mammalian methyl-CpG binding proteins. *Molecular and Cellular Biology, 18*(11), 6538–6547.
Heyn, P., Kircher, M., Dahl, A., Kelso, J., Tomancak, P., Kalinka, A. T., & Neugebauer, K. M. (2014). The earliest transcribed zygotic genes are short, newly evolved, and different across species. *Cell Reports, 6*(2), 285–292.
Hontelez, S., van Kruijsbergen, I., Georgiou, G., van Heeringen, S. J., Bogdanovic, O., Lister, R., & Veenstra, G. J. (2015). Embryonic transcription is controlled by maternally defined chromatin state. *Nature Communications, 6*, 10148.
Howe, J. A., Howell, M., Hunt, T., & Newport, J. W. (1995). Identification of a developmental timer regulating the stability of embryonic cyclin A and a new somatic A-type cyclin at gastrulation. *Genes & Development, 9*(10), 1164–1176.
Howe, J. A., & Newport, J. W. (1996). A developmental timer regulates degradation of cyclin E1 at the midblastula transition during *Xenopus* embryogenesis. *Proceedings of the National Academy of Sciences of the United States of America, 93*(5), 2060–2064.
Hu, G., Cui, K., Northrup, D., Liu, C., Wang, C., Tang, Q., … Zhao, K. (2013). H2A.Z facilitates access of active and repressive complexes to chromatin in embryonic stem cell self-renewal and differentiation. *Cell Stem Cell, 12*(2), 180–192.
Hughes, A. L., & Rando, O. J. (2014). Mechanisms underlying nucleosome positioning in vivo. *Annual Review of Biophysics, 43*, 41–63.
Hwang, W. Y., Fu, Y., Reyon, D., Maeder, M. L., Tsai, S. Q., Sander, J. D., … Joung, J. K. (2013). Efficient genome editing in zebrafish using a CRISPR-Cas system. *Nature Biotechnology, 31*(3), 227–229.
Jacobi, U. G., Akkers, R. C., Pierson, E. S., Weeks, D. L., Dagle, J. M., & Veenstra, G. J. (2007). TBP paralogs accommodate metazoan- and vertebrate-specific developmental gene regulation. *EMBO Journal, 26*(17), 3900–3909.
Jallow, Z., Jacobi, U. G., Weeks, D. L., Dawid, I. B., & Veenstra, G. J. C. (2004). Specialized and redundant roles of TBP and a vertebrate-specific TBP paralog in embryonic gene regulation in *Xenopus*. *Proceedings of the National Academy of Sciences of the United States of America, 101*(37), 13525–13530.
Jiang, C., & Pugh, B. F. (2009). Nucleosome positioning and gene regulation: advances through genomics. *Nature Reviews Genetics, 10*(3), 161–172.
Jiang, L., Zhang, J., Wang, J. J., Wang, L., Zhang, L., Li, G., … Liu, J. (2013). Sperm, but not oocyte, DNA methylome is inherited by zebrafish early embryos. *Cell, 153*(4), 773–784.
Jin, C., Zang, C., Wei, G., Cui, K., Peng, W., Zhao, K., & Felsenfeld, G. (2009). H3.3/H2A.Z double variant-containing nucleosomes mark 'nucleosome-free regions' of active promoters and other regulatory regions. *Nature Genetics, 41*(8), 941–945.
Juven-Gershon, T., & Kadonaga, J. T. (2010). Regulation of gene expression via the core promoter and the basal transcriptional machinery. *Developmental Biology, 339*(2), 225–229.
Kadonaga, J. T. (2012). Perspectives on the RNA polymerase II core promoter. *Wiley Interdisciplinary Reviews: Developmental Biology, 1*(1), 40–51.
Kalinka, A. T., Varga, K. M., Gerrard, D. T., Preibisch, S., Corcoran, D. L., Jarrells, J., … Tomancak, P. (2010). Gene expression divergence recapitulates the developmental hourglass model. *Nature, 468*(7325), 811–814.
Kane, D. A., Hammerschmidt, M., Mullins, M. C., Maischein, H. M., Brand, M., van Eeden, F. J., … Nüsslein-Volhard, C. (1996). The zebrafish epiboly mutants. *Development (Cambridge, England), 123*, 47–55.
Kane, D. A., & Kimmel, C. B. (1993). The zebrafish midblastula transition. *Development, 119*(2), 447–456.

Karr, T. L., Ali, Z., Drees, B., & Kornberg, T. (1985). The engrailed locus of *D. melanogaster* provides an essential zygotic function in precellular embryos. *Cell, 43*(3 Pt 2), 591–601.

Kedmi, A., Zehavi, Y., Glick, Y., Orenstein, Y., Ideses, D., Wachtel, C., … Juven-Gershon, T. (2014). *Drosophila* TRF2 is a preferential core promoter regulator. *Genes & Development, 28*(19), 2163–2174.

Kikyo, N., Wade, P. A., Guschin, D., Ge, H., & Wolffe, A. P. (2000). Active remodeling of somatic nuclei in egg cytoplasm by the nucleosomal ATPase ISWI. *Science, 289*(5488), 2360–2362.

Kimelman, D., Kirschner, M., & Scherson, T. (1987). The events of the midblastula transition in Xenopus are regulated by changes in the cell-cycle. *Cell, 48*(3), 399–407.

Kimmel, C. B., Ballard, W. W., Kimmel, S. R., Ullmann, B., & Schilling, T. F. (1995). Stages of embryonic development of the zebrafish. *Developmental Dynamics, 203*(3), 253–310.

Konermann, S., Brigham, M. D., Trevino, A. E., Hsu, P. D., Heidenreich, M., Cong, L., … Zhang, F. (2013). Optical control of mammalian endogenous transcription and epigenetic states. *Nature, 500*(7463), 472–476.

Konermann, S., Brigham, M. D., Trevino, A. E., Joung, J., Abudayyeh, O. O., Barcena, C., … Zhang, F. (2015). Genome-scale transcriptional activation by an engineered CRISPR-Cas9 complex. *Nature, 517*(7536), 583–588.

Konig, H., Matter, N., Bader, R., Thiele, W., & Muller, F. (2007). Splicing segregation: the minor spliceosome acts outside the nucleus and controls cell proliferation. *Cell, 131*(4), 718–729.

van Kruijsbergen, I., Hontelez, S., & Veenstra, G. J. (2015). Recruiting polycomb to chromatin. *International Journal of Biochemistry & Cell Biology, 67*, 177–187.

Kumari, P., Gilligan, P. C., Lim, S., Tran, L. D., Winkler, S., Philp, R., & Sampath, K. (2013). An essential role for maternal control of Nodal signaling. *eLife, 2*, e00683.

Lagha, M., Bothma, J. P., Esposito, E., Ng, S., Stefanik, L., Tsui, C., … Levine, M. S. (2013). Paused Pol II coordinates tissue morphogenesis in the *Drosophila* embryo. *Cell, 153*(5), 976–987.

Lee, M. T., Bonneau, A. R., & Giraldez, A. J. (2014). Zygotic genome activation during the maternal-to-zygotic transition. *Annual Review of Cell and Developmental Biology, 30*, 581–613.

Lee, M. T., Bonneau, A. R., Takacs, C. M., Bazzini, A. A., DiVito, K. R., Fleming, E. S., & Giraldez, A. J. (2013). Nanog, Pou5f1 and SoxB1 activate zygotic gene expression during the maternal-to-zygotic transition. *Nature, 503*(7476), 360–364.

Leichsenring, M., Maes, J., Mossner, R., Driever, W., & Onichtchouk, D. (2013). Pou5f1 transcription factor controls zygotic gene activation in vertebrates. *Science, 341*(6149), 1005–1009.

Lenhard, B., Sandelin, A., & Carninci, P. (2012). Metazoan promoters: emerging characteristics and insights into transcriptional regulation. *Nature Reviews Genetics, 13*(4), 233–235.

Leung, T., Soll, I., Arnold, S. J., Kemler, R., & Driever, W. (2003). Direct binding of Lef1 to sites in the boz promoter may mediate pre-midblastula-transition activation of boz expression. *Developmental Dynamics, 228*(3), 424–432.

Levine, M. (2011). Paused RNA polymerase II as a developmental checkpoint. *Cell, 145*(4), 502–511.

Levine, M., Cattoglio, C., & Tjian, R. (2014). Looping back to leap forward: transcription enters a new era. *Cell, 157*(1), 13–25.

Liang, H.-L., Nien, C.-Y., Liu, H.-Y., Metzstein, M. M., Kirov, N., & Rushlow, C. (2008). The zinc-finger protein Zelda is a key activator of the early zygotic genome in *Drosophila*. *Nature, 456*(7220), 400–403.

Lienert, F., Wirbelauer, C., Som, I., Dean, A., Mohn, F., & Schubeler, D. (2011). Identification of genetic elements that autonomously determine DNA methylation states. *Nature Genetics, 43*(11), 1091–1097.

Lindeman, L. C., Andersen, I. S., Reiner, A. H., Li, N., Aanes, H., Ostrup, O., ... Collas, P. (2011). Prepatterning of developmental gene expression by modified histones before zygotic genome activation. *Developmental Cell, 21*(6), 993–1004.

Long, H. K., Sims, D., Heger, A., Blackledge, N. P., Kutter, C., Wright, M. L., ... Klose, R. J. (2013). Epigenetic conservation at gene regulatory elements revealed by non-methylated DNA profiling in seven vertebrates. *eLife, 2*, e00348.

Lu, X., Li, J. M., Elemento, O., Tavazoie, S., & Wieschaus, E. F. (2009). Coupling of zygotic transcription to mitotic control at the *Drosophila* mid-blastula transition. *Development (Cambridge, England), 136*(12), 2101–2110.

Lund, E., & Dahlberg, J. E. (1992). Control of 4-8S RNA transcription at the midblastula transition in *Xenopus laevis* embryos. *Genes & Development, 6*(6), 1097–1106.

Mathavan, S., Lee, S. G. P., Mak, A., Miller, L. D., Murthy, K. R. K., Govindarajan, K. R., ... Lufkin, T. (2005). Transcriptome analysis of zebrafish embryogenesis using microarrays. *PLoS Genetics, 1*(2), 260–276.

Mavrich, T. N., Jiang, C., Ioshikhes, I. P., Li, X., Venters, B. J., Zanton, S. J., ... Pugh, B. F. (2008). Nucleosome organization in the *Drosophila* genome. *Nature, 453*(7193), 358–362.

McHale, P., Mizutani, C. M., Kosman, D., MacKay, D. L., Belu, M., Hermann, A., ... Hwa, T. (2011). Gene length may contribute to graded transcriptional responses in the *Drosophila* embryo. *Developmental Biology, 360*(1), 230–240.

Mikkelsen, T. S., Ku, M., Jaffe, D. B., Issac, B., Lieberman, E., Giannoukos, G., ... Bernstein, B. E. (2007). Genome-wide maps of chromatin state in pluripotent and lineage-committed cells. *Nature, 448*(7153), 553–560.

Müller, F., Demény, M. A., & Tora, L. (2007). New problems in RNA polymerase ii transcription initiation: matching the diversity of core promoters with a variety of promoter recognition factors. *Journal of Biological Chemistry, 282*(20), 14685–14689.

Muller, F., Lakatos, L., Dantonel, J., Strahle, U., & Tora, L. (2001). TBP is not universally required for zygotic RNA polymerase II transcription in zebrafish. *Current Biology, 11*(4), 282–287.

Muller, F., & Tora, L. (2014). Chromatin and DNA sequences in defining promoters for transcription initiation. *Biochimica et Biophysica Acta, 1839*(3), 118–128.

Muller, F., Zaucker, A., & Tora, L. (2010). Developmental regulation of transcription initiation: more than just changing the actors. *Current Opinion in Genetics & Development, 20*(5), 533–540.

Nair, S., Lindeman, R. E., & Pelegri, F. (2013). In vitro oocyte culture-based manipulation of zebrafish maternal genes. *Developmental Dynamics, 242*(1), 44–52. http://dx.doi.org/10.1002/dvdy.23894.

Neme, R., & Tautz, D. (2013). Phylogenetic patterns of emergence of new genes support a model of frequent de novo evolution. *BMC Genomics, 14*, 117.

Nepal, C., Hadzhiev, Y., Previti, C., Haberle, V., Li, N., Takahashi, H., ... Muller, F. (2013). Dynamic regulation of the transcription initiation landscape at single nucleotide resolution during vertebrate embryogenesis. *Genome Research, 23*(11), 1938–1950.

Newport, J., & Kirschner, M. (1982a). A major developmental transition in early *Xenopus* embryos: I. characterization and timing of cellular changes at the midblastula stage. *Cell, 30*(3), 675–686.

Newport, J., & Kirschner, M. (1982b). A major developmental transition in early *Xenopus* embryos: II. Control of the onset of transcription. *Cell, 30*(3), 687–696.

Ng, R. K., & Gurdon, J. B. (2008). Epigenetic memory of an active gene state depends on histone H3.3 incorporation into chromatin in the absence of transcription. *Nature Cell Biology, 10*(1), 102–109.

Ni, T., Corcoran, D. L., Rach, E. A., Song, S., Spana, E. P., Gao, Y., ... Zhu, J. (2010). A paired-end sequencing strategy to map the complex landscape of transcription initiation. *Nature Methods, 7*(7), 521–527.
Nien, C.-Y., Liang, H.-L., Butcher, S., Sun, Y., Fu, S., Gocha, T., ... Rushlow, C. (2011). Temporal coordination of gene networks by Zelda in the early *Drosophila* embryo. *PLoS Genetics, 7*(10), e1002339.
Nonchev, S., & Tsanev, R. (1990). Protamine-histone replacement and DNA replication in the male mouse pronucleus. *Molecular Reproduction and Development, 25*(1), 72–76.
Ohler, U., & Wassarman, D. A. (2009). Promoting developmental transcription. *Development (Cambridge, England), 137*(1), 15–26.
Pan, G., Tian, S., Nie, J., Yang, C., Ruotti, V., Wei, H., ... Thomson, J. A. (2007). Whole-genome analysis of histone H3 lysine 4 and lysine 27 methylation in human embryonic stem cells. *Cell Stem Cell, 1*(3), 299–312.
Paranjpe, S. S., Jacobi, U. G., van Heeringen, S. J., & Veenstra, G. J. (2013). A genome-wide survey of maternal and embryonic transcripts during *Xenopus* tropicalis development. *BMC Genomics, 14*, 762.
Plessy, C., Pascarella, G., Bertin, N., Akalin, A., Carrieri, C., Vassalli, A., ... Carninci, P. (2012). Promoter architecture of mouse olfactory receptor genes. *Genome Research, 22*(3), 486–497.
Poccia, D. (1986). Remodeling of nucleoproteins during gametogenesis, fertilization, and early development. *International Review of Cytology, 105*, 1–65.
Pointud, J. C., Mengus, G., Brancorsini, S., Monaco, L., Parvinen, M., Sassone-Corsi, P., & Davidson, I. (2003). The intracellular localisation of TAF7L, a paralogue of transcription factor TFIID subunit TAF7, is developmentally regulated during male germ-cell differentiation. *Journal of Cell Science, 116*(Pt 9), 1847–1858.
Ponjavic, J., Lenhard, B., Kai, C., Kawai, J., Carninci, P., Hayashizaki, Y., & Sandelin, A. (2006). Transcriptional and structural impact of TATA-initiation site spacing in mammalian core promoters. *Genome Biology, 7*(8), R78.
Potok, M. E., Nix, D. A., Parnell, T. J., & Cairns, B. R. (2013). Reprogramming the maternal zebrafish genome after fertilization to match the paternal methylation pattern. *Cell, 153*(4), 759–772.
Prioleau, M. N., Huet, J., Sentenac, A., & Méchali, M. (1994). Competition between chromatin and transcription complex assembly regulates gene expression during early development. *Cell, 77*(3), 439–449.
Rach, E. A., Winter, D. R., Benjamin, A. M., Corcoran, D. L., Ni, T., Zhu, J., & Ohler, U. (2011). Transcription initiation patterns indicate divergent strategies for gene regulation at the chromatin level. *PLoS Genetics, 7*(1), e1001274.
Ram, P. T., & Schultz, R. M. (1993). Reporter gene expression in G2 of the 1-cell mouse embryo. *Developmental Biology, 156*(2), 552–556.
Rotem, A., Ram, O., Shoresh, N., Sperling, R. A., Goren, A., Weitz, D. A., & Bernstein, B. E. (2015). Single-cell ChIP-seq reveals cell subpopulations defined by chromatin state. *Nature Biotechnology, 33*(11), 1165–1172.
Ruzov, A., Dunican, D. S., Prokhortchouk, A., Pennings, S., Stancheva, I., Prokhortchouk, E., & Meehan, R. R. (2004). Kaiso is a genome-wide repressor of transcription that is essential for amphibian development. *Development, 131*(24), 6185–6194.
Rybtsova, N., Leimgruber, E., Seguin-Estevez, Q., Dunand-Sauthier, I., Krawczyk, M., & Reith, W. (2007). Transcription-coupled deposition of histone modifications during MHC class II gene activation. *Nucleic Acids Research, 35*(10), 3431–3441.
Sandelin, A., Carninci, P., Lenhard, B., Ponjavic, J., Hayashizaki, Y., & Hume, D. A. (2007). Mammalian RNA polymerase II core promoters: insights from genome-wide studies. *Nature Reviews Genetics, 8*(6), 424–436.

Sasaki, S., Mello, C. C., Shimada, A., Nakatani, Y., Hashimoto, S., Ogawa, M., … Morishita, S. (2009). Chromatin-associated periodicity in genetic variation downstream of transcriptional start sites. *Science, 323*(5912), 401–404.

Sawai, T. (1979). Cyclic changes in the cortical layer of non-nucleated fragments of the newt's egg. *Journal of Embryology and Experimental Morphology, 51*, 183–193.

Schmidl, C., Rendeiro, A. F., Sheffield, N. C., & Bock, C. (2015). ChIPmentation: fast, robust, low-input ChIP-seq for histones and transcription factors. *Nature Methods, 12*(10), 963–965.

Schubeler, D. (2015). Function and information content of DNA methylation. *Nature, 517*(7534), 321–326.

Shechner, D. M., Hacisuleyman, E., Younger, S. T., & Rinn, J. L. (2015). Multiplexable, locus-specific targeting of long RNAs with CRISPR-Display. *Nature Methods, 12*(7), 664–670.

Shermoen, A. W., & O'Farrell, P. H. (1991). Progression of the cell cycle through mitosis leads to abortion of nascent transcripts. *Cell, 67*(2), 303–310.

Signoret, J., & Lefresne, J. (1975). Demonstration of RNA synthesis during segmentation in the Axolotl embryo. *Comptes Rendus Hebdomadaires des Seances de l'Academie des Sciences D, 281*(4), 301–303.

Siklenka, K., Erkek, S., Godmann, M., Lambrot, R., McGraw, S., Lafleur, C., … Kimmins, S. (2015). Disruption of histone methylation in developing sperm impairs offspring health transgenerationally. *Science, 350*(6261), aab2006.

Smale, S. T., & Kadonaga, J. T. (2003). The RNA polymerase II core promoter. - PubMed - NCBI. *Annual Review of Biochemistry, 72*(1), 449–479.

Stack, J. H., & Newport, J. W. (1997). Developmentally regulated activation of apoptosis early in *Xenopus* gastrulation results in cyclin A degradation during interphase of the cell cycle. *Development (Cambridge, England), 124*(16), 3185–3195.

Stancheva, I., El-Maarri, O., Walter, J., Niveleau, A., & Meehan, R. R. (2002). DNA methylation at promoter regions regulates the timing of gene activation in *Xenopus laevis* embryos. *Developmental Biology, 243*(1), 155–165.

Stein, P., & Schultz, R. M. (2000). Initiation of a chromatin-based transcriptionally repressive state in the preimplantation mouse embryo: lack of a primary role for expression of somatic histone H1. *Molecular Reproduction and Development, 55*(3), 241–248.

Stein, P., Worrad, D. M., Belyaev, N. D., Turner, B. M., & Schultz, R. M. (1997). Stage-dependent redistributions of acetylated histones in nuclei of the early preimplantation mouse embryo. *Molecular Reproduction and Development, 47*(4), 421–429.

Steinbach, O. C., Wolffe, A. P., & Rupp, R. A. (1997). Somatic linker histones cause loss of mesodermal competence in *Xenopus*. *Nature, 389*(6649), 395–399.

Struhl, K., & Segal, E. (2013). Determinants of nucleosome positioning. *Nature Structural & Molecular Bioliogy, 20*(3), 267–273.

Sun, Y., Nien, C.-Y., Chen, K., Liu, H.-Y., Johnston, J., Zeitlinger, J., & Rushlow, C. (2015). Zelda overcomes the high intrinsic nucleosome barrier at enhancers during *Drosophila* zygotic genome activation. *Genome Research, 25*(11), 1703–1714.

Tadros, W., & Lipshitz, H. D. (2009). The maternal-to-zygotic transition: a play in two acts. *Development, 136*(18), 3033–3042.

Tanaka, M., Hennebold, J. D., Macfarlane, J., & Adashi, E. Y. (2001). A mammalian oocyte-specific linker histone gene H1oo: homology with the genes for the oocyte-specific cleavage stage histone (cs-H1) of sea urchin and the B4/H1M histone of the frog. *Development (Cambridge, England), 128*(5), 655–664.

Thompson, E. M., Legouy, E., Christians, E., & Renard, J. P. (1995). Progressive maturation of chromatin structure regulates HSP70.1 gene expression in the preimplantation mouse embryo. *Development (Cambridge, England), 121*(10), 3425–3437.

Thomson, J. P., Skene, P. J., Selfridge, J., Clouaire, T., Guy, J., Webb, S., ... Bird, A. (2010). CpG islands influence chromatin structure via the CpG-binding protein Cfp1. *Nature, 464*(7291), 1082–1086.

Ulitsky, I., Shkumatava, A., Jan, C. H., Subtelny, A. O., Koppstein, D., Bell, G. W., ... Bartel, D. P. (2012). Extensive alternative polyadenylation during zebrafish development. *Genome Research, 22*(10), 2054–2066.

Valouev, A., Johnson, S. M., Boyd, S. D., Smith, C. L., Fire, A. Z., & Sidow, A. (2011). Determinants of nucleosome organization in primary human cells. *Nature, 474*(7352), 516–520.

Vastenhouw, N. L., & Schier, A. F. (2012). Bivalent histone modifications in early embryogenesis. *Current Opinion in Cell Biology, 24*(3), 374–386.

Vastenhouw, N. L., Zhang, Y., Woods, I. G., Imam, F., Regev, A., Liu, X. S., ... Schier, A. F. (2010). Chromatin signature of embryonic pluripotency is established during genome activation. *Nature, 464*(7290), 922–926.

Veenstra, G. J., Destree, O. H., & Wolffe, A. P. (1999). Translation of maternal TATA-binding protein mRNA potentiates basal but not activated transcription in *Xenopus* embryos at the midblastula transition. *Molecular and Cellular Biology, 19*(12), 7972–7982.

Veenstra, G. J., Weeks, D. L., & Wolffe, A. P. (2000). Distinct roles for TBP and TBP-like factor in early embryonic gene transcription in *Xenopus*. *Science, 290*(5500), 2312–2315.

Vermeulen, M., Mulder, K. W., Denissov, S., Pijnappel, W. W., van Schaik, F. M., Varier, R. A., ... Timmers, H. T. (2007). Selective anchoring of TFIID to nucleosomes by trimethylation of histone H3 lysine 4. *Cell, 131*(1), 58–69.

Ward, W. S., & Coffey, D. S. (1991). DNA packaging and organization in mammalian spermatozoa: comparison with somatic cells. *Biology of Reproduction, 44*(4), 569–574.

Wiekowski, M., Miranda, M., & DePamphilis, M. L. (1991). Regulation of gene expression in preimplantation mouse embryos: effects of the zygotic clock and the first mitosis on promoter and enhancer activities. *Developmental Biology, 147*(2), 403–414.

Wienholds, E., & Plasterk, R. H. (2004). Target-selected gene inactivation in zebrafish. *Methods in Cell Biology, 77*, 69–90.

Worrad, D. M., & Schultz, R. M. (1997). Regulation of gene expression in the preimplantation mouse embryo: temporal and spatial patterns of expression of the transcription factor Sp1. *Molecular Reproduction and Development, 46*(3), 268–277.

Wu, S. F., Zhang, H., & Cairns, B. R. (2011). Genes for embryo development are packaged in blocks of multivalent chromatin in zebrafish sperm. *Genome Research, 21*(4), 578–589.

Wykes, S. M., & Krawetz, S. A. (2003). The structural organization of sperm chromatin. *Journal of Molecular Chemistry, 278*(32), 29471–29477.

Xiao, L., Kim, M., & DeJong, J. (2006). Developmental and cell type-specific regulation of core promoter transcription factors in germ cells of frogs and mice. *Gene Expression Patterns, 6*(4), 409–419.

Xu, Z., Chen, H., Ling, J., Yu, D., Struffi, P., & Small, S. (2014). Impacts of the ubiquitous factor Zelda on Bicoid-dependent DNA binding and transcription in *Drosophila*. *Genes & Development, 28*(6), 608–621.

Xue, Z., Huang, K., Cai, C., Cai, L., Jiang, C.-Y., Feng, Y., ... Fan, G. (2013). Genetic programs in human and mouse early embryos revealed by single-cell RNA sequencing. *Nature, 500*(7464), 593–597.

Yamashita, R., Suzuki, Y., Sugano, S., & Nakai, K. (2005). Genome-wide analysis reveals strong correlation between CpG islands with nearby transcription start sites of genes and their tissue specificity. *Gene, 350*(2), 129–136.

Yanez-Cuna, J. O., Dinh, H. Q., Kvon, E. Z., Shlyueva, D., & Stark, A. (2012). Uncovering cis-regulatory sequence requirements for context specific transcription factor binding. *Genome Research, 22*. gr.132811.132111.

Yang, J., Tan, C., Darken, R. S., Wilson, P. A., & Klein, P. S. (2002). Beta-catenin/Tcf-regulated transcription prior to the midblastula transition. *Development, 129*(24), 5743–5752.

You, J. S., Kelly, T. K., De Carvalho, D. D., Taberlay, P. C., Liang, G., & Jones, P. A. (2011). OCT4 establishes and maintains nucleosome-depleted regions that provide additional layers of epigenetic regulation of its target genes. *Proceedings of the National Academy of Sciences of the United States of America, 108*(35), 14497–14502.

Zaret, K. S., & Carroll, J. S. (2011). Pioneer transcription factors: establishing competence for gene expression. *Genes & Development, 25*(21), 2227–2241.

Zehavi, Y., Kuznetsov, O., Ovadia-Shochat, A., & Juven-Gershon, T. (2014). Core promoter functions in the regulation of gene expression of *Drosophila* dorsal target genes. *Journal of Biological Chemistry, 289*(17), 11993–12004.

Zhang, Y., Vastenhouw, N. L., Feng, J., Fu, K., Wang, C., Ge, Y., ... Liu, X. S. (2014). Canonical nucleosome organization at promoters forms during genome activation. *Genome Research, 24*(2), 260–266.

Zhao, X. D., Han, X., Chew, J. L., Liu, J., Chiu, K. P., Choo, A., ... Wei, C.-L. (2007). Whole-genome mapping of histone H3 Lys4 and 27 trimethylations reveals distinct genomic compartments in human embryonic stem cells. *Cell Stem Cell, 1*(3), 286–298.

CHAPTER SEVEN

# The Zebrafish as Model for Deciphering the Regulatory Architecture of Vertebrate Genomes

**S. Rastegar*[,1] and U. Strähle*[,§,1]**
*Institute of Toxicology and Genetics, Karlsruhe Institute of Technology, Eggenstein-Leopoldshafen, Germany
[§]Ruprecht-Karls-Universität Heidelberg, Heidelberg, Germany
[1]Corresponding authors: E-mail: sepand.rastegar@kit.edu; uwe.straehle@kit.edu

## Contents



## Abstract

Despite enormous progress to map *cis*-regulatory modules (CRMs), like enhancers and promoters in genomes, elucidation of the regulatory landscape of the developing embryo remains a challenge. The zebrafish embryo with its experimental virtues has a great potential to contribute to this endeavor. However, so far progress remained behind expectation. We discuss here available methods and their limitations and how the zebrafish embryo could contribute in the future to unravel the wiring of the vertebrate genome.

## 1. INTRODUCTION

Spatially and temporally controlled retrieval of genetic information by means of differential gene transcription appears to be the most fundamental

*Advances in Genetics*, Volume 95
ISSN 0065-2660
http://dx.doi.org/10.1016/bs.adgen.2016.04.003

regulatory mechanism underlying vertebrate development and body homeostasis. This importance of transcription is reflected in the abundance of genes in the genome that have a function in gene transcription. Around 3303 genes in the zebrafish genome encode proteins with an inferred role as transcription factors, chromatin modifying factors, or proteins involved in the general process of transcription (Armant et al., 2013). Similar numbers of genes with a function in transcription regulation exist in the human genomes (Vaquerizas, Kummerfeld, Teichmann, & Luscombe, 2009).

The last 30 years, since the discovery of a transcriptional enhancer first in a virus and then in the beta-globin locus of rabbit (Banerji, Olson, & Schaffner, 1983; Banerji, Rusconi, & Schaffner, 1981) witnessed an explosion of our knowledge of the mechanisms underlying transcription regulation (Andersson, Sandelin, & Danko, 2015; Kolovos, Knoch, Grosveld, Cook, & Papantonis, 2012; Ong & Corces, 2011). It has become clear that noncoding DNA, in the past frequently also referred to as "junk" DNA of genomes (Ohno, 1972) is not at all irrelevant garbage but encodes additional information required to retrieve genetic information by regulating transcription. These so-called *cis*-regulatory modules (CRMs) comprise enhancers, silencers, and insulators and include gene proximal regulatory element, the promoter, from which transcription of the coding region of a gene starts (for review(s) Ferg et al., 2014; Kolovos et al., 2012). We will focus in this review on enhancers. Enhancers can act over long distances—in some cases over a megabase as, for example, a conserved enhancer in the *sonic hedgehog (shh)* locus driving transcription of *shh* mRNA in limbs and fins (Lettice et al., 2003, 2002; Sagai et al., 2009, 2004). In general, genes have multiple enhancers, for example, the zebrafish *shh* or *neurogenin1* (*neurog1*) genes harbor multiple CRMs that control transcription in spatially and temporally distinct but overlapping domains of expression of the gene in the embryo. These CRMs are hence partially redundant (Blader et al., 2004; Blader, Plessy, & Strahle, 2003; Ertzer et al., 2007). CRMs are characterized by a cluster of binding sites for transcription factors forming an open chromatin domain (Levine, 2010). Master or pioneer transcription factors open-up closed chromatin (Zaret & Carroll, 2011). Open chromatin which is transcriptionally active or which is poised for activation is accompanied by specific histone H3 modifications (for review see Maston, Landt, Snyder, & Green, 2012; Ong & Corces, 2011; Shlyueva, Stampfel, & Stark, 2014). It is increasingly appreciated that the nucleus has an inherent structure reflecting the transcriptional state of genes. Genes at the periphery tend to be repressed, whereas genes in the center are actively transcribed (Schneider & Grosschedl, 2007). By chromatin looping involving the Mediator complex and Cohesins

(Kagey et al., 2010), enhancers and promoters are brought together in domains of active transcription being also dubbed "transcription factories" (Cook, 1999; Rieder, Trajanoski, & McNally, 2012).

Changes in CRM sequences or emergence of new CRMs and concomitant changes in the patterns or responsiveness of expression of the linked genes are regarded as driving force of evolution (Carroll, 2008; Strahle & Rastegar, 2008; Wittkopp & Kalay, 2012). Moreover, many disease-linked single nucleotide polymorphisms (SNP) map to noncoding regions suggesting that changes in CRMs cause disease (Gordon & Lyonnet, 2014; Maurano et al., 2012; Smith & Shilatifard, 2014). Despite enormous progress in technologies to identify transcription factor binding, chromatin states, and interactions of CRMs at genome scale (Maston et al., 2012; Shlyueva et al., 2014), we are far from understanding the regulatory networks at the system's level. The latter will be a prerequisite to model and to predict how disturbances in regulatory networks can be compensated for and why this compensation fails and disease develops.

The zebrafish has been intensively used as genetic model to decipher the molecular pathways underlying development of the vertebrate body plan. The growth of its transparent embryos outside of the mother fostered initially in particular large-scale forward genetic screens (Driever et al., 1996; Golling et al., 2002; Haffter et al., 1996; Nusslein-Volhard, 2012). These screens uncovered many transcription factors such as FoxA2 (Norton et al., 2005), Pou2 (Belting et al., 2001), and tbx24 (Nikaido et al., 2002) as key regulators of developmental processes. When one types "transcription factor" into the search window of the Zebrafish Model Organism Database (ZFIN), one finds currently 971 mutant alleles of transcription factor genes (http://zfin.org/). However, while successful to pinpoint regulators of transcription, zebrafish genetics did not live up to expectations when it came to the identification of CRMs. Very few phenotype-causing mutations were reported in noncoding elements in the unbiased ENU screens (Montero-Balaguer et al., 2006; Schauerte et al., 1998). This is in striking contrast to the emerging evidence from human genome projects that many disease-causing mutations reside in noncoding regions of the human genome (Epstein, 2009; Khurana et al., 2013; Maurano et al., 2012; Weinhold, Jacobsen, Schultz, Sander, & Lee, 2014). We will address here this obvious discrepancy between the human data and the results of the genetic screens in the zebrafish. We will discuss available methods to analyze the regulatory architecture of the zebrafish genome at a global level and how new technological developments could help to overcome this knowledge deficit on the regulatory architecture of the zebrafish genome.

## 2. ENHANCERS ARE TRICKY TO FIND RELIABLY IN THE GENOME BY COMPUTATIONAL MEANS

Locating promoter(s) is guided by 5′-end mapping of the mRNA(s) encoded by a locus (Haberle et al., 2014; Nepal et al., 2013). However, identifying enhancers is much more challenging given their action over long distances and their flexible location both upstream, downstream and within transcribed regions (Strahle & Rastegar, 2008).

The DNA of enhancers is bound by transcription factors in a sequence-specific manner. This sequence-specific interaction constitutes the fundamental principle of the regulatory activity of enhancers. Computational approaches are very useful to identify potential transcription factor binding sites in short DNA regions with an already known enhancer function. However, the unbiased computational search of the entire genome sequence for binding sites of transcription factors was so far not too successful for the de novo identification of enhancers (Slattery et al., 2014). The code underlying transcription factor binding in the nucleus is more complex than just the mere presence of a recognition sequence in the naked DNA. Integration of determinants like accessibility of the site and the presence of competing or cooperating cofactors have to be taken into account (Slattery et al., 2014). In addition, the DNA-binding specificities of many transcription factors are still unknown and the recognition sequences are rather degenerate complicating such computational analyses further. However, new high-throughput methods for determining DNA—-protein interaction in vitro may help to overcome some of these limitations (Berger et al., 2006; Gubelmann et al., 2013; Hens et al., 2011; Jolma et al., 2010; Stormo & Zhao, 2010; White, Myers, Corbo, & Cohen, 2013). A critical point is how conserved DNA-binding specificities are. There is some evidence that the recognition sequences of specific transcription regulators evolve only slowly and are maintained over long evolutionary distances (Boyle et al., 2014; Weirauch & Hughes, 2010). However, use of mammalian DNA-binding specificities for computational searches of the zebrafish genome may still introduce variables due to evolutionary distance that blur the outcome of genome-wide analyses. Another strategy is the search for clusters of binding sites as multiple transcription factors interact usually with a single enhancer (Berman et al., 2002).

At some stage, straightforward sequence comparison of genomes of different species appeared highly promising as means to identify regulatory elements (Pennacchio & Rubin, 2001; Visel, Bristow, & Pennacchio, 2007; Wasserman & Sandelin, 2004). Some enhancers were indeed found to be extremely conserved between species. For example in zebrafish, most of the enhancers mapped functionally in the *neurog1 and shh* genes turned out to be highly conserved in mouse and human orthologues of the two genes (Blader et al., 2004; Ertzer et al., 2007; Strahle & Rastegar, 2008). However, it became also soon clear that this feature of conservation is predominantly linked to genes of transcription factors and developmental regulators (Bejerano et al., 2004; Plessy, Dickmeis, Chalmel, & Strahle, 2005; Sandelin et al., 2004; Woolfe et al., 2005). Furthermore, even among developmental regulators, the sequence of identified regulatory elements may not be conserved as, for example, in the case of an enhancer driving expression of the axon guidance protein Netrin1 in the floor plate (Rastegar et al., 2002). Why the sequence of enhancers linked to some developmental regulators is so extremely conserved remains a mystery (Harmston, Baresic, & Lenhard, 2013). Comparison of five enhancers all of which are highly conserved among orthologous genes and drive expression in the notochord combined with fine-grained mutational analysis of two of these enhancers suggested that high sequence conservation was not required for notochord expression (Rastegar et al., 2008). Only 47% of 528 conserved noncoding sequences tested in the mouse in a systematic transgenic study mediated expression (Visel et al., 2009). Even the function of conserved enhancers can have changed as in the case of the enhancers of the *shh* gene in mouse and fish (Ertzer et al., 2007; Strahle & Rastegar, 2008). Thus, conserved noncoding sequences are not a good predictor of enhancer activity. In general, regulatory elements show conservation in function but their flexible arrangement together with the variability of the recognition sequences can evolve in such a way that binding sites do not form easily recognizable blocks of conserved sequence as suggested by these conserved sequences fortuitously linked to enhancers of developmental regulators.

This is in line with the notion that the DNA structure of the regulatory architecture of genomes provides the playground of evolution (Arnold et al., 2014). Clearly, transcription factor genes are highly conserved over large evolutionary distances and they also use similar DNA recognition sequences (Boyle et al., 2014; Gehring, 1996; He, Holloway, Maerkl, & Kreitman, 2011; Villar, Flicek, & Odom, 2014). So it appears that re-arrangement and re-combination of transcription factor binding sites is the major mechanism of evolution of the regulatory genome (Carroll, 2008; Hare, Peterson, Iyer, Meier, & Eisen, 2008). Interestingly, when the conserved organization of genomic regulatory blocks was assayed, functional constraints possible at the level of chromatin conformation, selective enhancer/promoter interactions, and nuclear organization appear to exist (Naville et al., 2015). In summary, the genome-wide detection of enhancer sequences by computational means remains a challenge calling for new approaches (Yokoyama, Zhang, & Ma, 2014).

## 2.1 Chromatin Signatures as Guides to Enhancers on a Genome Scale

Over the years, several methods have emerged that allow identifying enhancer regions by particular structural features. Enhancers have an open chromatin structure resulting in discrete DNAse1 or micrococcal nuclease (MNase) hypersensitive islands in noncoding sequences. These regions can be mapped throughout the genome by deep sequencing (MNase-seq, DNAse-seq) of DNA fragments released from chromatin and subsequent alignment of the obtained sequences with the genome (Boyle et al., 2008; Boyle & Furey, 2009). However, also other functionally relevant regions like promoters and replication origins are bound by proteins and have an open chromatin structure. Thus, MNase-seq or DNAse-seq may not be standalone techniques, but in combination with techniques outlined below they provide very good ways to locate enhancer sequences.

Tremendous progress has been made by applying immunoprecipitation to enrich for specific features of enhancers followed by analysis of the pulled down DNA fragments on chips or by next generation sequencing. Distinct features can be probed in these approaches (Furey, 2012). With an antibody directed against a transcription factor, all sequences bound by this factor in the genome could in principal be pulled down and identified. It turned out, however, that transcription factors can bind unspecifically to chromatin or to sites that appear to have no functional relevance (Fisher et al., 2012; Li et al., 2008). For example, glucocorticoid receptor bound to more than 4000 sites throughout the genome of human lung carcinoma cells in response to the synthetic glucocorticoid dexamethasone while expression of only 236 genes was induced by this treatment (Reddy et al., 2009). A general feature of enhancers is the clustering of transcription factor binding sites. Reiterative immunoprecipitation of two transcription factors is thus a possibility to add some specificity to this approach. However, this requires antibodies with the exquisite specificity necessary to pull down the chromatin-bound transcription factor specifically. These are currently not available for each of the more than 1000 transcription factors encoded in vertebrate genomes. Another severe drawback is the fact that many of these antibodies are species-specific preventing cross-species applications. In zebrafish, few studies were carried out using this approach (Greenald et al., 2015; Morley et al., 2009; Nelson et al., 2014). For example, the Wardle lab used an antibody directed against the T-box transcription factor Notail/Brachyury to identify genes bound by the factor during gastrulation of the zebrafish embryo (Morley et al., 2009).

Distinct signatures of histone H3 posttranslational modifications are associated with the state of enhancer activity (Calo & Wysocka, 2013; Heintzman et al., 2007). Either the histone H3 modifications themselves (Heintzman et al., 2007) or the histone-modifying enzymes such as the histone acetylase CBP/p300 associated with the enhancers (Visel et al., 2009) were employed as epitopes to immunoprecipitate and thus identify enhancer sequences at a genome-wide scale. Enrichment of histone H3 monomethylated at lysin 4 (H3K4me1) over the trimethlyated state (H3K4me3) appears to be a reliable marker of enhancers (Heintzman et al., 2007). Acetylation of lysine 27 of histone H3 (H3K27ac) by CBP/p300 provides in addition a mark for active enhancers (Creyghton et al., 2010; Visel et al., 2009).

In zebrafish, chromatin signatures were intensively investigated in studies of the events linked to midblastula transition where the zygotic genome becomes active for the first time after fertilization (Kane & Kimmel, 1993). Bivalent histone marks both permissive (H3K4me3) and repressing (H3K27me3, H3K9me3) coexisted at enhancers of lineage regulators suggesting that enhancers in the zebrafish are kept in a state poised for activation as previously noted in pluripotent mouse ES cells (Aday, Zhu, Lakshmanan, Wang, & Lawson, 2011; Bernstein et al., 2006; Lindeman et al., 2011; Vastenhouw et al., 2010). It remains controversial whether these histone marks are deposited before or after activation of the zygotic genome (Lindeman et al., 2011; Vastenhouw et al., 2010). During subsequent stages, activity indicating acetylation marks (H3K27ac) increase and repressive marks (H3K27me3) are lost reflecting the activation of the zygotic genome and the emergence of the first cell types (Bogdanovic et al., 2012; Lindeman et al., 2011; Vastenhouw et al., 2010).

All of these methods have, however, severe limitations when applied to older embryos. Embryos do not consist of single cell types and even cells of the same type maybe in different states depending on cell cycle or external stimuli. These methods provide thus an average description of an enhancer region over all cell states and types encountered in the embryo. Only a subset of enhancers will be active in any given cell type/state (Buecker & Wysocka, 2012). Thus, the outcome will be at the best the potential of a DNA sequence to have enhancer activity, but it requires additional information/experiments to validate these results in the context of the cellular heterogeneity of an embryo. In addition, many enhancers may escape detection for reasons of sensitivity or latency (Ostuni et al., 2013).

## 2.2 Noncoding RNAs Linked to Enhancer Activity

It is increasingly appreciated that enhancers themselves drive expression of noncoding RNA so-called eRNA (Hah et al., 2011; Kim, Hemberg, & Gray, 2015; Kim et al., 2010; Wang et al., 2011). These eRNAs are around 2 kb long and are expressed about 10- to 100-fold less abundantly than other usually longer noncoding RNAs. eRNAs are neither spliced nor polyadenylated. eRNAs are transcribed by RNA polymerase II and require the integrator complex interacting with the C-terminus of RNA polymerase II for their synthesis (Kim, Lee, Yun, & Kim, 2015; Lai, Gardini, Zhang, & Shiekhattar, 2015). Enhancers and promoters may thus be much more similar than previously believed (Andersson, 2015; Andersson et al., 2014). However, there are some differences. Enhancers do not govern the direction of transcription; eRNAs are hence transcribed bidirectionally. Also enhancers do not show the H3K4me3 mark characteristic of promoters—a feature that may be linked with the low abundance of eRNAs in comparison to messenger RNAs initiated from promoters (De Santa et al., 2010; Kim et al., 2010). In a cap analysis of gene expression (CAGE) of RNAs isolated from a variety of human cells and tissues, more than 40,000 capped bidirectional RNAs were identified to arise from noncoding regions in the human genome suggesting widespread presence of these eRNAs (Andersson et al., 2014).

In a functional analysis in transgenic mouse embryos focusing on the limb and heart, 57% regions from which eRNAs were transcribed acted as tissue-specific enhancers. Most notably, the expression pattern of the eRNA matched that of the activity profile of the enhancer in 88% cases. In the same study only 36% and 46% eRNA expressing regions were also marked by p300 and H3K27ac (Wu et al., 2014). Thus, it remains an open question whether eRNA expressing regions represent a special subset of enhancer sequences. It was also noted that eRNAs were expressed from enhancers prior to expression from the linked promoter (Kim, Lee, et al., 2015). eRNAs may thus reflect a poised state similar to what was observed for certain histone marks in ES cells and zebrafish blastomeres (Creyghton et al., 2010; Lindeman et al., 2011; Vastenhouw et al., 2010).

A crucial question is what is the role of eRNAs? Are they transcriptional noise? Or do they have a specific function? Current models suggest enhancers are brought into close vicinity to the promoter and are held in a complex with the basic transcription machinery by Mediator and Cohesins (Andersson, 2015; Kagey et al., 2010). Whether eRNAs affect transcription from the nearby promoter that the enhancer controls remains to be addressed. With respect to eRNA, zebrafish seem to be uncharted territory despite previous efforts to identify noncoding RNAs (Pauli et al., 2012; Pauli, Valen, & Schier, 2015). However, given the fact that eRNAs are capped, the CAGE data available on the zebrafish transcriptome should provide a rich source of eRNAs (Haberle et al., 2014; Nepal et al., 2013). However, it remains to be seen how efficiently these methods can be tuned to detect the low levels of eRNAs efficiently.

## 2.3 Functional Methods of Enhancer Detection

The initial definition of an enhancer was based on its function ie, to increase transcription from promoters that it was linked to (Banerji et al., 1981). Since these early days methods to identify and functionally analyze enhancers such as FIREWACh (Murtha et al., 2014), SIF-Seq (Dickel et al., 2014), STARR-Seq (Arnold et al., 2013), and CRE-Seq (Mogno, Kwasnieski, & Cohen, 2013) both in tissue culture cells and animals have been drastically improved. They are mostly based on the original concept of enhancer testing by bringing an isolated piece of genomic regulatory DNA into the context of a heterologous promoter directing transcription of a reporter gene cassette.

In comparison to the structure-based methods of enhancer detection (Histone H3 marks, transcription factor binding, DNase-sensitivity), expression-based assays do not achieve the same throughput. The screening of enhancer activity on the basis of gain-of-function has also other drawbacks. There is increasing evidence that enhancer and promoters cooperate specifically: enhancers next to by-stander genes, for example, do not interact with the promoter of the bystander gene but with the remote promoter of their actual target gene (Kikuta et al., 2007; Navratilova & Becker, 2009). Only a minority of core promoters have TATA boxes and use other elements (Carninci et al., 2006; Fenouil et al., 2012; Yang, Bolotin, Jiang, Sladek, & Martinez, 2007). Usually the parallel enhancer testing methods do not take this diversity in core promoter structures into account. Moreover, chromatin structure is not only governed by the enhancer and the promoter but also by flanking sequences. In many of these expression systems, flanking sequences are heterologous sequences even from bacteria that may assemble chromatin in an aberrant way. Finally, testing a sequence out of its context may not reflect on the function of the sequence in its original location: the many nonfunctional binding sites of transcription factors in the genome may lead to positive scoring in the artificial assembly in the enhancer/reporter system but do not reflect on the actual regulatory function in the genome. Thus, these assays probe only for the potential of a sequence to have enhancer-like activity.

These simple systems can also not assess the interplay of multiple enhancers that are believed to integrate various regulatory inputs into the overall expression of a gene in time (developmental stage, response to external stimulus) and space (cell type, tissue). Probing enhancer function by mutating them in the natural context of the genome would be a much more telling approach. However, the only effective method available for these purposes were homologous recombination in the mouse or the fortuitous isolation of regulatory mutants in genetic screens. Besides the tedious technical issues, redundancy of enhancer function is another major obstacle in these approaches. For example, several enhancers were identified in the zebrafish and mouse *shh* gene to drive expression in the floor plate and notochord (Ertzer et al., 2007; Jeong & Epstein, 2003; Muller et al., 1999). Deletion of individual enhancers may not lead to a strong effect or any effect at all. In fact, deletion of highly conserved noncoding regions in the mouse by knockout did not lead to any visible effect; all mice survived and reproduce normally (Ahituv et al., 2007). Either the ultra-conserved sequences do not serve as enhancers or their function is redundantly encoded in multiple CRMs or their effects are so subtle that they escaped detection. In contrast, deletion studies of enhancers of the mouse *sox2* (Li et al., 2014) and *gata2* (Groschel et al., 2014) genes revealed strong defects in the expression of the cognate genes. Thus, such targeted mutational approaches can result in measurable effects on the expression of the linked genes. The challenge will be now to adapt the new CRISPR/Cas9-based genome-editing technologies (Sander & Joung, 2014; Varshney, Sood, & Burgess, 2015) to achieve the throughput to assess the function of the many enhancer-like sequences in vertebrate genomes. The zebrafish embryo with its technical advantages could be a fantastic model in this endeavor. Medium to high-throughput pipelines used previously to map enhancer—promoter interactions in the zebrafish (Gehrig et al., 2009) may facilitate the analysis of the resulting changes in expression patterns.

## 2.4 Limitations of Global Methods

The technologies which identify enhancers by their particular structural features (transcription factor binding, clustering of transcription factor binding sites, chromatin signatures, eRNA expression) or by their functional properties together with enormously grown data collections from ENCODE Project Consortium (Consortium, 2012) and improved bioinformatic search, analysis, and visualization tools put us nowadays into a position to unravel the regulatory networks of vertebrate genomes. While individual methods may have their drawbacks, the combination of methods should eventually lead to a complete description of the regulatory wiring of the genome at least in tissue culture cells.

The analysis of enhancers in a developing embryo or adult animal adds layers of complexity. The differences in enhancer activation/recruitment are fundamental determinants of the 200 cell types that constitute approximately the mature vertebrate body. Gene expression is highly dynamic both in development and during body homeostasis. In many cases, the combination of multiple enhancers determines the spatiotemporal pattern of expression and the responsiveness of a gene to a stimulus. The results from global genome analysis of intact embryos or organs are thus averaged over many different cell types. On the other hand, isolating individual cell types induces regeneration and repair programs in the cells torn out of their normal tissue context. Besides this, stem and progenitor cell populations are small in number in developing embryos and regenerating organs and thus do frequently not yield enough isolated cells for these global analysis methods. Moreover, the regulatory input of a gene will vary according to developmental stage and external stimuli: Latent enhancers are activated in response to signals that cells are exposed to (Heintzman et al., 2009; Hurtado, Holmes, Ross-Innes, Schmidt, & Carroll, 2011; Wang et al., 2011). Other enhancers become silenced or decommissioned when the next developmental stage is reached or stimuli have vanished (Ostuni et al., 2013). Thus, understanding the regulatory architecture of the vertebrate genome in the context of the entire organism remains a challenge. It calls for more sensitive techniques requiring fewer cells to detect enhancers in the genome. To this end, we need the possibility to assay the activity of individual enhancers in the intact genome in an embryo/tissue context with highest spatial and temporal resolution.

## 3. HOW CAN THE ZEBRAFISH MODEL CONTRIBUTE?

Clearly the global enhancer detection methods (Aday et al., 2011; Andersen, Reiner, Aanes, Alestrom, & Collas, 2012; Bogdanovic et al., 2012; Lindeman et al., 2011; Vastenhouw et al., 2010; Wardle et al., 2006) together with the CAGE sequencing efforts to map 5′ ends of capped RNAs (Haberle et al., 2014; Nepal et al., 2013) provides already a rich source of information on the regulatory landscape of the zebrafish genome. However, these studies focused on early stages of zebrafish development where cellular diversity has just started to emerge. It remains to be seen how efficiently CRMs can be detected by these methods at later stages, when the number of distinct cell types has significantly increased.

The zebrafish was introduced into the laboratories for its advantages to carry out forward genetic screens (Amsterdam & Hopkins, 2006; Granato & Nusslein-Volhard, 1996; Mullins, Hammerschmidt, Haffter, & Nusslein-Volhard, 1994; Strahle & Ingham, 1992). However, very few enhancer mutations in regulatory regions were reported so far after cloning the mutant loci. The *shh*$^{T4}$ allele targets most likely the promoter of *shh* resulting in lower levels of *shh* mRNAs (Chang, Blader, Fischer, Ingham, & Strahle, 1997; Schauerte et al., 1998). The *mothers superior*$^{m188}$ mutation leads to loss of expression of the transcription factor gene *foxd3* and concomitant loss of neural crest derivatives. The mutation was mapped outside of the coding region in the *foxd3* locus suggesting that the mutation altered an enhancer region required for expression of *foxd3* (Montero-Balaguer et al., 2006). Many of the retroviral and transposon insertion mutants carry integrations in noncoding regions including the first intron suggesting that the insertion may interfere with the regulation of the mutated gene (Amsterdam et al., 1999; Ellingsen et al., 2005; Kawakami et al., 2010; Kondrychyn, Garcia-Lecea, Emelyanov, Parinov, & Korzh, 2009; Kondrychyn et al., 2011; Parinov, Kondrichin, Korzh, & Emelyanov, 2004). Precise mapping of the insertions with respect to enhancer and promoter marks may thus uncover many more CRMs than currently described (S. Burgess, pers. comm.)

Despite these advances, it is puzzling why the forward genetic screens have not identified more regulatory mutants. A straightforward explanation is that the resulting phenotypes were too subtle and were hence overseen in the genetic screens: enhancers frequently control only expression in a subset of the cells in which the linked gene is normally expressed. Furthermore, redundancy of regulatory elements, while providing the playground for evolution with respect to gain of new function, may mask the phenotypic consequences of loss-of-function mutations in individual CRMs. Yet, increasing evidence suggests that many diseases causing polymorphisms in the human genome are noncoding and thus likely linked to regulatory sequences (Epstein, 2009; Gordon & Lyonnet, 2014).

With the introduction of the Crispr/Cas9 gene editing technology, we have now also in zebrafish the opportunity to alter the genome at specific regions (Hwang et al., 2013; Varshney, Pei, et al., 2015; Varshney, Sood, et al., 2015). This will allow us to remove individual or groups of enhancers and test their function in a systematic manner. A critical question is whether active enhancers can be identified in the developing embryo efficiently. If eRNAs indeed hold what they promise, they could potentially provide the tools to link enhancer activity with a specific cell type, a developmental stage, or external inducer (Cheng, Pan, Tsai, & Tsai, 2015; Yao et al., 2015). Techniques for detection of low level RNA expression are available (Li, Ni, Zhang, Ma, & Zhang, 2014; Ro, Park, Jin, Sanders, & Yan, 2006). These technical possibilities combined with the other virtues of the zebrafish embryo such as development outside of the mother, transparent embryos, and medium throughput analysis (Yang et al., 2009) will provide an outstanding platform to elucidate the regulatory architecture of the vertebrate genome.

## ACKNOWLEDGMENTS

Research in Uwe Strähle's laboratory is supported by the Deutsche Forschungsgemeinschaft (DFG Str439/5-1), the EU IP ZF-Health, FP7-HEALTH-2007-B2, Helmholtz Initiative on Synthetic Biology (HISynBio), the BMBF funded Molecular Interaction Engineering (MIE) network, and the ZENCODE-ITN, funded by the European Union under the H2020 programme.

## REFERENCES

Aday, A. W., Zhu, L. J., Lakshmanan, A., Wang, J., & Lawson, N. D. (2011). Identification of *cis* regulatory features in the embryonic zebrafish genome through large-scale profiling of H3K4me1 and H3K4me3 binding sites. *Developmental Biology, 357*(2), 450–462. http://dx.doi.org/10.1016/j.ydbio.2011.03.007.

Ahituv, N., Zhu, Y., Visel, A., Holt, A., Afzal, V., Pennacchio, L. A., & Rubin, E. M. (2007). Deletion of ultraconserved elements yields viable mice. *PLoS Biology, 5*(9), e234. http://dx.doi.org/10.1371/journal.pbio.0050234.

Amsterdam, A., Burgess, S., Golling, G., Chen, W., Sun, Z., Townsend, K., … Hopkins, N. (1999). A large-scale insertional mutagenesis screen in zebrafish. *Genes & Development, 13*(20), 2713–2724. Retrieved from: http://www.ncbi.nlm.nih.gov/pubmed/10541557.

Amsterdam, A., & Hopkins, N. (2006). Mutagenesis strategies in zebrafish for identifying genes involved in development and disease. *Trends in Genetics, 22*(9), 473–478. http://dx.doi.org/10.1016/j.tig.2006.06.011.

Andersen, I. S., Reiner, A. H., Aanes, H., Alestrom, P., & Collas, P. (2012). Developmental features of DNA methylation during activation of the embryonic zebrafish genome. *Genome Biology, 13*(7), R65. http://dx.doi.org/10.1186/gb-2012-13-7-r65.

Andersson, R. (2015). Promoter or enhancer, what's the difference? Deconstruction of established distinctions and presentation of a unifying model. *Bioessays, 37*(3), 314–323. http://dx.doi.org/10.1002/bies.201400162.

Andersson, R., Gebhard, C., Miguel-Escalada, I., Hoof, I., Bornholdt, J., Boyd, M., … Sandelin, A. (2014). An atlas of active enhancers across human cell types and tissues. *Nature, 507*(7493), 455–461. http://dx.doi.org/10.1038/nature12787.

Andersson, R., Sandelin, A., & Danko, C. G. (2015). A unified architecture of transcriptional regulatory elements. *Trends in Genetics, 31*(8), 426–433. http://dx.doi.org/10.1016/j.tig.2015.05.007.

Armant, O., Marz, M., Schmidt, R., Ferg, M., Diotel, N., Ertzer, R., … Rastegar, S. (2013). Genome-wide, whole mount in situ analysis of transcriptional regulators in zebrafish embryos. *Developmental Biology, 380*(2), 351–362. http://dx.doi.org/10.1016/j.ydbio.2013.05.006.

Arnold, C. D., Gerlach, D., Spies, D., Matts, J. A., Sytnikova, Y. A., Pagani, M., … Stark, A. (2014). Quantitative genome-wide enhancer activity maps for five *Drosophila* species show functional enhancer conservation and turnover during *cis*-regulatory evolution. *Nature Genetics, 46*(7), 685–692. http://dx.doi.org/10.1038/ng.3009.

Arnold, C. D., Gerlach, D., Stelzer, C., Boryn, L. M., Rath, M., & Stark, A. (2013). Genome-wide quantitative enhancer activity maps identified by STARR-seq. *Science, 339*(6123), 1074–1077. http://dx.doi.org/10.1126/science.1232542.

Banerji, J., Olson, L., & Schaffner, W. (1983). A lymphocyte-specific cellular enhancer is located downstream of the joining region in immunoglobulin heavy chain genes. *Cell, 33*(3), 729–740. Retrieved from: http://www.ncbi.nlm.nih.gov/pubmed/6409418.

Banerji, J., Rusconi, S., & Schaffner, W. (1981). Expression of a beta-globin gene is enhanced by remote SV40 DNA sequences. *Cell, 27*(2 Pt 1), 299–308. Retrieved from: http://www.ncbi.nlm.nih.gov/pubmed/6277502.

Bejerano, G., Pheasant, M., Makunin, I., Stephen, S., Kent, W. J., Mattick, J. S., & Haussler, D. (2004). Ultraconserved elements in the human genome. *Science, 304*(5675), 1321–1325. http://dx.doi.org/10.1126/science.1098119.

Belting, H. G., Hauptmann, G., Meyer, D., Abdelilah-Seyfried, S., Chitnis, A., Eschbach, C., … Driever, W. (2001). Spiel ohne grenzen/pou2 is required during establishment of the zebrafish midbrain-hindbrain boundary organizer. *Development, 128*(21), 4165–4176. Retrieved from: http://www.ncbi.nlm.nih.gov/pubmed/11684654.

Berger, M. F., Philippakis, A. A., Qureshi, A. M., He, F. S., Estep, P. W., 3rd, & Bulyk, M. L. (2006). Compact, universal DNA microarrays to comprehensively determine transcription-factor binding site specificities. *Nature Biotechnology, 24*(11), 1429–1435. http://dx.doi.org/10.1038/nbt1246.

Berman, B. P., Nibu, Y., Pfeiffer, B. D., Tomancak, P., Celniker, S. E., Levine, M., … Eisen, M. B. (2002). Exploiting transcription factor binding site clustering to identify *cis*-regulatory modules involved in pattern formation in the *Drosophila* genome. *Proceedings of the National Academy of Sciences of the United States of America, 99*(2), 757–762. http://dx.doi.org/10.1073/pnas.231608898.

Bernstein, B. E., Mikkelsen, T. S., Xie, X., Kamal, M., Huebert, D. J., Cuff, J., … Lander, E. S. (2006). A bivalent chromatin structure marks key developmental genes in embryonic stem cells. *Cell, 125*(2), 315–326. http://dx.doi.org/10.1016/j.cell.2006.02.041.

Blader, P., Lam, C. S., Rastegar, S., Scardigli, R., Nicod, J. C., Simplicio, N., … Strahle, U. (2004). Conserved and acquired features of neurogenin1 regulation. *Development, 131*(22), 5627–5637. http://dx.doi.org/10.1242/dev.01455.

Blader, P., Plessy, C., & Strahle, U. (2003). Multiple regulatory elements with spatially and temporally distinct activities control neurogenin1 expression in primary neurons of the zebrafish embryo. *Mechanisms of Development, 120*(2), 211–218. Retrieved from: http://www.ncbi.nlm.nih.gov/pubmed/12559493.

Bogdanovic, O., Fernandez-Minan, A., Tena, J. J., de la Calle-Mustienes, E., Hidalgo, C., van Kruysbergen, I., … Gomez-Skarmeta, J. L. (2012). Dynamics of enhancer chromatin signatures mark the transition from pluripotency to cell specification during embryogenesis. *Genome Research, 22*(10), 2043–2053. http://dx.doi.org/10.1101/gr.134833.111.

Boyle, A. P., Araya, C. L., Brdlik, C., Cayting, P., Cheng, C., Cheng, Y., … Snyder, M. (2014). Comparative analysis of regulatory information and circuits across distant species. *Nature, 512*(7515), 453–456. http://dx.doi.org/10.1038/nature13668.

Boyle, A. P., Davis, S., Shulha, H. P., Meltzer, P., Margulies, E. H., Weng, Z., … Crawford, G. E. (2008). High-resolution mapping and characterization of open chromatin across the genome. *Cell, 132*(2), 311–322. http://dx.doi.org/10.1016/j.cell.2007.12.014.

Boyle, A. P., & Furey, T. S. (2009). High-resolution mapping studies of chromatin and gene regulatory elements. *Epigenomics, 1*(2), 319–329. http://dx.doi.org/10.2217/epi.09.29.

Buecker, C., & Wysocka, J. (2012). Enhancers as information integration hubs in development: lessons from genomics. *Trends in Genetics, 28*(6), 276–284. http://dx.doi.org/10.1016/j.tig.2012.02.008.

Calo, E., & Wysocka, J. (2013). Modification of enhancer chromatin: what, how, and why? *Molecular Cell, 49*(5), 825–837. http://dx.doi.org/10.1016/j.molcel.2013.01.038.

Carninci, P., Sandelin, A., Lenhard, B., Katayama, S., Shimokawa, K., Ponjavic, J., … Hayashizaki, Y. (2006). Genome-wide analysis of mammalian promoter architecture and evolution. *Nature Genetics, 38*(6), 626–635. http://dx.doi.org/10.1038/ng1789.

Carroll, S. B. (2008). Evo-devo and an expanding evolutionary synthesis: a genetic theory of morphological evolution. *Cell, 134*(1), 25–36. http://dx.doi.org/10.1016/j.cell.2008.06.030.

Chang, B. E., Blader, P., Fischer, N., Ingham, P. W., & Strahle, U. (1997). Axial (HNF3beta) and retinoic acid receptors are regulators of the zebrafish sonic hedgehog promoter. *EMBO Journal, 16*(13), 3955–3964. http://dx.doi.org/10.1093/emboj/16.13.3955.

Cheng, J. H., Pan, D. Z., Tsai, Z. T., & Tsai, H. K. (2015). Genome-wide analysis of enhancer RNA in gene regulation across 12 mouse tissues. *Scientific Reports, 5*, 12648. http://dx.doi.org/10.1038/srep12648.

Consortium, Encode Project. (2012). An integrated encyclopedia of DNA elements in the human genome. *Nature, 489*(7414), 57–74. http://dx.doi.org/10.1038/nature11247.

Cook, P. R. (1999). The organization of replication and transcription. *Science, 284*(5421), 1790–1795. Retrieved from: http://www.ncbi.nlm.nih.gov/pubmed/10364545.

Creyghton, M. P., Cheng, A. W., Welstead, G. G., Kooistra, T., Carey, B. W., Steine, E. J., … Jaenisch, R. (2010). Histone H3K27ac separates active from poised enhancers and predicts developmental state. *Proceedings of the National Academy of Sciences of the United States of America, 107*(50), 21931–21936. http://dx.doi.org/10.1073/pnas.1016071107.

De Santa, F., Barozzi, I., Mietton, F., Ghisletti, S., Polletti, S., Tusi, B. K., … Natoli, G. (2010). A large fraction of extragenic RNA pol II transcription sites overlap enhancers. *PLoS Biology, 8*(5), e1000384. http://dx.doi.org/10.1371/journal.pbio.1000384.

Dickel, D. E., Zhu, Y., Nord, A. S., Wylie, J. N., Akiyama, J. A., Afzal, V., … Pennacchio, L. A. (2014). Function-based identification of mammalian enhancers using site-specific integration. *Nature Methods, 11*(5), 566–571. http://dx.doi.org/10.1038/nmeth.2886.

Driever, W., Solnica-Krezel, L., Schier, A. F., Neuhauss, S. C., Malicki, J., Stemple, D. L., … Boggs, C. (1996). A genetic screen for mutations affecting embryogenesis in zebrafish.

*Development, 123*, 37–46. Retrieved from: http://www.ncbi.nlm.nih.gov/pubmed/9007227.

Ellingsen, S., Laplante, M. A., Konig, M., Kikuta, H., Furmanek, T., Hoivik, E. A., & Becker, T. S. (2005). Large-scale enhancer detection in the zebrafish genome. *Development, 132*(17), 3799–3811. http://dx.doi.org/10.1242/dev.01951.

Epstein, D. J. (2009). *Cis*-regulatory mutations in human disease. *Briefings in Functional Genomics and Proteomics, 8*(4), 310–316. http://dx.doi.org/10.1093/bfgp/elp021.

Ertzer, R., Muller, F., Hadzhiev, Y., Rathnam, S., Fischer, N., Rastegar, S., & Strahle, U. (2007). Cooperation of sonic hedgehog enhancers in midline expression. *Developmental Biology, 301*(2), 578–589. http://dx.doi.org/10.1016/j.ydbio.2006.11.004.

Fenouil, R., Cauchy, P., Koch, F., Descostes, N., Cabeza, J. Z., Innocenti, C., … Andrau, J. C. (2012). CpG islands and GC content dictate nucleosome depletion in a transcription-independent manner at mammalian promoters. *Genome Research, 22*(12), 2399–2408. http://dx.doi.org/10.1101/gr.138776.112.

Ferg, M., Armant, O., Yang, L., Dickmeis, T., Rastegar, S., & Strahle, U. (2014). Gene transcription in the zebrafish embryo: regulators and networks. *Briefings in Functional Genomics, 13*(2), 131–143. http://dx.doi.org/10.1093/bfgp/elt044.

Fisher, W. W., Li, J. J., Hammonds, A. S., Brown, J. B., Pfeiffer, B. D., Weiszmann, R., … Celniker, S. E. (2012). DNA regions bound at low occupancy by transcription factors do not drive patterned reporter gene expression in *Drosophila*. *Proceedings of the National Academy of Sciences of the United States of America, 109*(52), 21330–21335. http://dx.doi.org/10.1073/pnas.1209589110.

Furey, T. S. (2012). ChIP-seq and beyond: new and improved methodologies to detect and characterize protein-DNA interactions. *Nature Reviews Genetics, 13*(12), 840–852. http://dx.doi.org/10.1038/nrg3306.

Gehrig, J., Reischl, M., Kalmar, E., Ferg, M., Hadzhiev, Y., Zaucker, A., … Muller, F. (2009). Automated high-throughput mapping of promoter-enhancer interactions in zebrafish embryos. *Nature Methods, 6*(12), 911–916. http://dx.doi.org/10.1038/nmeth.1396.

Gehring, W. J. (1996). The master control gene for morphogenesis and evolution of the eye. *Genes to Cells, 1*(1), 11–15. Retrieved from: http://www.ncbi.nlm.nih.gov/pubmed/9078363.

Golling, G., Amsterdam, A., Sun, Z., Antonelli, M., Maldonado, E., Chen, W., … Hopkins, N. (2002). Insertional mutagenesis in zebrafish rapidly identifies genes essential for early vertebrate development. *Nature Genetics, 31*(2), 135–140. http://dx.doi.org/10.1038/ng896.

Gordon, C. T., & Lyonnet, S. (2014). Enhancer mutations and phenotype modularity. *Nature Genetics, 46*(1), 3–4. http://dx.doi.org/10.1038/ng.2861.

Granato, M., & Nusslein-Volhard, C. (1996). Fishing for genes controlling development. *Current Opinion in Genetics & Development, 6*(4), 461–468. Retrieved from: http://www.ncbi.nlm.nih.gov/pubmed/8791528.

Greenald, D., Jeyakani, J., Pelster, B., Sealy, I., Mathavan, S., & van Eeden, F. J. (2015). Genome-wide mapping of Hif-1alpha binding sites in zebrafish. *BMC Genomics, 16*(1), 923. http://dx.doi.org/10.1186/s12864-015-2169-x.

Groschel, S., Sanders, M. A., Hoogenboezem, R., de Wit, E., Bouwman, B. A., Erpelinck, C., … Delwel, R. (2014). A single oncogenic enhancer rearrangement causes concomitant EVI1 and GATA2 deregulation in leukemia. *Cell, 157*(2), 369–381. http://dx.doi.org/10.1016/j.cell.2014.02.019.

Gubelmann, C., Waszak, S. M., Isakova, A., Holcombe, W., Hens, K., Iagovitina, A., … Deplancke, B. (2013). A yeast one-hybrid and microfluidics-based pipeline to map mammalian gene regulatory networks. *Molecular Systems Biology, 9*, 682. http://dx.doi.org/10.1038/msb.2013.38.

Haberle, V., Li, N., Hadzhiev, Y., Plessy, C., Previti, C., Nepal, C., ... Lenhard, B. (2014). Two independent transcription initiation codes overlap on vertebrate core promoters. *Nature, 507*(7492), 381–385. http://dx.doi.org/10.1038/nature12974.

Haffter, P., Granato, M., Brand, M., Mullins, M. C., Hammerschmidt, M., Kane, D. A., ... Nusslein-Volhard, C. (1996). The identification of genes with unique and essential functions in the development of the zebrafish, *Danio rerio*. *Development, 123*, 1–36. Retrieved from: http://www.ncbi.nlm.nih.gov/pubmed/9007226.

Hah, N., Danko, C. G., Core, L., Waterfall, J. J., Siepel, A., Lis, J. T., & Kraus, W. L. (2011). A rapid, extensive, and transient transcriptional response to estrogen signaling in breast cancer cells. *Cell, 145*(4), 622–634. http://dx.doi.org/10.1016/j.cell.2011.03.042.

Hare, E. E., Peterson, B. K., Iyer, V. N., Meier, R., & Eisen, M. B. (2008). Sepsid even-skipped enhancers are functionally conserved in *Drosophila* despite lack of sequence conservation. *PLoS Genetics, 4*(6), e1000106. http://dx.doi.org/10.1371/journal.pgen.1000106.

Harmston, N., Baresic, A., & Lenhard, B. (2013). The mystery of extreme non-coding conservation. *Philosophical Transactions of the Royal Society of London B Biological Sciences, 368*(1632), 20130021. http://dx.doi.org/10.1098/rstb.2013.0021.

He, B. Z., Holloway, A. K., Maerkl, S. J., & Kreitman, M. (2011). Does positive selection drive transcription factor binding site turnover? A test with *Drosophila cis*-regulatory modules. *PLoS Genetics*, 7(4), e1002053. http://dx.doi.org/10.1371/journal.pgen.1002053.

Heintzman, N. D., Hon, G. C., Hawkins, R. D., Kheradpour, P., Stark, A., Harp, L. F., ... Ren, B. (2009). Histone modifications at human enhancers reflect global cell-type-specific gene expression. *Nature, 459*(7243), 108–112. http://dx.doi.org/10.1038/nature07829.

Heintzman, N. D., Stuart, R. K., Hon, G., Fu, Y., Ching, C. W., Hawkins, R. D., ... Ren, B. (2007). Distinct and predictive chromatin signatures of transcriptional promoters and enhancers in the human genome. *Nature Genetics, 39*(3), 311–318. http://dx.doi.org/10.1038/ng1966.

Hens, K., Feuz, J. D., Isakova, A., Iagovitina, A., Massouras, A., Bryois, J., ... Deplancke, B. (2011). Automated protein-DNA interaction screening of *Drosophila* regulatory elements. *Nature Methods, 8*(12), 1065–1070. http://dx.doi.org/10.1038/nmeth.1763.

Hurtado, A., Holmes, K. A., Ross-Innes, C. S., Schmidt, D., & Carroll, J. S. (2011). FOXA1 is a key determinant of estrogen receptor function and endocrine response. *Nature Genetics, 43*(1), 27–33. http://dx.doi.org/10.1038/ng.730.

Hwang, W. Y., Fu, Y., Reyon, D., Maeder, M. L., Tsai, S. Q., Sander, J. D., ... Joung, J. K. (2013). Efficient genome editing in zebrafish using a CRISPR-Cas system. *Nature Biotechnology, 31*(3), 227–229. http://dx.doi.org/10.1038/nbt.2501.

Jeong, Y., & Epstein, D. J. (2003). Distinct regulators of Shh transcription in the floor plate and notochord indicate separate origins for these tissues in the mouse node. *Development, 130*(16), 3891–3902. Retrieved from: http://www.ncbi.nlm.nih.gov/pubmed/12835403.

Jolma, A., Kivioja, T., Toivonen, J., Cheng, L., Wei, G., Enge, M., ... Taipale, J. (2010). Multiplexed massively parallel SELEX for characterization of human transcription factor binding specificities. *Genome Research, 20*(6), 861–873. http://dx.doi.org/10.1101/gr.100552.109.

Kagey, M. H., Newman, J. J., Bilodeau, S., Zhan, Y., Orlando, D. A., van Berkum, N. L., ... Young, R. A. (2010). Mediator and cohesin connect gene expression and chromatin architecture. *Nature, 467*(7314), 430–435. http://dx.doi.org/10.1038/nature09380.

Kane, D. A., & Kimmel, C. B. (1993). The zebrafish midblastula transition. *Development, 119*(2), 447–456. Retrieved from: http://www.ncbi.nlm.nih.gov/pubmed/8287796.

Kawakami, K., Abe, G., Asada, T., Asakawa, K., Fukuda, R., Ito, A., … Yoshida, M. (2010). zTrap: zebrafish gene trap and enhancer trap database. *BMC Developmental Biology, 10*, 105. http://dx.doi.org/10.1186/1471-213X-10-105.
Khurana, E., Fu, Y., Colonna, V., Mu, X. J., Kang, H. M., Lappalainen, T., … Gerstein, M. (2013). Integrative annotation of variants from 1092 humans: application to cancer genomics. *Science, 342*(6154), 1235587. http://dx.doi.org/10.1126/science.1235587.
Kikuta, H., Laplante, M., Navratilova, P., Komisarczuk, A. Z., Engstrom, P. G., Fredman, D., … Becker, T. S. (2007). Genomic regulatory blocks encompass multiple neighboring genes and maintain conserved synteny in vertebrates. *Genome Research, 17*(5), 545–555. http://dx.doi.org/10.1101/gr.6086307.
Kim, T. K., Hemberg, M., & Gray, J. M. (2015). Enhancer RNAs: a class of long noncoding RNAs synthesized at enhancers. *Cold Spring Harbor Perspectives in Biology*, 7(1), a018622. http://dx.doi.org/10.1101/cshperspect.a018622.
Kim, T. K., Hemberg, M., Gray, J. M., Costa, A. M., Bear, D. M., Wu, J., … Greenberg, M. E. (2010). Widespread transcription at neuronal activity-regulated enhancers. *Nature, 465*(7295), 182–187. http://dx.doi.org/10.1038/nature09033.
Kim, Y. W., Lee, S., Yun, J., & Kim, A. (2015). Chromatin looping and eRNA transcription precede the transcriptional activation of gene in the beta-globin locus. *Bioscience Reports, 35*(2). http://dx.doi.org/10.1042/BSR20140126.
Kolovos, P., Knoch, T. A., Grosveld, F. G., Cook, P. R., & Papantonis, A. (2012). Enhancers and silencers: an integrated and simple model for their function. *Epigenetics Chromatin, 5*(1), 1. http://dx.doi.org/10.1186/1756-8935-5-1.
Kondrychyn, I., Garcia-Lecea, M., Emelyanov, A., Parinov, S., & Korzh, V. (2009). Genome-wide analysis of Tol2 transposon reintegration in zebrafish. *BMC Genomics, 10*, 418. http://dx.doi.org/10.1186/1471-2164-10-418.
Kondrychyn, I., Teh, C., Garcia-Lecea, M., Guan, Y., Kang, A., & Korzh, V. (2011). Zebrafish Enhancer TRAP transgenic line database ZETRAP 2.0. *Zebrafish, 8*(4), 181–182. http://dx.doi.org/10.1089/zeb.2011.0718.
Lai, F., Gardini, A., Zhang, A., & Shiekhattar, R. (2015). Integrator mediates the biogenesis of enhancer RNAs. *Nature, 525*(7569), 399–403. http://dx.doi.org/10.1038/nature14906.
Lettice, L. A., Heaney, S. J., Purdie, L. A., Li, L., de Beer, P., Oostra, B. A., … de Graaff, E. (2003). A long-range Shh enhancer regulates expression in the developing limb and fin and is associated with preaxial polydactyly. *Human Molecular Genetics, 12*(14), 1725–1735. Retrieved from: http://www.ncbi.nlm.nih.gov/pubmed/12837695.
Lettice, L. A., Horikoshi, T., Heaney, S. J., van Baren, M. J., van der Linde, H. C., Breedveld, G. J., … Noji, S. (2002). Disruption of a long-range *cis*-acting regulator for Shh causes preaxial polydactyly. *Proceedings of the National Academy of Sciences of the United States of America, 99*(11), 7548–7553. http://dx.doi.org/10.1073/pnas.112212199.
Levine, M. (2010). Transcriptional enhancers in animal development and evolution. *Current Biology, 20*(17), R754–R763. http://dx.doi.org/10.1016/j.cub.2010.06.070.
Li, X., Ni, M., Zhang, C., Ma, W., & Zhang, Y. (2014). A convenient system for highly specific and sensitive detection of miRNA expression. *RNA, 20*(2), 252–259. http://dx.doi.org/10.1261/rna.040220.113.
Li, X. Y., MacArthur, S., Bourgon, R., Nix, D., Pollard, D. A., Iyer, V. N., … Biggin, M. D. (2008). Transcription factors bind thousands of active and inactive regions in the *Drosophila* blastoderm. *PLoS Biology, 6*(2), e27. http://dx.doi.org/10.1371/journal.pbio.0060027.
Li, Y., Rivera, C. M., Ishii, H., Jin, F., Selvaraj, S., Lee, A. Y., … Ren, B. (2014). CRISPR reveals a distal super-enhancer required for Sox2 expression in mouse embryonic stem cells. *PLoS One, 9*(12), e114485. http://dx.doi.org/10.1371/journal.pone.0114485.
Lindeman, L. C., Andersen, I. S., Reiner, A. H., Li, N., Aanes, H., Ostrup, O., … Collas, P. (2011). Prepatterning of developmental gene expression by modified histones before

zygotic genome activation. *Developmental Cell, 21*(6), 993–1004. http://dx.doi.org/10.1016/j.devcel.2011.10.008.
Maston, G. A., Landt, S. G., Snyder, M., & Green, M. R. (2012). Characterization of enhancer function from genome-wide analyses. *Annual Review of Genomics and Human Genetics, 13*, 29–57. http://dx.doi.org/10.1146/annurev-genom-090711-163723.
Maurano, M. T., Humbert, R., Rynes, E., Thurman, R. E., Haugen, E., Wang, H., … Stamatoyannopoulos, J. A. (2012). Systematic localization of common disease-associated variation in regulatory DNA. *Science, 337*(6099), 1190–1195. http://dx.doi.org/10.1126/science.1222794.
Mogno, I., Kwasnieski, J. C., & Cohen, B. A. (2013). Massively parallel synthetic promoter assays reveal the in vivo effects of binding site variants. *Genome Research, 23*(11), 1908–1915. http://dx.doi.org/10.1101/gr.157891.113.
Montero-Balaguer, M., Lang, M. R., Sachdev, S. W., Knappmeyer, C., Stewart, R. A., De La Guardia, A., … Knapik, E. W. (2006). The mother superior mutation ablates foxd3 activity in neural crest progenitor cells and depletes neural crest derivatives in zebrafish. *Developmental Dynamics, 235*(12), 3199–3212. http://dx.doi.org/10.1002/dvdy.20959.
Morley, R. H., Lachani, K., Keefe, D., Gilchrist, M. J., Flicek, P., Smith, J. C., & Wardle, F. C. (2009). A gene regulatory network directed by zebrafish No tail accounts for its roles in mesoderm formation. *Proceedings of the National Academy of Sciences of the United States of America, 106*(10), 3829–3834. http://dx.doi.org/10.1073/pnas.0808382106.
Muller, F., Chang, B., Albert, S., Fischer, N., Tora, L., & Strahle, U. (1999). Intronic enhancers control expression of zebrafish sonic hedgehog in floor plate and notochord. *Development, 126*(10), 2103–2116. Retrieved from: http://www.ncbi.nlm.nih.gov/pubmed/10207136.
Mullins, M. C., Hammerschmidt, M., Haffter, P., & Nusslein-Volhard, C. (1994). Large-scale mutagenesis in the zebrafish: in search of genes controlling development in a vertebrate. *Current Biology, 4*(3), 189–202. Retrieved from: http://www.ncbi.nlm.nih.gov/pubmed/7922324.
Murtha, M., Tokcaer-Keskin, Z., Tang, Z., Strino, F., Chen, X., Wang, Y., … Dailey, L. (2014). FIREWACh: high-throughput functional detection of transcriptional regulatory modules in mammalian cells. *Nature Methods, 11*(5), 559–565. http://dx.doi.org/10.1038/nmeth.2885.
Naville, M., Ishibashi, M., Ferg, M., Bengani, H., Rinkwitz, S., Krecsmarik, M., … Roest Crollius, H. (2015). Long-range evolutionary constraints reveal *cis*-regulatory interactions on the human X chromosome. *Nature Communications, 6*, 6904. http://dx.doi.org/10.1038/ncomms7904.
Navratilova, P., & Becker, T. S. (2009). Genomic regulatory blocks in vertebrates and implications in human disease. *Briefings in Functional Genomics and Proteomics, 8*(4), 333–342. http://dx.doi.org/10.1093/bfgp/elp019.
Nelson, A. C., Cutty, S. J., Niini, M., Stemple, D. L., Flicek, P., Houart, C., … Wardle, F. C. (2014). Global identification of Smad2 and Eomesodermin targets in zebrafish identifies a conserved transcriptional network in mesendoderm and a novel role for Eomesodermin in repression of ectodermal gene expression. *BMC Biology, 12*, 81. http://dx.doi.org/10.1186/s12915-014-0081-5.
Nepal, C., Hadzhiev, Y., Previti, C., Haberle, V., Li, N., Takahashi, H., … Muller, F. (2013). Dynamic regulation of the transcription initiation landscape at single nucleotide resolution during vertebrate embryogenesis. *Genome Research, 23*(11), 1938–1950. http://dx.doi.org/10.1101/gr.153692.112.
Nikaido, M., Kawakami, A., Sawada, A., Furutani-Seiki, M., Takeda, H., & Araki, K. (2002). Tbx24, encoding a T-box protein, is mutated in the zebrafish somite-segmentation

mutant fused somites. *Nature Genetics, 31*(2), 195–199. http://dx.doi.org/10.1038/ng899.
Norton, W. H., Mangoli, M., Lele, Z., Pogoda, H. M., Diamond, B., Mercurio, S., … Wilson, S. W. (2005). Monorail/Foxa2 regulates floorplate differentiation and specification of oligodendrocytes, serotonergic raphe neurones and cranial motoneurones. *Development, 132*(4), 645–658. http://dx.doi.org/10.1242/dev.01611.
Nusslein-Volhard, C. (2012). The zebrafish issue of development. *Development, 139*(22), 4099–4103. http://dx.doi.org/10.1242/dev.085217.
Ohno, S. (1972). So much "junk" DNA in our genome. *Brookhaven Symposium in Biology, 23*, 366–370. Retrieved from: http://www.ncbi.nlm.nih.gov/pubmed/5065367.
Ong, C. T., & Corces, V. G. (2011). Enhancer function: new insights into the regulation of tissue-specific gene expression. *Nature Reviews Genetics, 12*(4), 283–293. http://dx.doi.org/10.1038/nrg2957.
Ostuni, R., Piccolo, V., Barozzi, I., Polletti, S., Termanini, A., Bonifacio, S., … Natoli, G. (2013). Latent enhancers activated by stimulation in differentiated cells. *Cell, 152*(1–2), 157–171. http://dx.doi.org/10.1016/j.cell.2012.12.018.
Parinov, S., Kondrichin, I., Korzh, V., & Emelyanov, A. (2004). Tol2 transposon-mediated enhancer trap to identify developmentally regulated zebrafish genes in vivo. *Developmental Dynamics, 231*(2), 449–459. http://dx.doi.org/10.1002/dvdy.20157.
Pauli, A., Valen, E., Lin, M. F., Garber, M., Vastenhouw, N. L., Levin, J. Z., … Schier, A. F. (2012). Systematic identification of long noncoding RNAs expressed during zebrafish embryogenesis. *Genome Research, 22*(3), 577–591. http://dx.doi.org/10.1101/gr.133009.111.
Pauli, A., Valen, E., & Schier, A. F. (2015). Identifying (non-)coding RNAs and small peptides: challenges and opportunities. *Bioessays, 37*(1), 103–112. http://dx.doi.org/10.1002/bies.201400103.
Pennacchio, L. A., & Rubin, E. M. (2001). Genomic strategies to identify mammalian regulatory sequences. *Nature Reviews Genetics, 2*(2), 100–109. http://dx.doi.org/10.1038/35052548.
Plessy, C., Dickmeis, T., Chalmel, F., & Strahle, U. (2005). Enhancer sequence conservation between vertebrates is favoured in developmental regulator genes. *Trends in Genetics, 21*(4), 207–210. http://dx.doi.org/10.1016/j.tig.2005.02.006.
Rastegar, S., Albert, S., Le Roux, I., Fischer, N., Blader, P., Muller, F., & Strahle, U. (2002). A floor plate enhancer of the zebrafish netrin1 gene requires Cyclops (Nodal) signalling and the winged helix transcription factor FoxA2. *Developmental Biology, 252*(1), 1–14. Retrieved from: http://www.ncbi.nlm.nih.gov/pubmed/12453456.
Rastegar, S., Hess, I., Dickmeis, T., Nicod, J. C., Ertzer, R., Hadzhiev, Y., … Strahle, U. (2008). The words of the regulatory code are arranged in a variable manner in highly conserved enhancers. *Developmental Biology, 318*(2), 366–377. http://dx.doi.org/10.1016/j.ydbio.2008.03.034.
Reddy, T. E., Pauli, F., Sprouse, R. O., Neff, N. F., Newberry, K. M., Garabedian, M. J., & Myers, R. M. (2009). Genomic determination of the glucocorticoid response reveals unexpected mechanisms of gene regulation. *Genome Research, 19*(12), 2163–2171. http://dx.doi.org/10.1101/gr.097022.109.
Rieder, D., Trajanoski, Z., & McNally, J. G. (2012). Transcription factories. *Frontiers in Genetics, 3*, 221. http://dx.doi.org/10.3389/fgene.2012.00221.
Ro, S., Park, C., Jin, J., Sanders, K. M., & Yan, W. (2006). A PCR-based method for detection and quantification of small RNAs. *Biochemical and Biophysical Research Communications, 351*(3), 756–763. http://dx.doi.org/10.1016/j.bbrc.2006.10.105.
Sagai, T., Amano, T., Tamura, M., Mizushina, Y., Sumiyama, K., & Shiroishi, T. (2009). A cluster of three long-range enhancers directs regional Shh expression in the epithelial linings. *Development, 136*(10), 1665–1674. http://dx.doi.org/10.1242/dev.032714.

Sagai, T., Masuya, H., Tamura, M., Shimizu, K., Yada, Y., Wakana, S., … Shiroishi, T. (2004). Phylogenetic conservation of a limb-specific, *cis*-acting regulator of Sonic hedgehog (Shh). *Mammalian Genome, 15*(1), 23–34. http://dx.doi.org/10.1007/s00335-033-2317-5.

Sandelin, A., Bailey, P., Bruce, S., Engstrom, P. G., Klos, J. M., Wasserman, W. W., … Lenhard, B. (2004). Arrays of ultraconserved non-coding regions span the loci of key developmental genes in vertebrate genomes. *BMC Genomics, 5*(1), 99. http://dx.doi.org/10.1186/1471-2164-5-99.

Sander, J. D., & Joung, J. K. (2014). CRISPR-Cas systems for editing, regulating and targeting genomes. *Nature Biotechnology, 32*(4), 347–355. http://dx.doi.org/10.1038/nbt.2842.

Schauerte, H. E., van Eeden, F. J., Fricke, C., Odenthal, J., Strahle, U., & Haffter, P. (1998). Sonic hedgehog is not required for the induction of medial floor plate cells in the zebrafish. *Development, 125*(15), 2983–2993. Retrieved from: http://www.ncbi.nlm.nih.gov/pubmed/9655820.

Schneider, R., & Grosschedl, R. (2007). Dynamics and interplay of nuclear architecture, genome organization, and gene expression. *Genes & Development, 21*(23), 3027–3043. http://dx.doi.org/10.1101/gad.1604607.

Shlyueva, D., Stampfel, G., & Stark, A. (2014). Transcriptional enhancers: from properties to genome-wide predictions. *Nature Reviews Genetics, 15*(4), 272–286. http://dx.doi.org/10.1038/nrg3682.

Slattery, M., Zhou, T., Yang, L., Dantas Machado, A. C., Gordan, R., & Rohs, R. (2014). Absence of a simple code: how transcription factors read the genome. *Trends in Biochemical Sciences, 39*(9), 381–399. http://dx.doi.org/10.1016/j.tibs.2014.07.002.

Smith, E., & Shilatifard, A. (2014). Enhancer biology and enhanceropathies. *Nature Structural & Molecular Biology, 21*(3), 210–219. http://dx.doi.org/10.1038/nsmb.2784.

Stormo, G. D., & Zhao, Y. (2010). Determining the specificity of protein-DNA interactions. *Nature Reviews Genetics, 11*(11), 751–760. http://dx.doi.org/10.1038/nrg2845.

Strahle, U., & Ingham, P. W. (1992). Flight of fancy or a major new school? *Current Biology, 2*(3), 135–138. Retrieved from: http://www.ncbi.nlm.nih.gov/pubmed/15335988.

Strahle, U., & Rastegar, S. (2008). Conserved non-coding sequences and transcriptional regulation. *Brain Research Bulletin, 75*(2–4), 225–230. http://dx.doi.org/10.1016/j.brainresbull.2007.11.010.

Vaquerizas, J. M., Kummerfeld, S. K., Teichmann, S. A., & Luscombe, N. M. (2009). A census of human transcription factors: function, expression and evolution. *Nature Reviews Genetics, 10*(4), 252–263. http://dx.doi.org/10.1038/nrg2538.

Varshney, G. K., Pei, W., LaFave, M. C., Idol, J., Xu, L., Gallardo, V., … Burgess, S. M. (2015). High-throughput gene targeting and phenotyping in zebrafish using CRISPR/Cas9. *Genome Research, 25*(7), 1030–1042. http://dx.doi.org/10.1101/gr.186379.114.

Varshney, G. K., Sood, R., & Burgess, S. M. (2015). Understanding and editing the zebrafish genome. *Advances in Genetics, 92*, 1–52. http://dx.doi.org/10.1016/bs.adgen.2015.09.002.

Vastenhouw, N. L., Zhang, Y., Woods, I. G., Imam, F., Regev, A., Liu, X. S., … Schier, A. F. (2010). Chromatin signature of embryonic pluripotency is established during genome activation. *Nature, 464*(7290), 922–926. http://dx.doi.org/10.1038/nature08866.

Villar, D., Flicek, P., & Odom, D. T. (2014). Evolution of transcription factor binding in metazoans - mechanisms and functional implications. *Nature Reviews Genetics, 15*(4), 221–233. http://dx.doi.org/10.1038/nrg3481.

Visel, A., Blow, M. J., Li, Z., Zhang, T., Akiyama, J. A., Holt, A., … Pennacchio, L. A. (2009). ChIP-seq accurately predicts tissue-specific activity of enhancers. *Nature, 457*(7231), 854–858. http://dx.doi.org/10.1038/nature07730.

Visel, A., Bristow, J., & Pennacchio, L. A. (2007). Enhancer identification through comparative genomics. *Seminars in Cell & Developmental Biology, 18*(1), 140–152. http://dx.doi.org/10.1016/j.semcdb.2006.12.014.

Wang, D., Garcia-Bassets, I., Benner, C., Li, W., Su, X., Zhou, Y., … Fu, X. D. (2011). Reprogramming transcription by distinct classes of enhancers functionally defined by eRNA. *Nature, 474*(7351), 390–394. http://dx.doi.org/10.1038/nature10006.

Wardle, F. C., Odom, D. T., Bell, G. W., Yuan, B., Danford, T. W., Wiellette, E. L., … Smith, J. C. (2006). Zebrafish promoter microarrays identify actively transcribed embryonic genes. *Genome Biology, 7*(8), R71. http://dx.doi.org/10.1186/gb-2006-7-8-R71.

Wasserman, W. W., & Sandelin, A. (2004). Applied bioinformatics for the identification of regulatory elements. *Nature Reviews Genetics, 5*(4), 276–287. http://dx.doi.org/10.1038/nrg1315.

Weinhold, N., Jacobsen, A., Schultz, N., Sander, C., & Lee, W. (2014). Genome-wide analysis of noncoding regulatory mutations in cancer. *Nature Genetics, 46*(11), 1160–1165. http://dx.doi.org/10.1038/ng.3101.

Weirauch, M. T., & Hughes, T. R. (2010). Conserved expression without conserved regulatory sequence: the more things change, the more they stay the same. *Trends in Genetics, 26*(2), 66–74. http://dx.doi.org/10.1016/j.tig.2009.12.002.

White, M. A., Myers, C. A., Corbo, J. C., & Cohen, B. A. (2013). Massively parallel in vivo enhancer assay reveals that highly local features determine the *cis*-regulatory function of ChIP-seq peaks. *Proceedings of the National Academy of Sciences of the United States of America, 110*(29), 11952–11957. http://dx.doi.org/10.1073/pnas.1307449110.

Wittkopp, P. J., & Kalay, G. (2012). *Cis*-regulatory elements: molecular mechanisms and evolutionary processes underlying divergence. *Nature Reviews Genetics, 13*(1), 59–69. http://dx.doi.org/10.1038/nrg3095.

Woolfe, A., Goodson, M., Goode, D. K., Snell, P., McEwen, G. K., Vavouri, T., … Elgar, G. (2005). Highly conserved non-coding sequences are associated with vertebrate development. *PLoS Biology, 3*(1), e7. http://dx.doi.org/10.1371/journal.pbio.0030007.

Wu, H., Nord, A. S., Akiyama, J. A., Shoukry, M., Afzal, V., Rubin, E. M., … Visel, A. (2014). Tissue-specific RNA expression marks distant-acting developmental enhancers. *PLoS Genetics, 10*(9), e1004610. http://dx.doi.org/10.1371/journal.pgen.1004610.

Yang, C., Bolotin, E., Jiang, T., Sladek, F. M., & Martinez, E. (2007). Prevalence of the initiator over the TATA box in human and yeast genes and identification of DNA motifs enriched in human TATA-less core promoters. *Gene, 389*(1), 52–65. http://dx.doi.org/10.1016/j.gene.2006.09.029.

Yang, L., Ho, N. Y., Alshut, R., Legradi, J., Weiss, C., Reischl, M., … Strahle, U. (2009). Zebrafish embryos as models for embryotoxic and teratological effects of chemicals. *Reproductive Toxicology, 28*(2), 245–253. http://dx.doi.org/10.1016/j.reprotox.2009.04.013.

Yao, P., Lin, P., Gokoolparsadh, A., Assareh, A., Thang, M. W., & Voineagu, I. (2015). Coexpression networks identify brain region-specific enhancer RNAs in the human brain. *Nature Neuroscience, 18*(8), 1168–1174. http://dx.doi.org/10.1038/nn.4063.

Yokoyama, K. D., Zhang, Y., & Ma, J. (2014). Tracing the evolution of lineage-specific transcription factor binding sites in a birth-death framework. *PLoS Computational Biology, 10*(8), e1003771. http://dx.doi.org/10.1371/journal.pcbi.1003771.

Zaret, K. S., & Carroll, J. S. (2011). Pioneer transcription factors: establishing competence for gene expression. *Genes & Development, 25*(21), 2227–2241. http://dx.doi.org/10.1101/gad.176826.111.

CHAPTER EIGHT

# Transcriptomic Approaches in the Zebrafish Model for Tuberculosis—Insights Into Host- and Pathogen-specific Determinants of the Innate Immune Response

**E.L. Benard, J. Rougeot, P.I. Racz, H.P. Spaink and A.H. Meijer**[1]
Institute of Biology, Leiden University, Leiden, The Netherlands
[1]Corresponding author: E-mail: a.h.meijer@biology.leidenuniv.nl

## Contents



*Advances in Genetics*, Volume 95
ISSN 0065-2660
http://dx.doi.org/10.1016/bs.adgen.2016.04.004

## Abstract

*Mycobacterium marinum* infection in zebrafish has become a well-established model of tuberculosis. Both embryonic and adult zebrafish infection studies have contributed to our knowledge of the development and function of tuberculous granulomas, which are typical of mycobacterial pathogenesis. In this review we discuss how transcriptome profiling studies have helped to characterize this infection process. We illustrate this using new RNA sequencing (RNA-Seq) data that reveals three main phases in the host response to *M. marinum* during the early stages of granuloma development in zebrafish embryos and larvae. The early phase shows induction of complement and transcription factors, followed by a relatively minor induction of pro-inflammatory cytokines within hours following phagocytosis of *M. marinum*. A minimal response is observed in the mid-phase, between 6 hours and 1 day post infection, when the tissue dissemination of *M. marinum* begins. During subsequent larval development the granulomas expand and a late-phase response is apparent, which is characterized by progressively increasing induction of complement, transcription factors, pro-inflammatory cytokines, matrix metalloproteinases, and other defense and inflammation-related gene groups. This late-phase response shares common components with the strong and acute host transcriptome response that has previously been reported for *Salmonella typhimurium* infection in zebrafish embryos. In contrast, the early/mid-phase response to *M. marinum* infection, characterized by suppressed pro-inflammatory signaling, is strikingly different from the acute response to *S. typhimurium* infection. Furthermore, *M. marinum* infection shows a collective and strongly fluctuating regulation of lipoproteins, while *S. typhimurium* infection has pronounced effects on amino acid metabolism and glycolysis.


## 1. INTRODUCTION

*Mycobacterium tuberculosis* is a highly successful human pathogen, estimated to have currently infected about one-third of the human world population (WHO Global Tuberculosis report 2014). In the majority of infected individuals the immune system can effectively contain the infection, but there remains a 5–10% lifetime risk of developing the active and contagious form of tuberculosis disease (Russell, 2011). Host cells express several innate immune pattern recognition receptors (PRRs), such as Toll-like receptors (TLRs), NOD-like receptors (NLRs), mannose receptors, scavenger receptors, and complement receptors, that function in the recognition of pathogen-associated molecular patterns (PAMPs) of *M. tuberculosis* and in the recognition of danger-associated molecular patterns (DAMPs) released during the death of infected cells (Ottenhoff, 2012). This recognition function by the innate immune system represents the first protective barrier of the host against *M. tuberculosis* and it functions in concert with the adaptive immune response to control infection at later stages. The immune response to *M. tuberculosis* infection is not only regulated by the host itself but also manipulated by mycobacterial virulence factors, which provide a suitable environment for proliferation and also promote the immune-mediated responses required for the transmission of disease (Cooper & Torrado, 2012).

The BCG vaccine does not provide adequate protection against the latent or reactivated forms of tuberculosis and there is an increasing occurrence of multidrug-resistant strains of *M. tuberculosis* (Chiang, Centis, & Migliori, 2010). This highlights the need for developing novel therapeutic approaches requiring a better understanding of the mechanistic basis of the host–pathogen interactions underlying tuberculosis (Hawn, Matheson, Maley, & Vandal, 2013). In order to address this topic, a combination of functional genomics approaches and modern cell biology is required in animal models that mimic the human disease (O'Toole, 2010). The zebrafish has become a valuable addition to the animal models used for tuberculosis research (Cronan & Tobin, 2014; van Leeuwen, van der Sar, & Bitter, 2015; Meijer, 2015; Ramakrishnan, 2013). Zebrafish can be infected with *Mycobacterium marinum*, which is the closest genetic relative of the *M. tuberculosis* complex and a natural pathogen of fish and other cold-blooded animals. The genomes of both strains show that there is an *M. marinum* orthologue for 80% of all coding DNA sequences in *M. tuberculosis*, that the averaged shared amino acid identity is 85% (Stinear et al., 2008), and that the majority of virulence factors found in *M. tuberculosis* have homologous genes in *M. marinum* (Ruley et al., 2004; Tobin & Ramakrishnan, 2008). Both *M. marinum* and *M. tuberculosis* are capable of surviving and replicating within macrophages. They avoid the bactericidal mechanisms of the macrophage by blocking phagosome–lysosome fusion and both species are also capable of escaping from the phagosome into the cytosol (Simeone et al., 2012, 2015; Stamm et al., 2003; Vergne, Chua, Singh, & Deretic, 2004; van der Wel et al., 2007). Additionally, the hallmark of infection with both pathogens is the development of latency with survival of the pathogen in caseating granulomas, which are highly structured aggregates of infected and noninfected immune cells (Parikka et al., 2012; Russell, 2011; Swaim et al., 2006). Historically these granulomas were strictly regarded as host defense structures containing the infection, but recent insight that for a major part has resulted from studies in zebrafish, implicate granulomas also in the expansion and dissemination of mycobacterial infection, especially at the early stages of granuloma development (Ramakrishnan, 2012).

Many examples of how studies using the zebrafish *M. marinum* model have contributed to tuberculosis research are discussed in recent reviews (Cronan & Tobin, 2014; van Leeuwen et al., 2015; Meijer, 2015; Ramakrishnan, 2013). Here we focus specifically on the results of transcriptome profiling studies. Following a brief introduction to the zebrafish immune system, we give an overview of how embryonic, larval, and adult zebrafish models are used to study the development and function of granulomas in mycobacterial infection. Subsequently, we illustrate how transcriptome profiling studies represent a vital tool for characterizing these models by considering RNA sequencing (RNA-Seq) data from a detailed time course of the early stages of granuloma development in zebrafish embryos and larvae. Finally, a comparison of the RNA-Seq profiles of *M. marinum* and *Salmonella typhimurium* infections shows the specific and common components in the innate immune response to different pathogens.

## 2. ZEBRAFISH AS A MODEL FOR STUDYING IMMUNITY AND INFECTION

Zebrafish embryos are fertilized externally and develop an innate immune system within the first days following fertilization. The development and differentiation of zebrafish immune cells is controlled by a transcriptional program similar to that found in mammals (Ellett & Lieschke, 2010). During the primitive hematopoiesis, mesoderm cells differentiate into myeloid cells under control of transcription factor Spi1/Pu.1 or into erythroid cells, controlled by Gata1 (Belele et al., 2009; Rhodes et al., 2005). During the definitive hematopoiesis, hematopoietic stem cells (HSCs) are formed and erythroid and myeloid lineage cells derive from a common erythromyeloid progenitor (Bertrand et al., 2007; Davidson & Zon, 2004). The first myeloid cell type to develop is the primitive macrophage, which is already functional in 1-day-old zebrafish embryos (Herbomel, Thisse, & Thisse, 1999). Macrophages can be distinguished by specific marker genes, such as *csf1r*, *mpeg1.1*, and *mfap4* (Benard et al., 2015; Gray et al., 2011; Walton, Cronan, Beerman, & Tobin, 2015). At this stage, macrophages are capable of phagocytosing pathogens and tissue debris. Neutrophils are the second innate immune cell type to develop. They are detectable by *mpx* gene expression on day 1 and produce antimicrobial proteins within their characteristic granules by 2 days post fertilization (Le Guyader et al., 2008; Renshaw et al., 2006). These cells are the first responders to inflammatory stimuli and are also efficient scavengers of bacteria in infected tissues. The first immature

T-cells can be detected in 4-day-old larvae (Langenau, 2004; Langenau et al., 2003), but it takes several weeks before zebrafish also have antibody-producing B-cells (Page et al., 2013) and a fully functional adaptive immune system. Using the zebrafish as a host model has several advantages. Firstly, the temporal separation of innate and adaptive immunity allows us to study the isolated function of innate immune factors during infection in embryos and early larval stages. Secondly, various transgenic lines are available that have fluorescently labeled macrophages, neutrophils, or other immune cell types, and the behavior of these cells can be imaged in real-time and with excellent resolution during the early life stages when zebrafish are transparent. Reporter lines for innate immune response genes have also recently been developed showing that subsets of macrophages are differentially activated during infection (Nguyen-Chi et al., 2015; Sanderson et al., 2015). Thirdly, the published sequenced genome of the zebrafish allows us to apply state-of-the-art RNA sequencing techniques to study gene expression profiles. Fourthly, many genetic tools are available to create temporary knockdowns or permanent mutations of genes of interest. All these tools combined have led to the development of zebrafish models for a wide variety of pathogenic infections (Goody, Sullivan, & Kim, 2014; Gratacap & Wheeler, 2014; Levraud, Palha, Langevin, & Boudinot, 2014; Palha et al., 2013; Torraca, Masud, Spaink, & Meijer, 2014).

## 3. MYCOBACTERIAL GRANULOMA DEVELOPMENT IN ZEBRAFISH

Infecting the optically transparent zebrafish embryo with fluorescent *M. marinum* allows real-time visualization of the early stages of a forming granuloma in vivo (Davis et al., 2002; Ramakrishnan, 2013). *M. marinum* is recognized and phagocytosed extremely rapidly by primitive macrophages following injection into the blood circulation of the zebrafish embryo (Benard, Roobol, Spaink, & Meijer, 2014; Davis et al., 2002). Phagocytosis by macrophages has proven to be an important mechanism of innate immunity, as zebrafish embryos lacking functional leukocytes are not capable of restricting mycobacterial growth (Clay et al., 2007). Despite the controlling mechanism of macrophages on the growth of *M. marinum*, this pathogen is still capable of growing restrictedly within these cells. Furthermore, *M. marinum* can direct macrophages to aggregate into granulomas by producing the virulence determinant RD1 (region of deletion 1)-dependent signal (Volkman et al., 2004). Subsequently, infected macrophages can be seen to leave primary granulomas, either via tissue migration or via the vasculature, and can then initiate secondary granulomas (Davis & Ramakrishnan, 2009). These events take place during the first few days of zebrafish embryo and early larval development (Traver et al., 2003). As these early life stages lack an adaptive immune system, the context of innate immunity appears to be sufficient for granuloma formation (Davis et al., 2002). The importance of innate immune control is also well demonstrated in the zebrafish embryo *M. marinum* infection model during knockdown of tumor necrosis factor receptor 1 (tnfrsf1a) or in null mutants of myeloid differentiation factor 88 (*myd88*). In the absence of TNF signaling, *M. marinum* intracellular growth and granuloma formation are accelerated and followed by necrotic death of overladen macrophages and breakdown of granulomas (Clay, Volkman, & Ramakrishnan, 2008). Similarly, mutation of *myd88* accelerates formation of granuloma-like aggregates and increases the bacterial burden, which was shown to be associated with lower induction of genes central to innate immunity (van der Vaart, van Soest, Spaink, & Meijer, 2013).

In the adult zebrafish model, *M. marinum* is injected intraperitoneally into the fish to initiate either acute symptoms or a chronic progressive disease with multiple organ involvement and formation of granuloma-like lesions (Parikka et al., 2012; Prouty, Correa, Barker, Jagadeeswaran, & Klose, 2003; van der Sar et al., 2004; Swaim et al., 2006). The granulomas of adult fish are comprised of an outer layer of dense epithelial cells and a central region containing mainly macrophages, granulocytes, and lymphocytes. Similar to the human tuberculous granuloma, this central region of the zebrafish granuloma is often necrotic and is hypoxic (Oehlers et al., 2015; van der Sar et al., 2004; Swaim et al., 2006). The adult zebrafish model has also proven to be extremely valuable because it provides a means to address latency and reactivation, mechanisms that are difficult to study due to lack of standardized animal models (Hammaren et al., 2014; Parikka et al., 2012). Intraperitoneal injection of a low dose of *M. marinum* leads to a latent disease in most fish which relies on *rag1*-dependent adaptive immunity (Parikka et al., 2012). The reactivation of dormant *M. marinum* can be efficiently induced by gamma irradiation that transiently depletes granulo/monocyte and lymphocyte pools (Parikka et al., 2012).

The *M. marinum* zebrafish infection model thus contributes to our knowledge of tuberculosis by providing two important features that are challenging to observe in other animal models: the ability to visualize the stages of early granuloma formation in the zebrafish embryo in vivo and the possibility to study latency in adults (Meijer, 2015).

## 4. TIMING OF EVENTS DURING THE *M. MARINUM* INFECTION PROCESS IN ZEBRAFISH EMBRYOS AND LARVAE

Microinjection of mycobacteria intravenously at the blood island at 30 hpf has been shown to yield a robust systemic infection in which various aspects of the granuloma formation can be studied in a noninvasive manner (Davis et al., 2002). As soon as 40 min post infection (mpi) after intravenous injection of a single cell suspension of *M. marinum*, the bacteria are completely internalized predominantly by macrophages (Benard et al., 2014; Clay et al., 2007) (Fig. 1A) and occasionally by neutrophils around the site of injection (Fig. 1B). Between 6 h post injection (hpi) and 1 day post injection (dpi) infected cells travel through the vascular system (Fig. 1C–G) and can occasionally be observed outside the blood circulation, adjacent to the duct of Cuvier (Fig. 1E) and posterior to the transition of the caudal vein and artery (Fig. 1G). However, the majority of the infected cells remain within the vasculature during these stages. From 2 dpi onward *M. marinum*-infected macrophages can be observed within the caudal fin fold of the embryo (Fig. 1H). In addition, infected cells travel through the intersegmental vessels, the infection spreads from the caudal vein to the dorsal longitudinal lymphatic vessel, and *M. marinum* can also be seen outside the caudal vein (Fig. 1I) (Fenaroli et al., 2014). Between 2 and 5 dpi clusters of *M. marinum* tend to stay in the same position, where they continue to grow and form tight infection structures. During this growth process the structures extend from inside the blood circulation to outside the vessels (Fig. 1I and J).

The formation of granuloma-like structures becomes evident by comparing uninfected embryos (Fig. 1K–O) with *M. marinum*-infected embryos (Fig. 1P–Y) obtained from double transgenic zebrafish carrying the macrophage marker *mpeg1.1:mCherry-F* (Bernut et al., 2014) and the neutrophil marker *mpx:GFP* (Renshaw et al., 2006). At 1 dpi the infected macrophages and uninfected neutrophils are still spread evenly and no clusters of cells can be seen (Fig. 1P), quite similar to the macrophage and neutrophil distribution of uninfected embryos (Fig. 1K). At 2 dpi the first clusters can be observed consisting of multiple infected and uninfected macrophages and a few uninfected neutrophils (Fig. 1Q). At 3 dpi neutrophils are attracted to the developing tight aggregates of infected macrophages (Fig. 1R) and infected neutrophils can be observed for the first time within the granuloma-like structures. At 4 dpi the morphology of the infected

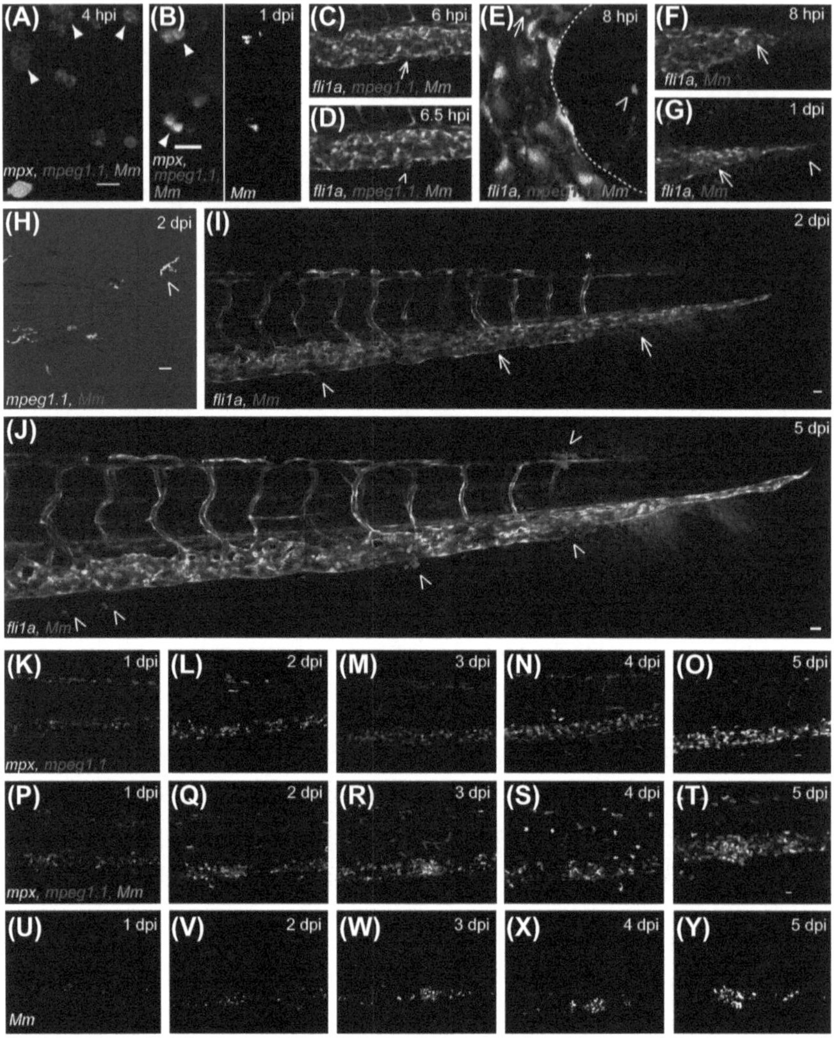


**Figure 1** ***Overview of the*** **M. marinum** ***infection process imaged in zebrafish embryos.*** Zebrafish embryos were injected at 28 hpf with 250 cfu of *M. marinum* in 2% PVP or with 2% PVP alone into the blood island. (A) E2-Crimson-expressing *M. marinum* (Swaim et al., 2006) bacterial infection in *Tg(mpeg1.1::mCherry-F)/Tg(mpx:GFP)* embryos at 4 hpi imaged at Duct of Cuvier site. *Solid arrowheads* indicate infected macrophages. (B) E2-Crimson-expressing *M. marinum* bacterial infection in *Tg(mpeg1.1::mCherry-F)/ Tg(mpx:GFP)* embryos at caudal vein site. Left panel, overlay of all three fluorescence channels; right panel, E2-Crimson fluorescence channel with *M. marinum* fluorescence depicted as white pixels. *Solid arrowheads* indicate infected neutrophils. (C–E) E2-Crimson-expressing *M. marinum* bacterial infection in *Tg(fli1a:EGFP)/Tg(mpeg1.1:: mCherry-F)* embryo imaged over time at (C and D) the caudal vein at 6 and 6.5 hpi

macrophages changes to a more rounded shape (Fig. 1S) and these cells contain a high level of bacterial fluorescence (Fig. 1X), indicating that they are severely infected. This phenomenon is even more apparent at 5 dpi (Fig. 1T) when the *mpeg1.1* driven-macrophage fluorescence markedly diminishes (Benard et al., 2015). The use of a recently developed transgenic line using *mfap4* as an embryonic macrophage-specific promoter will allow stable fluorescence throughout several days of infection because unlike *mpeg1.1*, *mfap4* is not downregulated during infection (Walton et al., 2015). At 4 and 5 dpi the number of neutrophils surrounding the granuloma-like structure is further increased (Fig. 1S and T) and the granuloma-like structures start protruding the epithelial layer of the embryo. It has been shown that granuloma-activated genes of *M. marinum* start to be expressed at these stages, consistent with the immune cell aggregates in zebrafish larvae representing the early stages of tuberculous granulomas (Davis et al., 2002; Stoop,

and (E) the Duct of Cuvier site at 8 hpi. *M. marinum*-infected macrophages (purple color: double positive for mCherry and E2-Crimson) are observed both (*arrows*) inside and (*pointed arrowheads*) outside the vasculature. *Dotted line* in (E) indicates border of vasculature in Duct of Cuvier. (F and G) mCherry-expressing *M. marinum* (van der Sar et al., 2004) bacterial infection in a *Tg(fli1a:EGFP)* (Lawson & Weinstein, 2002) embryo at the transition of the caudal vein and artery at (F) 8 hpi and (G) 1 dpi. *M. marinum* can be observed (*arrow*) inside and (*pointed arrowhead*) outside the vasculature. (H) mCherry-expressing *M. marinum* bacterial infection in a *Tg(mpeg1.1:EGFP)* embryo imaged at the posterior tail area. Merged image of bright field, GFP, and mCherry channels show an infected macrophage (*pointed arrowhead*) located in the caudal fin. (I and J) mCherry-expressing *M. marinum* bacterial infection in a *Tg(fli1a:EGFP)* embryo imaged at the tail region at (I) 2 dpi and (J) 5 dpi. An *M. marinum* cluster can be seen (*asterisk*) in the dorsal longitudinal lymphatic vessel and various *M. marinum* clusters can be seen residing (*arrow*) within the caudal vein and (*pointed arrowhead*) outside the caudal vein at (I) 2 dpi. These clusters increase in size at (J) 5 dpi and clusters previously inside the caudal vein at 2 dpi are now observed (*pointed arrowhead*) outside the caudal vein. (K–O) Uninfected (mock injected) *Tg(mpeg1.1::mCherry-F)/Tg(mpx:GFP)* embryo tail region posterior to the urogenital opening imaged over a time course of (K) 1 dpi, (L) 2 dpi, (MI) 3 dpi, (N) 4 dpi, and (O) 5 dpi. Images are *z*-stack projections of the merged channels of GFP and mCherry. (P–Y) E2-Crimson-labeled *M. marinum* bacterial infection in a *Tg(mpeg1.1::mCherry-F)/Tg(mpx:GFP)* embryo imaged in identical area as (K–O) uninfected embryo over time course of (P) 1 dpi, (Q) 2 dpi, (R) 3 dpi, (S) 4 dpi, and (T) 5 dpi. Confocal images were taken on a Leica TCS SPE with an HC APO 20× 0.7 NA objective and an HCX APO 40× 0.8 NA objective or on a Zeiss LSM5 Exciter/AxioObserver with an HC APO 20× 0.7 NA objective. Images are *z*-stack projections of the merged channels of GFP, mCherry, and E2-Crimson. (U–Y) *z*-stack projections of the E2-Crimson channel isolated from P-T and depicted as white pixels. Scale bars: 20 μm. (See color plate)

Bitter, & van der Sar, 2012). Genetic screens using the larval model have revealed multiple host as well as bacterial factors that are required for granuloma formation (Stoop et al., 2012; Tobin et al., 2010; Torraca et al., 2015).

## 5. ZEBRAFISH TRANSCRIPTOMICS

To strengthen the utility of the zebrafish model for *M. marinum* infection, an increasing number of transcriptome studies have been performed to provide a detailed expression analysis of the immune response toward this infection. Initially, a multiplatform microarray study was performed in *M. marinum*-infected zebrafish adults which showed that zebrafish express many homologs of human immune response genes and genes that have previously been implicated in the response to mycobacterial infection (Meijer et al., 2005). Microarrays, however, have a low dynamic range and are limited in content because the probe design is based on inference from prior genomic sequence data and fluorescence intensity is used as a surrogate for gene expression. As a result of this, exon junctions for novel expressed regions and RNA editing events will be missed and allele-specific differences in gene expression will not be detected (Malone & Oliver, 2011). Therefore, a transcriptome study has been performed using Digital Gene Expression (DGE), which is a tagged-based transcriptome sequencing method derived from serial analysis of gene expression (SAGE) (Hegedus et al., 2009). Like SAGE, DGE generates short signature sequences (tags) for the 3′-end regions of mRNA transcripts. When the results of microarray and DGE were compared, a large overlap was found in differentially expressed transcripts as well as technology-dependent differences, indicating the value of using two complementary transcriptome analysis methods. In addition, a great benefit of using DGE over microarray was that the DGE tag data could be mapped onto transcript databases and onto the genomic sequence, which revealed that many alternative transcript forms and novel transcripts are regulated in a disease-specific fashion (Hegedus et al., 2009).

In a different microarray study, infections of *M. marinum* acute and chronic strains in adult zebrafish were compared (van der Sar, Spaink, Zakrzewska, Bitter, & Meijer, 2009). The results show that both infections evoke partially overlapping host transcriptome signatures, despite them inducing profoundly different disease phenotypes. Infection with the acute *M. marinum* strain led to high expression levels of genes encoding MHC class

I proteins, matrix metalloproteinases, transcription factors, cytokines, and other common immune response proteins. In contrast, infection with the chronic *M. marinum* strain led to high expression levels of small GTPase and histone gene groups. Moreover, there was a large overlap of mycobacterium-responsive genes between infected zebrafish adults and embryos at various stages of granuloma formation. The embryo response consists solely of an innate immune response, whereas the adult zebrafish response consists of both innate and adaptive response which is similar to mammals. The large overlap between these two groups indicates how important the contribution of the innate immune response is during mycobacterial infection.

RNA-Seq, which involves sequencing of the whole transcriptome, has clear advantages over microarray and DGE approaches and is currently the preferred method for transcriptome analysis (Wang, Gerstein, & Snyder, 2009). To illustrate the utility of this approach, below we compare new RNA-Seq profiles of *M. marinum*-infected and *M. marinum*-noninfected embryos throughout a detailed time course of infection.

## 6. THE PROFILES OF THE GENE EXPRESSION RESPONSE TO *M. MARINUM* INFECTION CAN BE DIVIDED INTO AN EARLY, MID, AND LATE PHASE

For a detailed characterization of the host innate immune response during the course of *M. marinum* infection, infected and noninfected embryos were sampled every 2 h from 2 hpi to 8 hpi and daily from 1 dpi to 5 dpi. At 2 hpi, when all *M. marinum* bacteria have been phagocytosed (Benard et al., 2014) (Fig. 1A), an immediate response to the infection was observed with 170 upregulated genes and 64 downregulated genes. Subsequently, over the time period between 4 hpi and 1 dpi, the total number of differently regulated genes was at least threefold lower (55 up and 33 down at 4 hpi), with a further decline between 6 hpi and 1 dpi (Fig. 2A, Table 1). At this time point, *M. marinum* is predominantly contained by macrophages (Fig. 1A), although uninfected neutrophils have mounted a systemic response by upregulating nitrotyrosine in a TLR/IL1R-MyD88 signaling-dependent manner (Elks et al., 2014). Afterward, an increase in up- and downregulated genes was seen at 2 dpi (310 up and 459 down), the time point when the first clusters can be observed consisting of multiple infected and uninfected macrophages and a few uninfected neutrophils (Fig. 1Q). The number of differentially regulated genes decreased slightly at 3 dpi but continued to increase remarkably at 4 (1165

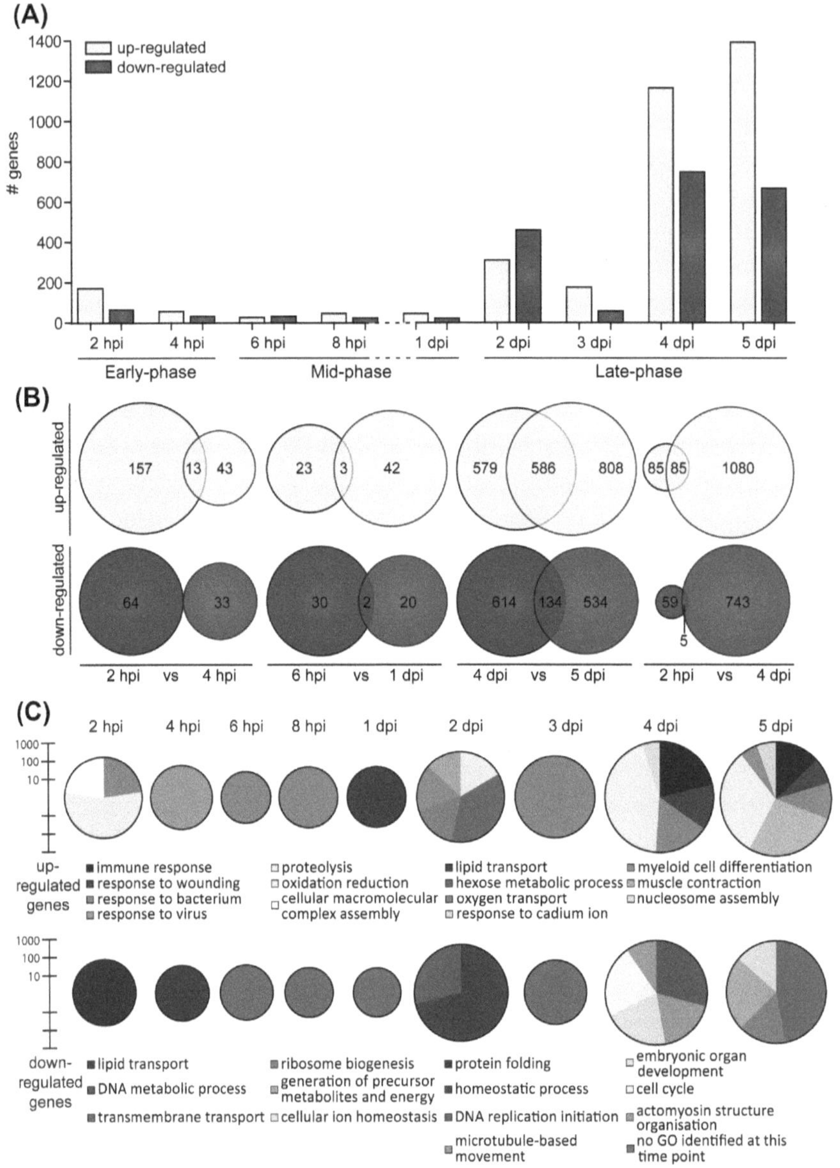


**Figure 2** ***Overview of total number of genes differentially regulated in*** **M. marinum*-infected embryos, their overlap between various time points, and gene ontology terms enriched per time point.*** AB/TL embryos were infected with *M. marinum* Mma20 strain at 1 day post fertilization (dpf) and RNA was isolated at 2, 4, 6, and 8 hpi and 1, 2, 3, 4, and 5 dpi. Four biological replicates each containing 30 embryos of each time point were used for RNA sequencing and DeSeq analysis of transcript count data was used for statistical comparison with mock-injected control embryos sampled at the same

**Table 1** Overlap of genes up- and downregulated during the time course of *M. marinum* infection

| **# Overlap up** | **2 hpi** | **4 hpi** | **6 hpi** | **8 hpi** | **1 dpi** | **2 dpi** | **3 dpi** | **4 dpi** | **5 dpi** |
|---|---|---|---|---|---|---|---|---|---|
| 2 hpi | 170 | 13 | 6 | 23 | 18 | 28 | 18 | 85 | 88 |
| 4 hpi | | 56 | 7 | 9 | 3 | 29 | 23 | 37 | 38 |
| 6 hpi | | | 26 | 9 | 3 | 16 | 12 | 17 | 17 |
| 8 hpi | | | | 46 | 21 | 22 | 14 | 20 | 24 |
| 1 dpi | | | | | 45 | 20 | 6 | 17 | 14 |
| 2 dpi | | | | | | 310 | 41 | 70 | 69 |
| 3 dpi | | | | | | | 176 | 83 | 98 |
| 4 dpi | | | | | | | | 1165 | 586 |
| 5 dpi | | | | | | | | | 1394 |
| **# Overlap down** | **2 hpi** | **4 hpi** | **6 hpi** | **8 hpi** | **1 dpi** | **2 dpi** | **3 dpi** | **4 dpi** | **5 dpi** |
| 2 hpi | 64 | 0 | 2 | 3 | 1 | 8 | 4 | 5 | 6 |
| 4 hpi | | 33 | 1 | 2 | 2 | 1 | 0 | 6 | 1 |
| 6 hpi | | | 32 | 2 | 2 | 1 | 0 | 2 | 1 |
| 8 hpi | | | | 23 | 3 | 7 | 0 | 2 | 3 |
| 1 dpi | | | | | 22 | 9 | 5 | 3 | 3 |
| 2 dpi | | | | | | 459 | 6 | 40 | 29 |
| 3 dpi | | | | | | | 57 | 14 | 7 |
| 4 dpi | | | | | | | | 748 | 134 |
| 5 dpi | | | | | | | | | 668 |

time points (* = fold change > 2, $p < 0.05$) (Anders & Huber, 2010). (A) Overview of the total number of genes either significantly upregulated (yellow) or downregulated (blue) per time point in the *M. marinum*-infected samples compared to the uninfected control embryos. (B) Area-proportionate Venn-diagrams representing the overlap of both upregulated and downregulated genes during *M. marinum* infection. The time points compared are 2 hpi versus 4 hpi, 6 hpi versus 1 dpi, 4 dpi versus 5 dpi, and 2 hpi versus 4 dpi (respectively left versus right circle for each comparison). (C) Gene ontology terms identified per time point represented as pie charts. The diameter of each chart is proportional to the total number of genes upregulated (top) or downregulated (bottom) per time point. Colors illustrate the relative proportion of each identified gene ontology term. Genes without associated gene ontology terms are not represented in the pie charts (except for time points with no gene ontologies identified). Proportional Venn diagrams were created with a Venn Diagram Generator (http://jura.wi.mit.edu/bioc/tools/venn.php). Gene ontology analysis was performed using DAVID v6.7 (http://david.abcc.ncifcrf.gov/home.jsp) (Huang da et al., 2009). Gene ontology terms are considered significant when the false-discovery rate was below 5%. Gene Expression Omnibus (GEO) database accession for RNA-Seq: GSE76499. (See color plate) up and 748 down) and 5 dpi (1394 up and 668 down) when the embryos were highly infected (Fig. 1S and T).

Based on the apparent trend in gene regulation, the transcriptional response can be divided into three main phases: an immediate early phase with a clear response at 2 hpi and rapid decline at 4 hpi, a mid-phase response from 4 hpi to 1 dpi when there is low differential gene regulation, and a late-phase response, which starts from 2 dpi but is mainly characterized by a striking increase in differential gene expression at 4 and 5 dpi (Fig. 2A). When we look at the overlap between different time points, we find that there is remarkable little overlap between consecutive time points of the series up to 4 dpi (Table 1). For example, in the early phase only 13 upregulated genes were overlapping between 2 hpi and 4 hpi, and there was no overlap of the downregulated genes (Fig. 2B). The only genes that were significantly upregulated for each time point during 2 hpi until 8 hpi are the matrix metalloproteinase gene *mmp9* and a noncoding RNA (*si: dkey-97i18.5*). Hardly any overlap can be found when comparing two mid-phase stages (6 hpi with 1 dpi), during which a very low number of genes are differentially regulated (Fig. 2B). This is quite the opposite when the final two time points of the late-phase time points are compared (4 dpi with 5 dpi), when a high overlap is evident between the upregulated genes (Fig. 2B and Table 1). Furthermore, there was also higher overlap between the earliest time point (2 hpi) and the latest time points (4 and 5 dpi). This is reflected in the results of gene ontology (GO) analysis using DAVID (Huang da, Sherman, & Lempicki, 2009), which shows upregulated biological process categories that are identified during both the early-phase and late-phase responses: response to bacterium and proteolysis (Fig. 2C). In contrast, during the mid-phase response the category of lipid transport is exclusively detected at 1 dpi. Hexose metabolic process, oxygen transport, and muscle contraction are exclusively upregulated at the beginning of the late phase at 2 dpi (Fig. 2C). Notable at this time point was also the downregulation of genes related to the GO-terms homeostatic process and DNA replication initiation. The up- and downregulated gene groups at 2 dpi show that the transition between the mid-phase and late phase in the infection process is associated with strong effects on the host metabolism. There was no overlap between the time points in downregulated biological process categories (Fig. 2C). In addition to the strong downregulation response at 2 dpi, many downregulated gene groups were detected at 4 and 5 dpi, with gene ontology terms mainly related to DNA and protein metabolism and transmembrane transport.

## 7. EARLY AND LATE REGULATION OF TRANSCRIPTION FACTOR GENES

Following a global gene ontology analysis, expression patterns of different functional groups of genes were studied over the infection time course. A significant change in expression levels was found within a group of transcription factor genes. These included members of the activating transcription factor (ATF) family, the interferon regulatory factor (IRF) family, the CCAAT/enhancer binding protein family (CEBP), the AP-1 (FOS and JUN family), and the NFκB family. ATF3 is a member of the CREB/ATF family of basic leucine zipper transcription factors. It has been shown to act primarily as a repressor, but splice forms that are unable to bind DNA can also stimulate transcription presumably by sequestration of inhibitory cofactors (Chen, Liang, Whelan, & Hai, 1994). In mice, ATF3 is used as an ER stress marker that is expressed in the macrophage-rich area of *M. tuberculosis*-induced granulomas (Seimon et al., 2010). GEO profiles in the NCBI database show that *Atf3* expression is gradually induced in mice during early lung immune response to *M. tuberculosis* (GDS4257) and that *ATF3* expression is induced in human macrophages in response to hypoxia (GDS2036). Hypoxia occurs in the environment of the human tuberculous granulomas and it may be responsible for *M. tuberculosis* dormancy (Boshoff & Barry, 2005). It was recently shown in mice that ROS-mediated superinduction of *ATF3* causes high susceptibility to bacterial infections through the suppression of interleukin 6 (IL-6) and that *ATF3−/−* mice are protected against bacterial infections (Hoetzenecker et al., 2012). *ATF3* can also be induced by LPS in bone marrow-derived macrophages from mice and this induction was suggested to regulate TLR-stimulated inflammatory responses as part of a negative-feedback loop (Gilchrist et al., 2006). Previous research has shown that genetic disruption of IRF-1 in mice leads to an increased susceptibility to tuberculosis (Cooper, Pearl, Brooks, Ehlers, & Orme, 2000; Kamijo et al., 1994; Pine, 2002). This can be explained by the necessity of IRF-1 in inducible nitric oxide synthase (iNOS) regulation (Pine, 2002). In turn, the iNOS-dependent production of nitric oxide and reactive nitrogen species is important for killing of *M. tuberculosis* in murine models of tuberculosis (Chan, Xing, Magliozzo, & Bloom, 1992) and iNOS signaling has also been shown to have an antimicrobial function in neutrophils against *M. marinum* (Elks et al., 2013). Expression of both C/EBP beta and C/EBP delta have been shown to be induced during inflammation (Akira et al., 1990; Alam, An, & Papaconstantinou, 1992; Kinoshita, Akira,

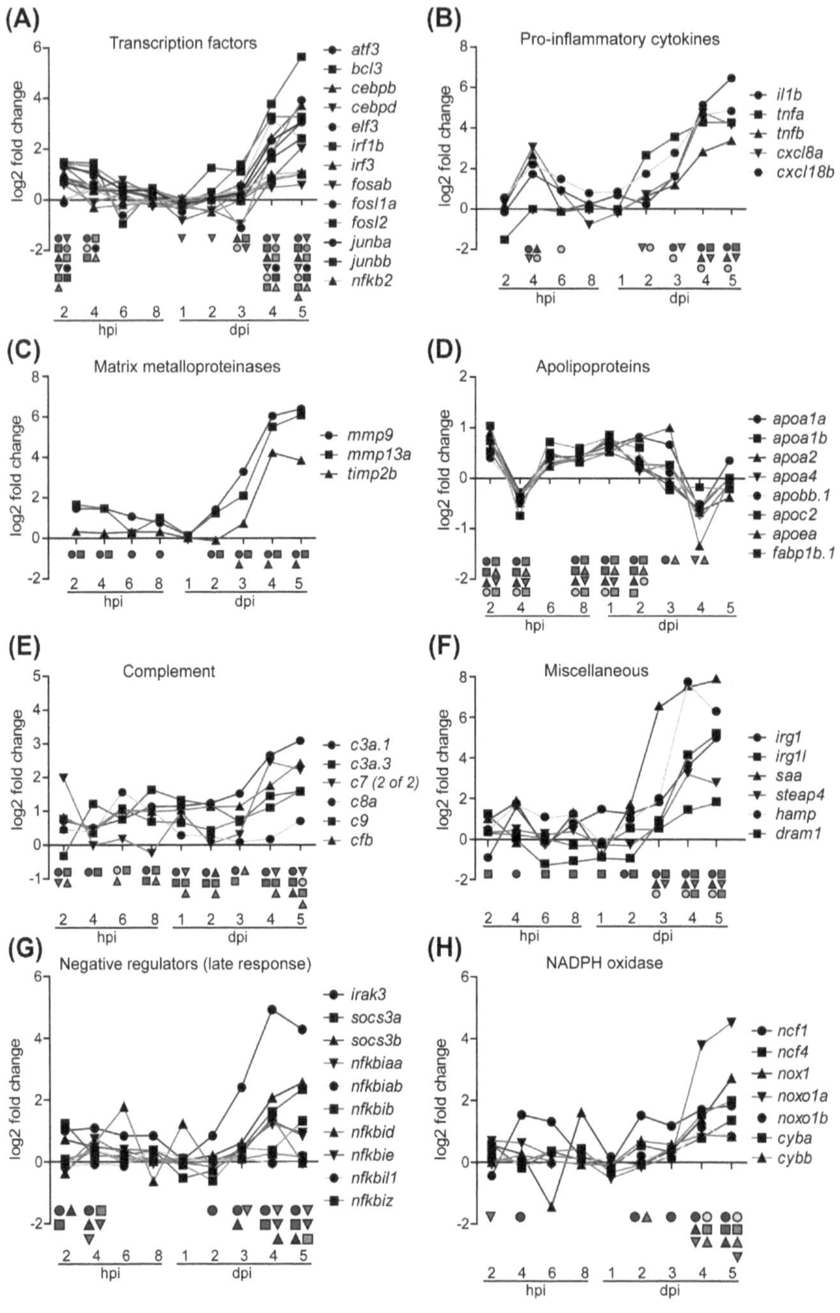


**Figure 3** ***Overview of gene expression profiles during*** **M. marinum** ***infection.*** Gene expression levels of groups of selected genes during a time course of *M. marinum*

& Kishimoto, 1992) and are similarly induced in murine macrophages in response to exposure to 19-kDa lipoprotein (LpqH) of *M. tuberculosis* (Pennini et al., 2007). Interestingly, both AP-1 and NF-κB play critical roles in the induction of *IL8* promoter activity during the response to the early secreted antigenic target of 6 kDa (ESAT-6) of *M. tuberculosis* (Boggaram, Gottipati, Wang, & Samten, 2013). In addition, both NF-κB and AP-1 subunits contribute to the regulation of MMP secretion in vitro in human microglial cells and in vivo in CNS TB patients (Green et al., 2010). In zebrafish, genes belonging to the NF-κB, AP-1, and MMP families (see below) are also upregulated during *M. marinum* infection (Fig. 3C). Directly after recognition and phagocytosis of *M. marinum* (Fig. 1A), an immediate early response to *M. marinum* infection at 2 hpi consists for a major part of the upregulation of these transcription factor genes (Fig. 3A). About half of these genes are still upregulated at 4 hpi, after which their expression returns to the uninfected control levels between 6 hpi to approximately 3 dpi only to become significantly induced again during the final stage of the late phase response at 4–5 dpi (Fig. 3A).

## 8. REGULATION OF PRO-INFLAMMATORY CYTOKINE GENES

Pro-inflammatory cytokines produced by phagocytes, such as tumor necrosis factor alpha (TNF-α) and interleukin 1 beta (IL-1β), play an important role in the control of mycobacterial infection (Mayer-Barber & Sher, 2015; O'Garra et al., 2013). IL-1β production is tightly controlled by a signal that can induce the expression of pro-IL-1β and a signal that results in the maturation of IL-1β via inflammasome-mediated cleavage by caspase-1 (Mariathasan & Monack, 2007). During in vivo *M. tuberculosis*

infection are represented as log2 fold change values. Groups of genes represented encode: (A) transcription factors; (B) pro-inflammatory cytokines and chemokines; (C) matrix metalloproteinases and inhibitor; (D) apolipoproteins; (E) complement components; (F) miscellaneous late-phase responses; (G) negative regulators; (H) NADPH oxidase. Statistically significant data points (adjusted *P*-value $< 0.1$) are indicated below the *x*-axis as the shape filled with the color of the specific gene (see legend for gene shape and color). Gene symbols are according to ZFIN nomenclature, except for the following genes that are here identified by their ENSEMBL IDs: *c3a.1*, ENSDARG0000012694; *c3a.3*, ENSDARG00000052207; *c7 (2 of 2)*, ENSDARG00000057121. Only genes are shown for which differential expression was significant in at least one time point. (See color plate) infection in mice, it was shown that IL-1β is essential for MyD88-dependent host resistance (Mayer-Barber et al., 2010). In zebrafish, the presence of a MyD88-dependent signaling route leading to *il1b* induction has been demonstrated during *M. marinum* infection (van der Vaart et al., 2013). The zebrafish genome contains two homologs of the TNF-α gene (named *tnfa* and *tnfb*) (Kinoshita, Biswas, Kono, Hikima, & Sakai, 2014) and it was shown that a strict balance of TNF-α signaling is essential for the control of *M. marinum* infection in zebrafish embryos (Tobin et al., 2010). During the *M. marinum* infection, upregulation of *il1b* and both *tnf-α* genes can be observed as well as upregulation of two chemokine genes, *cxcl8a* (also referred to as *il8*) and *cxcl18b* (previously referred to as *cxcl-c1c*). Similar to its human counterpart, zebrafish *cxcl8a* has been shown to mediate neutrophil recruitment (Deng et al., 2013; de Oliveira et al., 2013) and *cxcl18b* is an IL8-related chemokine that is also induced during *S. typhimurium* infection (Stockhammer, Zakrzewska, Hegedus, Spaink, & Meijer, 2009). Analysis of the *M. marinum* infection time course shows that there are two apparent waves of significant upregulation of pro-inflammatory cytokines, the first at 4 hpi when *il1b*, *tnfb* (tumor necrosis factor alpha b), *cxcl8a*, and *cxcl18b* peak and the second at 3–5 dpi when *cxcl8a* and *cxcl18b* are first upregulated at 3 dpi followed by the additional upregulation of *il1b* at 4 dpi and *tnfa* (tumor necrosis factor alpha a) and *tnfb* at 4 dpi onward (Fig. 2B). The transient peak of pro-inflammatory cytokine gene induction during the early phase of *M. marinum* infection was found to depend on scavenger receptor Marco (macrophage receptor with collagenous structure), which is required for the rapid phagocytosis of *M. marinum* and is thought to function as a co-receptor of TLR2 (Benard et al., 2014).

## 9. INDUCTION OF MATRIX METALLOPROTEINASE GENES

Matrix metalloproteinases (MMPs) are capable of degrading components of the extracellular matrix (Sternlicht & Werb, 2001), thereby allowing immune cells to migrate to sites of inflammation from the bloodstream. In addition, MMPs are capable of cleaving cytokines and chemokines to augment or to inhibit their functional activity (Ito et al., 1996; Schonbeck, Mach, & Libby, 1998). In our infection model, *mmp9* and *mmp13a* are significantly upregulated at most time points during the course of infection (Fig. 3C). A similar pattern can be observed for the expression of *tissue inhibitor of metalloproteinase 2b* (*timp2b*), presumably functioning to inhibit the strongly induced MMPs to avoid excess tissue damage that can occur due to an imbalance in MMP production during mycobacterial infection (Chang et al., 1996; Hrabec, Strek, Zieba, Kwiatkowska, & Hrabec, 2002; Izzo, Izzo, Kasimos, & Majka, 2004; Price et al., 2001; Price, Gilman, Uddin, Recavarren & Friedland, 2003; Quiding-Jarbrink, Smith, & Bancroft, 2001; Rivera-Marrero, Schuyler, Roser, & Roman, 2000). TIMPs can bind in a noncovalent manner to the catalytic domain of MMPs, thus regulating their activity (Brew, Dinakarpandian, & Nagase, 2000). Other metalloproteinases, such as members of the ADAM (a disintegrin and metalloproteinase) family (*adam8a, adam8b, adamts1, adamts8a*), are only upregulated during the late phase of infection (data not shown). Macrophages do not need to migrate toward the infection site in this infection model because *M. marinum* is injected directly into the blood circulation where they are phagocytosed by macrophages. However, at later stages the infected immune cells do migrate outside the vascular system by traversing the endothelial barrier and migrating into surrounding tissues (Fig. 1C–E) (Clay et al., 2007) which requires proteolysis of the basement membrane by metalloproteinases. At 2 dpi and 3 dpi, when uninfected macrophages (Fig. 1M) and neutrophils (Fig. 1N) are first attracted to the developing granulomas (which requires cytokine dependent leukocyte attraction and extracellular matrix degradation), *mmp9* and *mmp13a* (Fig. 3C) are once again upregulated. This induction increases over time until 5 dpi, the time point when the granuloma structures start protruding the epithelial layer and infected macrophages can leave the granuloma to form new granulomas (Davis & Ramakrishnan, 2009). The increasing expression of *mmp9* during this period is consistent with its proposed function in facilitating the recruitment of macrophages during granuloma formation (Volkman et al., 2010).

## 10. COLLECTIVE AND STRONGLY FLUCTUATING REGULATION OF LIPID METABOLISM AND TRANSPORT GENES

When mycobacteria are intracellular, they can release mycobacterial lipids that are incorporated into exocytosing vesicles of macrophages (Beatty, Ullrich, & Russell, 2001). This allows the mycobacterial lipids to be transferred between cells and this process is mediated by host apolipoprotein E (van den Elzen et al., 2005). By transferring antigenic lipids from cell to cell, such as cord factor trehalose 6,6′-dimycolate (TDM), the pathogen can extend its area of influence beyond its own host cell (Geisel, Sakamoto, Russell, & Rhoades, 2005). Such lipid transfer has been found to attract cells to form a granuloma-like structure in a myd88-dependent manner in mice (Geisel et al., 2005), similar to the effect that secreted virulence factor ESAT-6 has on enhancing recruitment of macrophages in a Mmp9-dependent manner in zebrafish (Volkman et al., 2010). When studying the response of apolipoprotein genes during *M. marinum* infection, a collective regulation of a group of eight genes related to lipid metabolism and transport is observed, including the gene coding for apolipoprotein E (*apoea*). This group of seven apolipoprotein genes and a fatty acid binding protein gene (*fabp1b.1*) is first upregulated at 2 hpi and subsequently downregulated at 4 hpi (Fig. 3D), exactly coinciding with the transient peak in pro-inflammatory cytokine induction (Fig. 3B). After the downregulation at 4 hpi, the gene group is notably upregulated again between 8 hpi and 2 dpi, and subsequently shows another trend toward downregulation at 4 dpi, but only significant for some genes in the group. The way these genes are regulated is quite remarkable because it is the only group of genes, besides complement genes, that are upregulated during the mid-phase of infection.

## 11. EARLY INDUCTION OF COMPLEMENT COMPONENT GENES

Numerous complement components in zebrafish are transferred from mother to egg as proteins or mRNA (Wang, Zhang, Tong, Li, & Wang, 2009), thus protecting the embryo before it is capable of synthesizing immune-relevant molecules. This is essential because complement can be immediately activated in response to an invading pathogen. Here, we have observed that the expression levels of complement component genes such as *c3a.1*, *c7*, *c9*, and *cfb* are rapidly upregulated by *M. marinum* infection at 2 hpi (Fig. 3E). This indicates that already at 1 dpf, the zebrafish embryo itself is capable of inducing complement components in response to *M. marinum* infection. It has been shown that complement component C3 on the surface of *M. tuberculosis* mediates phagocytosis of this bacterium (Schlesinger, Bellinger-Kawahara, Payne, & Horwitz, 1990) and recently all the homologs of the mammalian complement fundamental components were found to be present in zebrafish as revealed by phylogenetic analysis (Zhang & Cui, 2014). Therefore, it is possible that zebrafish C3a.1 and C3a.3 function in the same manner to assist phagocytosis of *M. marinum*, a process which occurs progressively during granuloma maturation. In the subsequent stages of infection different complement components continue to show upregulated expression. Thus, this is one of the few upregulated gene groups during the mid-phase, similar to the apolipoprotein gene group. However, in contrast to the apolipoprotein genes, expression of complement genes strongly increased during the late phase of infection (4–5 dpi) (Fig. 3E).

## 12. LATE-PHASE RESPONSE GENES

A miscellaneous group of genes that were previously shown to be important during infection were also identified to be significantly regulated during *M. marinum* infection (Fig. 3F). The following genes belong to this group: the genes for mitochondria-localizing enzymes encoded by *immunoresponsive gene 1-like* (*irg1l*) and *immunoresponsive gene 1* (*irg1*), which were shown to augment bactericidal activity of macrophage-lineage cells by regulating β-oxidation-dependent mitochondrial ROS production (Hall et al., 2013); the acute phase response gene *serum amyloid A* (*saa*), an apolipoprotein family member highly expressed in response to inflammation and tissue injury (Urieli-Shoval, Linke, & Matzner, 2000); the metalloreductase gene *six transmembrane epithelial antigen of prostate 4* (*steap4*) (also known as *tumor necrosis factor alpha-induced protein 9* (*tnfaip9*) or *six-transmembrane protein of prostate 2* (*stamp2*)), which was shown to play a role in the integration of inflammatory and metabolic responses and previously linked with *M. marinum* expression in zebrafish (Hegedus et al., 2009; Stockhammer et al., 2010; Wellen et al., 2007); the *hepcidin antimicrobial peptide 2* (*hamp*) gene involved in microbial killing (Zhang, Yu, Li, & Sun, 2014), and the autophagy modulator gene *DNA-damage regulated autophagy modulator 1* (*dram1*), which was recently shown to protect against mycobacterial infection in zebrafish (van der Vaart et al., 2014). All these genes are strongly upregulated during the late phase of infection at 4 and 5 dpi, while *irg1* and *irg1l* additionally show up- or downregulated expression at earlier time points (Fig. 3F). Concomitant with the upregulation of this miscellaneous group of genes during the late phase of infection, three other groups of genes are upregulated, namely a group of prostaglandin genes (*ptgis*, *ptgs1*, *ptgs2a*, *ptgs2b*), a group of genes encoding various negative regulators of the immune response (Fig. 3G), and a group of NADPH oxidase complex genes (Fig. 3H). Susceptibility to *M. marinum* in zebrafish can result from either inadequate or excessive acute inflammation (Kanwal et al., 2013; Tobin et al., 2012); therefore it is not surprising that concomitant with the induction of pro-inflammatory genes and transcriptional activators, numerous negative regulators of the TLR pathway (*irak3*, *socs3a*, and a group of NFκB inhibitors) are also upregulated during early- and late-phase infection (Fig. 3G). A similar induction of negative regulators occurs during *S. typhimurium* infection (Kanwal et al., 2013; Stockhammer et al., 2009). During granuloma maturation, infected cells containing high numbers of *M. marinum* undergo cell death and are phagocytosed by multiple newly recruited macrophages (Davis & Ramakrishnan, 2009). This process seems beneficial for the mycobacteria because the granuloma expands in this manner. However, dead macrophages also attract neutrophils which are capable of exerting antimicrobial functions against *M. marinum* via inducible iNOS signaling (Elks et al., 2013) and can phagocytose dying, infected macrophages after which they can kill internalized mycobacteria through NADPH-oxidase dependent mechanisms (Yang et al., 2012). Neutrophils containing *M. marinum* can be observed as soon as 1 dpi around the injection site (Fig. 1B) and at 3 dpi within the granuloma-like structures (Fig. 1R). In addition, genes of the phagocyte NADPH oxidase complex are induced significantly at 4 and 5 dpi (Fig. 3H), which correlates with the infection stage where infected macrophages start dying and increasing numbers of neutrophils are attracted to the maturing granuloma (Fig. 1S and T). These observations support the hypothesis that neutrophils play an important role during this stage of infection when macrophages are no longer capable of containing mycobacterial growth.

## 13. HOST-SPECIFIC DETERMINANTS OF THE INNATE IMMUNE RESPONSE

Disruption of TLR/IL1R signaling in *myd88* mutant zebrafish leads to an increased susceptibility to *M. marinum* and other infections (van der Vaart et al., 2013). RNA-Seq analysis has shown that the immunodeficiency of *myd88* mutants is associated with a strongly reduced expression of most of the genes that display an upregulated response during infection in wildtype zebrafish. This includes the pro-inflammatory marker genes that are characteristic for the late phase of infection, such as *il1b* and *mmp9*. In contrast, these pro-inflammatory genes and many other *M. marinum*-induced genes show a hyperinduced expression in the absence of the host factor Ptpn6, a protein tyrosine phosphatase that is required for feedback control of the innate immune response (Kanwal et al., 2013). When *ptpn6* is knocked down, zebrafish embryos are more susceptible to infection similar to *myd88* mutants. These studies have revealed that increased susceptibility to *M. marinum* infection can have different underlying causes, which are reflected in the transcriptome signature either by a hypoinduction or by a hyperinduction of the innate immune response. TLR2 is one of the receptors functioning upstream of the MYD88-dependent response to *M. tuberculosis* infection, mediating the recognition of mycobacterial lipoproteins (Drage et al., 2009). In zebrafish embryos, the TLR2 ligand Pam3CSK4 induces a gene set distinct from that induced by the TLR5 ligand flagellin. The regulation of these Pam3CSK4- and flagellin-specific gene sets is inhibited by knockdown of *tlr2* or *tlr5*, respectively, and the Pam3CSK4 gene set includes several transcription factor genes, including *cebpb*, *fosb*, *nr4a1*, and *egr3* (Yang, Marin-Juez, Meijer, & Spaink, 2015).

## 14. COMMON AND SPECIFIC PATHOGEN-DEPENDENT INNATE IMMUNE RESPONSES

To determine specific and common components of the innate immune response to various pathogens, a comparison can be made between the RNA-Seq profiles of two well-characterized infection processes in zebrafish embryos: the acute *S. typhimurium* infection (Ordas et al., 2013) and the chronically progressive *M. marinum* infection (Fig. 2). This comparison demonstrates that the host response to *M. marinum* at 8 hpi is minimal, in sharp contrast with the strong pro-inflammatory response of zebrafish embryos to *S. typhimurium* infection at the same time point (Fig. 4). This conclusion is also supported by the comparison of microarray data of *M. marinum*, *S. typhimurium* and *Edwardsiella tarda* infections where *E. tarda* causes an acute response similar to that caused by *S. typhimurium* (van der Vaart, Spaink, & Meijer, 2012). However, the response to *M. marinum* at 5 dpi shows a much stronger inflammatory signature than the response to *S. typhimurium* at 8 hpi (Fig. 4). Genes belonging to the gene ontology group called "response to bacteria" can be seen as a common component of the innate immune response because they are differentially regulated during both *M. marinum* and *S. typhimurium* infection. Interestingly, both infections lead to their own specific responses, for example *S. typhimurium* shows upregulation of genes related to aminoacyl-tRNA biosynthesis, amino acid and nucleotide metabolism, and downregulation of genes related to glycolysis and spliceosome. *M. marinum* infection also induces a specific response at 5 dpi which is represented by upregulation of genes related to erythrocyte

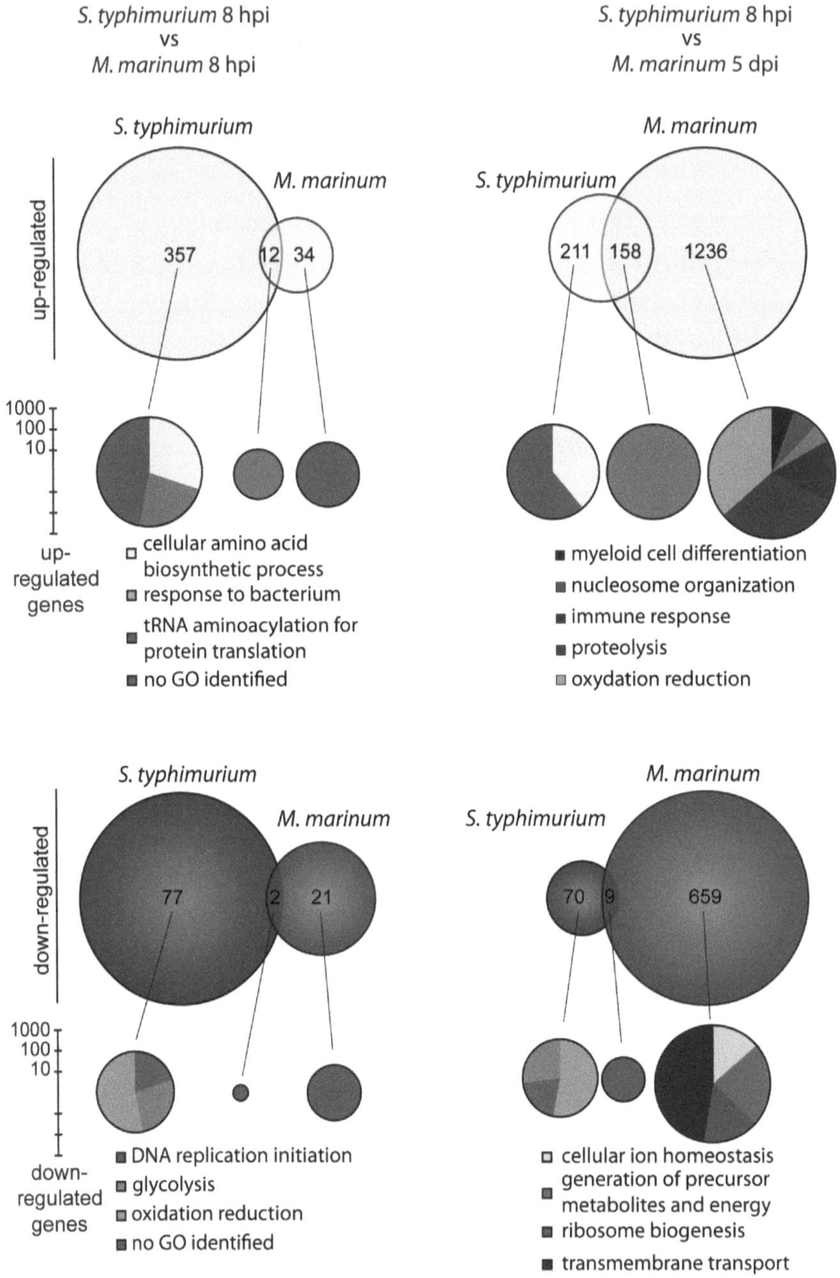


**Figure 4** ***Overview of total number of genes differentially regulated in* S. typhimurium- *and* M. marinum-*infected embryos and their overlap between various time points and gene ontology terms enriched per time point.*** RNA-Seq data from *M. marinum* infection

homeostasis, tight junction, and cell adhesion molecules, while other adhesion-related pathways, such as ECM–receptor interaction and focal adhesion, are downregulated. A striking difference between the two responses is that no effect on lipid metabolism or transport was observed in response to *S. typhimurium* infection, while a group of apolipoprotein genes showed a common and fluctuating regulation during *M. marinum* infection (Fig. 3D). It is not yet known how these coordinately regulated apolipoprotein genes function during the infection process or whether *M. marinum* might be capable of modulating cell function via apolipoproteins. While no differential expression of apolipoprotein genes has been observed in wildtype zebrafish embryos infected with *S. typhimurium*, six members of this gene family (*apoa1a*, *apoa4*, *apoba*, *apobb*, *apoc1l*, and *apoeb*) were found to be infection inducible under knockdown conditions of miR-146a, a microRNA that is highly conserved between zebrafish and human (Ordas et al., 2013). miR-146a is known as a negative regulator of TLR signaling, targeting the downstream signaling intermediates IRAK1 and TRAF6 (Taganov, Boldin, Chang, & Baltimore, 2006). In addition, miR-146a has been implicated in the control of lipid accumulation (Yang et al., 2011). Changes in lipoprotein composition are common to the acute phase response and might protect the host against infection and inflammation (Khovidhunkit et al., 2004). Alternatively, changes in lipid metabolism in the host might alter the bacterial microenvironment to the advantage of the pathogen (Lovewell, Sassetti, & VanderVen, 2015). A better understanding of lipid metabolism and trafficking processes is an important challenge for identifying new potential therapeutic targets for tuberculosis and other infectious diseases (Lovewell et al., 2015).

(Fig. 2) was compared to RNA-Seq data from *S. typhimurium*-infected AB/TL embryos (Ordas et al., 2013) that has been re-analyzed with genome assembly ZV9, annotation version 71. Area proportionate Venn diagrams are shown representing significantly upregulated (yellow) and downregulated (blue) genes during *S. typhimurium* and *M. marinum* infection in zebrafish embryos. The *S. typhimurium* 8 hpi time point was compared to *M. marinum* 8 hpi (left) and 5 dpi (right). Gene ontology terms identified per time point per comparison are represented as pie charts below the Venn diagrams. Colors in the pie charts illustrate the relative proportion of each identified gene ontology term. Genes without associated gene ontology term are not represented in the pie charts (except for time points with no gene ontologies identified). Gene ontology analysis was performed using DAVID v6.7 (http://david.abcc.ncifcrf.gov/home.jsp) (Huang da et al., 2009). Gene ontology terms are considered significant when the false-discovery rate was below 5%. (See color plate)

## 15. CONCLUDING REMARKS

The zebrafish *M. marinum* infection model has provided significant new insight into the development and function of granuloma-like structures, which are the hallmark of infection with pathogenic mycobacteria, including the human pathogen *M. tuberculosis* (Cronan & Tobin, 2014; van Leeuwen et al., 2015; Meijer, 2015; Ramakrishnan, 2013). Despite the widespread use of this model, the innate immune response of zebrafish embryos to *M. marinum* infection, especially during the early stages of infection, had not yet been well characterized. Previous microarray data had

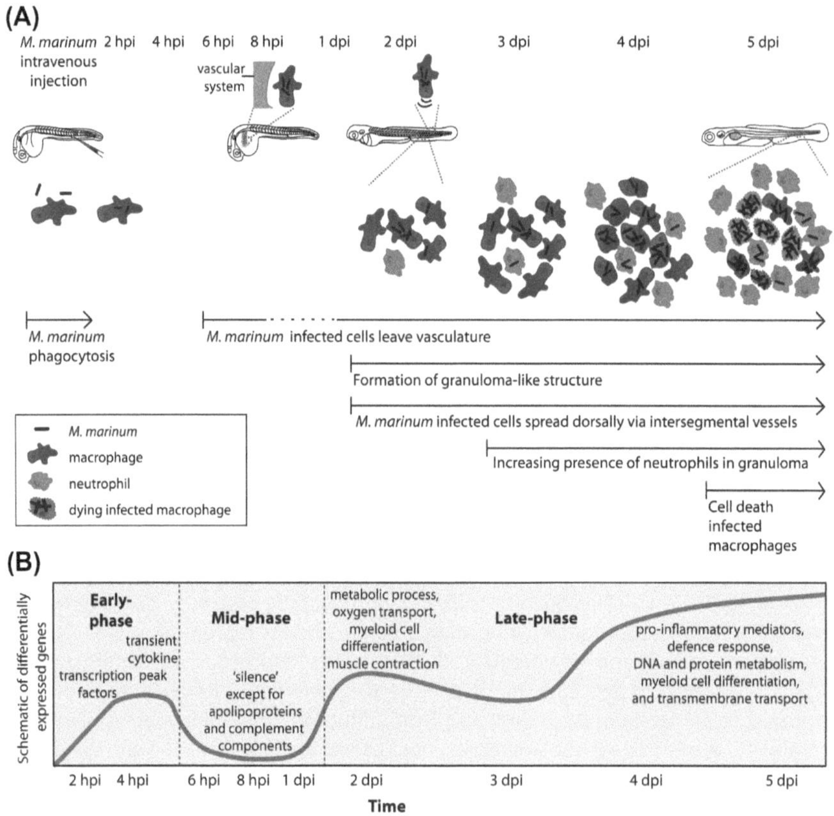


**Figure 5** ***Summary of the infection process of*** **M. marinum** ***in zebrafish embryos.*** A schematic overview representing (A) host–pathogen interactions and (B) the three distinct phases observed in gene expression profile responses (*orange line* represents total genes up- and downregulated during *M. marinum* infection). (See color plate)

shown that early after infection (8 hpi), the host response to *M. marinum* is minimal, in sharp contrast with the strong pro-inflammatory response of zebrafish embryos to other pathogens like *E. tarda* and *S. typhimurium*, which cause acute disease (van der Vaart et al., 2012). As reported here, detection of more subtle changes in gene expression during *M. marinum* infection requires RNA deep sequencing. The RNA-Seq analysis presented here over a time course from 2 hpi up to 5 dpi visualization of the host–pathogen interactions throughout the infection process until granuloma formation and the interrelation of the infection stages with the host gene expression profiles. As summarized in Fig. 5, the gene expression profiles reveal three main phases in the host response: early phase: induction of complement and transcription factors followed by induction of pro-inflammatory cytokines within hours after phagocytosis of *M. marinum*; mid-phase: minimal response between 6 hpi to 1 dpi when the tissue dissemination of *M. marinum* has begun, except for the induction of complement and apolipoprotein genes; late phase: progressively increasing induction of complement, transcription factors, pro-inflammatory cytokines, matrix metalloproteinases, and other defense- and inflammation-related gene groups, along with the expansion of granulomas. In future work it will be important to extend these whole embryo/larvae transcriptome profiling studies toward analysis of the transcriptome changes that specifically occur in the infected macrophages or neutrophils. This can be achieved by fluorescence-activated cell sorting using the growing collection zebrafish reporter lines for different leukocyte subtypes (Rougeot et al., 2014).

## ACKNOWLEDGMENTS

The authors would like to thank Georges Lutfalla (University Montpellier 2) for the kind gift of *Tg(mpeg1.1::mCherry-F)* fish, all members of the aquarium facility for fish care, and other lab members for helpful discussions. EB, HPS, and AHM were supported by the Smart Mix Program of the Netherlands Ministry of Economic Affairs and the Ministry of Education, Culture and Science, JR by a Marie Curie Fellowship of the European Seventh Framework Initial Training Network FishForPharma (PITG-GA-2011-289209), and PIR by a Marie Curie Intra-European Fellowship (PIEF-GA-2010-274389).

## REFERENCES

Akira, S., Isshiki, H., Sugita, T., Tanabe, O., Kinoshita, S., Nishio, Y., … Kishimoto, T. (1990). A nuclear factor for IL-6 expression (NF-IL6) is a member of a C/EBP family. *The EMBO Journal, 9*(6), 1897–1906.

Alam, T., An, M. R., & Papaconstantinou, J. (1992). Differential expression of three C/EBP isoforms in multiple tissues during the acute phase response. *The Journal of Biological Chemistry, 267*(8), 5021–5024.

Anders, S., & Huber, W. (2010). Differential expression analysis for sequence count data. *Genome Biology, 11*(10), R106.

Beatty, W. L., Ullrich, H. J., & Russell, D. G. (2001). Mycobacterial surface moieties are released from infected macrophages by a constitutive exocytic event. *European Journal of Cell Biology, 80*(1), 31–40.

Belele, C. L., English, M. A., Chahal, J., Burnetti, A., Finckbeiner, S. M., Gibney, G., … Liu, P. P. (2009). Differential requirement for Gata1 DNA binding and transactivation between primitive and definitive stages of hematopoiesis in zebrafish. *Blood, 114*(25), 5162–5172.

Benard, E. L., Racz, P. I., Rougeot, J., Nezhinsky, A. E., Verbeek, F. J., Spaink, H. P., & Meijer, A. H. (2015). Macrophage-expressed perforins mpeg1 and mpeg1.2 have an anti-bacterial function in zebrafish. *Journal of Innate Immunity, 7*(2), 136–152.

Benard, E. L., Roobol, S. J., Spaink, H. P., & Meijer, A. H. (2014). Phagocytosis of mycobacteria by zebrafish macrophages is dependent on the scavenger receptor Marco, a key control factor of pro-inflammatory signalling. *Developmental and Comparative Immunology, 47*(2), 223–233.

Bernut, A., Herrmann, J. L., Kissa, K., Dubremetz, J. F., Gaillard, J. L., Lutfalla, G., & Kremer, L. (2014). *Mycobacterium abscessus* cording prevents phagocytosis and promotes abscess formation. *Proceedings of the National Academy of Sciences of the United States of America, 111*(10), E943–E952.

Bertrand, J. Y., Kim, A. D., Violette, E. P., Stachura, D. L., Cisson, J. L., & Traver, D. (2007). Definitive hematopoiesis initiates through a committed erythromyeloid progenitor in the zebrafish embryo. *Development (Cambridge, England), 134*(23), 4147–4156.

Boggaram, V., Gottipati, K. R., Wang, X., & Samten, B. (2013). Early secreted antigenic target of 6 kDa (ESAT-6) protein of *Mycobacterium tuberculosis* induces interleukin-8 (IL-8) expression in lung epithelial cells via protein kinase signaling and reactive oxygen species. *The Journal of Biological Chemistry, 288*(35), 25500–25511.

Boshoff, H. I., & Barry, C. E., 3rd (2005). Tuberculosis – metabolism and respiration in the absence of growth (Review) *Nature Reviews. Microbiology, 3*(1), 70–80.

Brew, K., Dinakarpandian, D., & Nagase, H. (2000). Tissue inhibitors of metalloproteinases: evolution, structure and function. *Biochimica et Biophysica Acta, 1477*(1–2), 267–283.

Chan, J., Xing, Y., Magliozzo, R. S., & Bloom, B. R. (1992). Killing of virulent *Mycobacterium tuberculosis* by reactive nitrogen intermediates produced by activated murine macrophages. *The Journal of Experimental Medicine, 175*(4), 1111–1122.

Chang, J. C., Wysocki, A., Tchou-Wong, K. M., Moskowitz, N., Zhang, Y., & Rom, W. N. (1996). Effect of *Mycobacterium tuberculosis* and its components on macrophages and the release of matrix metalloproteinases. *Thorax, 51*(3), 306–311.

Chen, B. P., Liang, G., Whelan, J., & Hai, T. (1994). ATF3 and ATF3 delta Zip. Transcriptional repression versus activation by alternatively spliced isoforms. *The Journal of Biological Chemistry, 269*(22), 15819–15826.

Chiang, C. Y., Centis, R., & Migliori, G. B. (2010). Drug-resistant tuberculosis: past, present, future (Review) *Respirology: Official Journal of the Asian Pacific Society of Respirology (Carlton, VIC), 15*(3), 413–432.

Clay, H., Davis, J. M., Beery, D., Huttenlocher, A., Lyons, S. E., & Ramakrishnan, L. (2007). Dichotomous role of the macrophage in early *Mycobacterium marinum* infection of the zebrafish. *Cell Host & Microbe, 2*(1), 29–39.

Clay, H., Volkman, H. E., & Ramakrishnan, L. (2008). Tumor necrosis factor signaling mediates resistance to mycobacteria by inhibiting bacterial growth and macrophage death. *Immunity, 29*(2), 283–294.

Cooper, A. M., Pearl, J. E., Brooks, J. V., Ehlers, S., & Orme, I. M. (2000). Expression of the nitric oxide synthase 2 gene is not essential for early control of *Mycobacterium tuberculosis* in the murine lung. *Infection and Immunity, 68*(12), 6879–6882.

Cooper, A. M., & Torrado, E. (2012). Protection versus pathology in tuberculosis: recent insights. *Current Opinion in Immunology, 24*(4), 431–437.

Cronan, M. R., & Tobin, D. M. (2014). Fit for consumption: zebrafish as a model for tuberculosis. *Disease Models & Mechanisms, 7*(7), 777–784.

Davidson, A. J., & Zon, L. I. (2004). The 'definitive' (and 'primitive') guide to zebrafish hematopoiesis. *Oncogene, 23*(43), 7233–7246.

Davis, J. M., Clay, H., Lewis, J. L., Ghori, N., Herbomel, P., & Ramakrishnan, L. (2002). Real-time visualization of mycobacterium-macrophage interactions leading to initiation of granuloma formation in zebrafish embryos. *Immunity, 17*(6), 693–702.

Davis, J. M., & Ramakrishnan, L. (2009). The role of the granuloma in expansion and dissemination of early tuberculous infection. *Cell, 136*(1), 37–49.

Deng, Q., Sarris, M., Bennin, D. A., Green, J. M., Herbomel, P., & Huttenlocher, A. (2013). Localized bacterial infection induces systemic activation of neutrophils through Cxcr2 signaling in zebrafish. *Journal of Leukocyte Biology, 93*(5), 761–769.

Drage, M. G., Pecora, N. D., Hise, A. G., Febbraio, M., Silverstein, R. L., Golenbock, D. T., … Harding, C. V. (2009). TLR2 and its co-receptors determine responses of macrophages and dendritic cells to lipoproteins of *Mycobacterium tuberculosis*. *Cellular Immunology, 258*(1), 29–37.

Elks, P. M., Brizee, S., van der Vaart, M., Walmsley, S. R., van Eeden, F. J., Renshaw, S. A., & Meijer, A. H. (2013). Hypoxia inducible factor signaling modulates susceptibility to mycobacterial infection via a nitric oxide dependent mechanism. *PLoS Pathogens, 9*(12), e1003789.

Elks, P. M., van der Vaart, M., van Hensbergen, V., Schutz, E., Redd, M. J., Murayama, E., … Meijer, A. H. (2014). Mycobacteria counteract a TLR-mediated nitrosative defense mechanism in a zebrafish infection model. *PLoS One, 9*(6), e100928.

Ellett, F., & Lieschke, G. J. (2010). Zebrafish as a model for vertebrate hematopoiesis. *Current Opinion in Pharmacology, 10*(5), 563–570.

van den Elzen, P., Garg, S., Leon, L., Brigl, M., Leadbetter, E. A., Gumperz, J. E., … Brenner, M. B. (2005). Apolipoprotein-mediated pathways of lipid antigen presentation. *Nature, 437*(7060), 906–910.

Fenaroli, F., Westmoreland, D., Benjaminsen, J., Kolstad, T., Skjeldal, F. M., Meijer, A. H., … Griffiths, G. (2014). Nanoparticles as drug delivery system against tuberculosis in zebrafish embryos: direct visualization and treatment. *ACS Nano, 8*(7), 7014–7026.

Geisel, R. E., Sakamoto, K., Russell, D. G., & Rhoades, E. R. (2005). In vivo activity of released cell wall lipids of *Mycobacterium bovis* bacillus Calmette-Guerin is due principally to trehalose mycolates. *Journal of Immunology, 174*(8), 5007–5015.

Gilchrist, M., Thorsson, V., Li, B., Rust, A. G., Korb, M., Roach, J. C., … Aderem, A. (2006). Systems biology approaches identify ATF3 as a negative regulator of Toll-like receptor 4. *Nature, 441*(7090), 173–178.

Goody, M. F., Sullivan, C., & Kim, C. H. (2014). Studying the immune response to human viral infections using zebrafish (Review) *Developmental and Comparative Immunology, 46*(1), 84–95.

Gratacap, R. L., & Wheeler, R. T. (2014). Utilization of zebrafish for intravital study of eukaryotic pathogen-host interactions. *Developmental and Comparative Immunology, 46*(1), 108–115.

Gray, C., Loynes, C. A., Whyte, M. K., Crossman, D. C., Renshaw, S. A., & Chico, T. J. (2011). Simultaneous intravital imaging of macrophage and neutrophil behaviour during inflammation using a novel transgenic zebrafish. *Thrombosis and Haemostasis, 105*(5), 811–819.

Green, J. A., Elkington, P. T., Pennington, C. J., Roncaroli, F., Dholakia, S., Moores, R. C., … Friedland, J. S. (2010). *Mycobacterium tuberculosis* upregulates microglial matrix metalloproteinase-1 and -3 expression and secretion via NF-kappaB- and

activator protein-1-dependent monocyte networks. *Journal of Immunology, 184*(11), 6492–6503.
Hall, C. J., Boyle, R. H., Astin, J. W., Flores, M. V., Oehlers, S. H., Sanderson, L. E., … Crosier, P. S. (2013). Immunoresponsive gene 1 augments bactericidal activity of macrophage-lineage cells by regulating beta-oxidation-dependent mitochondrial ROS production. *Cell Metabolism, 18*(2), 265–278.
Hammaren, M. M., Oksanen, K. E., Nisula, H. M., Luukinen, B. V., Pesu, M., Ramet, M., & Parikka, M. (2014). Adequate Th2-type response associates with restricted bacterial growth in latent mycobacterial infection of zebrafish. *PLoS Pathogens, 10*(6), e1004190.
Hawn, T. R., Matheson, A. I., Maley, S. N., & Vandal, O. (2013). Host-directed therapeutics for tuberculosis: can we harness the host? *Microbiology and Molecular Biology Reviews: MMBR,* 77(4), 608–627.
Hegedus, Z., Zakrzewska, A., Agoston, V. C., Ordas, A., Racz, P., Mink, M., … Meijer, A. H. (2009). Deep sequencing of the zebrafish transcriptome response to *Mycobacterium* infection. *Molecular Immunology, 46*(15), 2918–2930.
Herbomel, P., Thisse, B., & Thisse, C. (1999). Ontogeny and behaviour of early macrophages in the zebrafish embryo. *Development (Cambridge, England), 126*(17), 3735–3745.
Hoetzenecker, W., Echtenacher, B., Guenova, E., Hoetzenecker, K., Woelbing, F., Bruck, J., … Rocken, M. (2012). ROS-induced ATF3 causes susceptibility to secondary infections during sepsis-associated immunosuppression. *Nature Medicine, 18*(1), 128–134.
Hrabec, E., Strek, M., Zieba, M., Kwiatkowska, S., & Hrabec, Z. (2002). Circulation level of matrix metalloproteinase-9 is correlated with disease severity in tuberculosis patients. *The International Journal of Tuberculosis and Lung Disease: The Official Journal of the International Union Against Tuberculosis and Lung Disease, 6*(8), 713–719.
Huang da, W., Sherman, B. T., & Lempicki, R. A. (2009). Systematic and integrative analysis of large gene lists using DAVID bioinformatics resources (Research Support, N.I.H., Extramural) *Nature Protocols, 4*(1), 44–57.
Ito, A., Mukaiyama, A., Itoh, Y., Nagase, H., Thogersen, I. B., Enghild, J. J., … Mori, Y. (1996). Degradation of interleukin 1beta by matrix metalloproteinases. *The Journal of Biological Chemistry, 271*(25), 14657–14660.
Izzo, A. A., Izzo, L. S., Kasimos, J., & Majka, S. (2004). A matrix metalloproteinase inhibitor promotes granuloma formation during the early phase of *Mycobacterium tuberculosis* pulmonary infection. *Tuberculosis, 84*(6), 387–396.
Kamijo, R., Harada, H., Matsuyama, T., Bosland, M., Gerecitano, J., Shapiro, D., et al. (1994). Requirement for transcription factor IRF-1 in NO synthase induction in macrophages. *Science, 263*(5153), 1612–1615.
Kanwal, Z., Zakrzewska, A., den Hertog, J., Spaink, H. P., Schaaf, M. J., & Meijer, A. H. (2013). Deficiency in hematopoietic phosphatase ptpn6/Shp1 hyperactivates the innate immune system and impairs control of bacterial infections in zebrafish embryos. *The Journal of Immunology: Official Journal of the American Association of Immunologists, 190*(4), 1631–1645.
Khovidhunkit, W., Kim, M. S., Memon, R. A., Shigenaga, J. K., Moser, A. H., Feingold, K. R., & Grunfeld, C. (2004). Effects of infection and inflammation on lipid and lipoprotein metabolism: mechanisms and consequences to the host. *Journal of Lipid Research, 45*(7), 1169–1196.
Kinoshita, S., Akira, S., & Kishimoto, T. (1992). A member of the C/EBP family, NF-IL6 beta, forms a heterodimer and transcriptionally synergizes with NF-IL6. *Proceedings of the National Academy of Sciences of the United States of America, 89*(4), 1473–1476.
Kinoshita, S., Biswas, G., Kono, T., Hikima, J., & Sakai, M. (2014). Presence of two tumor necrosis factor (tnf)-alpha homologs on different chromosomes of zebrafish (*Danio rerio*) and medaka (*Oryzias latipes*). *Marine Genomics, 13*, 1–9.

Langenau, D. M. (2004). In vivo tracking of T cell development, ablation, and engraftment in transgenic zebrafish. *Proceedings of the National Academy of Sciences of the United States of America, 101*(19), 7369–7374.

Langenau, D. M., Traver, D., Ferrando, A. A., Kutok, J. L., Aster, J. C., Kanki, J. P., … Look, A. T. (2003). Myc-induced T cell leukemia in transgenic zebrafish. *Science, 299*(5608), 887–890.

Lawson, N. D., & Weinstein, B. M. (2002). In vivo imaging of embryonic vascular development using transgenic zebrafish. *Developmental Biology, 248*(2), 307–318.

Le Guyader, D., Redd, M. J., Colucci-Guyon, E., Murayama, E., Kissa, K., Briolat, V., … Herbomel, P. (2008). Origins and unconventional behavior of neutrophils in developing zebrafish. *Blood, 111*(1), 132–141.

van Leeuwen, L. M., van der Sar, A. M., & Bitter, W. (2015). Animal models of tuberculosis: zebrafish (Review) *Cold Spring Harbor Perspectives in Medicine, 5*(3), a018580.

Levraud, J. P., Palha, N., Langevin, C., & Boudinot, P. (2014). Through the looking glass: witnessing host-virus interplay in zebrafish. *Trends in Microbiology, 22*(9), 490–497.

Lovewell, R. R., Sassetti, C. M., & VanderVen, B. C. (2015). Chewing the fat: lipid metabolism and homeostasis during *M. tuberculosis* infection (Review) *Current Opinion in Microbiology, 29*, 30–36.

Malone, J. H., & Oliver, B. (2011). Microarrays, deep sequencing and the true measure of the transcriptome. *BMC Biology, 9*, 34.

Mariathasan, S., & Monack, D. M. (2007). Inflammasome adaptors and sensors: intracellular regulators of infection and inflammation (Review) *Nature Reviews. Immunology, 7*(1), 31–40.

Mayer-Barber, K. D., Barber, D. L., Shenderov, K., White, S. D., Wilson, M. S., Cheever, A., … Sher, A. (2010). Caspase-1 independent IL-1beta production is critical for host resistance to *Mycobacterium tuberculosis* and does not require TLR signaling in vivo. *Journal of Immunology, 184*(7), 3326–3330.

Mayer-Barber, K. D., & Sher, A. (2015). Cytokine and lipid mediator networks in tuberculosis. *Immunological Reviews, 264*(1), 264–275.

Meijer, A. H. (2015). Protection and pathology in TB: learning from the zebrafish model. *Seminars in Immunopathology, 38*(2).

Meijer, A. H., Verbeek, F. J., Salas-Vidal, E., Corredor-Adamez, M., Bussman, J., van der Sar, A. M., … Spaink, H. P. (2005). Transcriptome profiling of adult zebrafish at the late stage of chronic tuberculosis due to *Mycobacterium marinum* infection. *Molecular Immunology, 42*(10), 1185–1203.

Nguyen-Chi, M., Laplace-Builhe, B., Travnickova, J., Luz-Crawford, P., Tejedor, G., Phan, Q. T., … Djouad, F. (2015). Identification of polarized macrophage subsets in zebrafish. *eLife, 4*, e07288.

Oehlers, S. H., Cronan, M. R., Scott, N. R., Thomas, M. I., Okuda, K. S., Walton, E. M., … Tobin, D. M. (2015). Interception of host angiogenic signalling limits mycobacterial growth. *Nature, 517*(7536), 612–615.

O'Garra, A., Redford, P. S., McNab, F. W., Bloom, C. I., Wilkinson, R. J., & Berry, M. P. (2013). The immune response in tuberculosis. *Annual Review of Immunology, 31*, 475–527.

de Oliveira, S., Reyes-Aldasoro, C. C., Candel, S., Renshaw, S. A., Mulero, V., & Calado, A. (2013). Cxcl8 (IL-8) mediates neutrophil recruitment and behavior in the zebrafish inflammatory response. *Journal of Immunology, 190*(8), 4349–4359.

Ordas, A., Kanwal, Z., Lindenberg, V., Rougeot, J., Mink, M., Spaink, H. P., & Meijer, A. H. (2013). MicroRNA-146 function in the innate immune transcriptome response of zebrafish embryos to *Salmonella typhimurium* infection. *BMC Genomics, 14*, 696.

O'Toole, R. (2010). Experimental models used to study human tuberculosis. *Advances in Applied Microbiology, 71*, 75–89.

Ottenhoff, T. H. (2012). New pathways of protective and pathological host defense to mycobacteria. *Trends in Microbiology, 20*(9), 419–428.

Page, D. M., Wittamer, V., Bertrand, J. Y., Lewis, K. L., Pratt, D. N., Delgado, N., … Traver, D. (2013). An evolutionarily conserved program of B-cell development and activation in zebrafish. *Blood, 122*(8), e1–11.

Palha, N., Guivel-Benhassine, F., Briolat, V., Lutfalla, G., Sourisseau, M., Ellett, F., … Levraud, J. P. (2013). Real-time whole-body visualization of Chikungunya virus infection and host interferon response in zebrafish. *PLoS Pathogens, 9*(9), e1003619.

Parikka, M., Hammaren, M. M., Harjula, S. K., Halfpenny, N. J., Oksanen, K. E., Lahtinen, M. J., … Ramet, M. (2012). *Mycobacterium marinum* causes a latent infection that can be reactivated by gamma irradiation in adult zebrafish. *PLoS Pathogens, 8*(9), e1002944.

Pennini, M. E., Liu, Y., Yang, J., Croniger, C. M., Boom, W. H., & Harding, C. V. (2007). CCAAT/enhancer-binding protein beta and delta binding to CIITA promoters is associated with the inhibition of CIITA expression in response to *Mycobacterium tuberculosis* 19-kDa lipoprotein. *Journal of Immunology, 179*(10), 6910–6918.

Pine, R. (2002). IRF and tuberculosis (Review) *Journal of Interferon & Cytokine Research: The Official Journal of the International Society for Interferon and Cytokine Research, 22*(1), 15–25.

Price, N. M., Farrar, J., Tran, T. T., Nguyen, T. H., Tran, T. H., & Friedland, J. S. (2001). Identification of a matrix-degrading phenotype in human tuberculosis in vitro and in vivo. *Journal of Immunology, 166*(6), 4223–4230.

Price, N. M., Gilman, R. H., Uddin, J., Recavarren, S., & Friedland, J. S. (2003). Unopposed matrix metalloproteinase-9 expression in human tuberculous granuloma and the role of TNF-alpha-dependent monocyte networks. *Journal of Immunology, 171*(10), 5579–5586.

Prouty, M. G., Correa, N. E., Barker, L. P., Jagadeeswaran, P., & Klose, K. E. (2003). Zebrafish-*Mycobacterium marinum* model for mycobacterial pathogenesis. *FEMS Microbiology Letters, 225*(2), 177–182.

Quiding-Jarbrink, M., Smith, D. A., & Bancroft, G. J. (2001). Production of matrix metalloproteinases in response to mycobacterial infection. *Infection and Immunity, 69*(9), 5661–5670.

Ramakrishnan, L. (2012). Revisiting the role of the granuloma in tuberculosis. *Nature Reviews. Immunology, 12*(5), 352–366.

Ramakrishnan, L. (2013). The zebrafish guide to tuberculosis immunity and treatment. *Cold Spring Harbor Symposia on Quantitative Biology, 78*, 179–192.

Renshaw, S. A., Loynes, C. A., Trushell, D. M., Elworthy, S., Ingham, P. W., & Whyte, M. K. (2006). A transgenic zebrafish model of neutrophilic inflammation. *Blood, 108*(13), 3976–3978.

Rhodes, J., Hagen, A., Hsu, K., Deng, M., Liu, T. X., Look, A. T., & Kanki, J. P. (2005). Interplay of Pu.1 and gata1 determines myelo-erythroid progenitor cell fate in zebrafish. *Developmental Cell, 8*(1), 97–108.

Rivera-Marrero, C. A., Schuyler, W., Roser, S., & Roman, J. (2000). Induction of MMP-9 mediated gelatinolytic activity in human monocytic cells by cell wall components of *Mycobacterium tuberculosis*. *Microbial Pathogenesis, 29*(4), 231–244.

Rougeot, J., Zakrzewska, A., Kanwal, Z., Jansen, H. J., Spaink, H. P., & Meijer, A. H. (2014). RNA sequencing of FACS-sorted immune cell populations from zebrafish infection models to identify cell specific responses to intracellular pathogens. *Methods in Molecular Biology, 1197*, 261–274.

Ruley, K. M., Ansede, J. H., Pritchett, C. L., Talaat, A. M., Reimschuessel, R., & Trucksis, M. (2004). Identification of *Mycobacterium marinum* virulence genes using signature-tagged mutagenesis and the goldfish model of mycobacterial pathogenesis. *FEMS Microbiology Letters, 232*(1), 75–81.

Russell, D. G. (2011). *Mycobacterium tuberculosis* and the intimate discourse of a chronic infection. *Immunological Reviews, 240*(1), 252–268.

Sanderson, L. E., Chien, A. T., Astin, J. W., Crosier, K. E., Crosier, P. S., & Hall, C. J. (2015). An inducible transgene reports activation of macrophages in live zebrafish larvae. *Developmental and Comparative Immunology, 53*(1), 63–69.

van der Sar, A. M., Abdallah, A. M., Sparrius, M., Reinders, E., Vandenbroucke-Grauls, C. M. J. E., & Bitter, W. (2004). *Mycobacterium marinum* strains can Be divided into two distinct types based on genetic diversity and virulence. *Infection and Immunity, 72*(11), 6306–6312.

van der Sar, A. M., Spaink, H. P., Zakrzewska, A., Bitter, W., & Meijer, A. H. (2009). Specificity of the zebrafish host transcriptome response to acute and chronic mycobacterial infection and the role of innate and adaptive immune components. *Molecular Immunology, 46*(11–12), 2317–2332.

Schlesinger, L. S., Bellinger-Kawahara, C. G., Payne, N. R., & Horwitz, M. A. (1990). Phagocytosis of *Mycobacterium tuberculosis* is mediated by human monocyte complement receptors and complement component C3. *The Journal of Immunology: Official Journal of the American Association of Immunologists, 144*(7), 2771–2780.

Schonbeck, U., Mach, F., & Libby, P. (1998). Generation of biologically active IL-1 beta by matrix metalloproteinases: a novel caspase-1-independent pathway of IL-1 beta processing. *Journal of Immunology, 161*(7), 3340–3346.

Seimon, T. A., Kim, M. J., Blumenthal, A., Koo, J., Ehrt, S., Wainwright, H., … Russell, D. G. (2010). Induction of ER stress in macrophages of tuberculosis granulomas. *PLoS One, 5*(9), e12772.

Simeone, R., Bobard, A., Lippmann, J., Bitter, W., Majlessi, L., Brosch, R., & Enninga, J. (2012). Phagosomal rupture by *Mycobacterium tuberculosis* results in toxicity and host cell death. *PLoS Pathogens, 8*(2), e1002507.

Simeone, R., Sayes, F., Song, O., Groschel, M. I., Brodin, P., Brosch, R., & Majlessi, L. (2015). Cytosolic access of *Mycobacterium tuberculosis*: critical impact of phagosomal acidification control and demonstration of occurrence in vivo. *PLoS Pathogens, 11*(2), e1004650.

Stamm, L. M., Morisaki, J. H., Gao, L. Y., Jeng, R. L., McDonald, K. L., Roth, R., … Brown, E. J. (2003). *Mycobacterium marinum* escapes from phagosomes and is propelled by actin-based motility. *The Journal of Experimental Medicine, 198*(9), 1361–1368.

Sternlicht, M. D., & Werb, Z. (2001). How matrix metalloproteinases regulate cell behavior. *Annual Review of Cell and Developmental Biology, 17*, 463–516.

Stinear, T. P., Seemann, T., Harrison, P. F., Jenkin, G. A., Davies, J. K., Johnson, P. D., … Cole, S. T. (2008). Insights from the complete genome sequence of *Mycobacterium marinum* on the evolution of *Mycobacterium tuberculosis*. *Genome Research, 18*(5), 729–741.

Stockhammer, O. W., Rauwerda, H., Wittink, F. R., Breit, T. M., Meijer, A. H., & Spaink, H. P. (2010). Transcriptome analysis of Traf6 function in the innate immune response of zebrafish embryos. *Molecular Immunology, 48*(1–3), 179–190.

Stockhammer, O. W., Zakrzewska, A., Hegedus, Z., Spaink, H. P., & Meijer, A. H. (2009). Transcriptome profiling and functional analyses of the zebrafish embryonic innate immune response to *Salmonella* infection. *The Journal of Immunology: Official Journal of the American Association of Immunologists, 182*(9), 5641–5653.

Stoop, E. J., Bitter, W., & van der Sar, A. M. (2012). Tubercle bacilli rely on a type VII army for pathogenicity. *Trends in Microbiology, 20*(10), 477–484.

Swaim, L. E., Connolly, L. E., Volkman, H. E., Humbert, O., Born, D. E., & Ramakrishnan, L. (2006). *Mycobacterium marinum* infection of adult zebrafish causes caseating granulomatous tuberculosis and is moderated by adaptive immunity. *Infection and Immunity, 74*(11), 6108–6117.

Taganov, K. D., Boldin, M. P., Chang, K. J., & Baltimore, D. (2006). NF-kappaB-dependent induction of microRNA miR-146, an inhibitor targeted to signaling proteins of innate immune responses. *Proceedings of the National Academy of Sciences of the United States of America, 103*(33), 12481–12486.

Tobin, D. M., & Ramakrishnan, L. (2008). Comparative pathogenesis of *Mycobacterium marinum* and *Mycobacterium tuberculosis*. *Cellular Microbiology, 10*(5), 1027–1039.

Tobin, D. M., Roca, F. J., Oh, S. F., McFarland, R., Vickery, T. W., Ray, J. P., … Ramakrishnan, L. (2012). Host genotype-specific therapies can optimize the inflammatory response to mycobacterial infections. *Cell, 148*(3), 434–446.

Tobin, D. M., Vary, J. C., Ray, J. P., Walsh, G. S., Dunstan, S. J., Bang, N. D., … Ramakrishnan, L. (2010). The lta4h locus modulates susceptibility to mycobacterial infection in zebrafish and humans. *Cell, 140*(5), 717–730.

Torraca, V., Cui, C., Boland, R., Bebelman, J. P., van der Sar, A. M., Smit, M. J., … Meijer, A. H. (2015). The CXCR3-CXCL11 signaling axis mediates macrophage recruitment and dissemination of mycobacterial infection. *Disease Models & Mechanisms, 8*(3), 253–269.

Torraca, V., Masud, S., Spaink, H. P., & Meijer, A. H. (2014). Macrophage-pathogen interactions in infectious diseases: new therapeutic insights from the zebrafish host model. *Disease Models & Mechanisms, 7*(7), 785–797.

Traver, D., Herbomel, P., Patton, E. E., Murphey, R. D., Yoder, J. A., Litman, G. W., … Trede, N. S. (2003). The zebrafish as a model organism to study development of the immune system. *Advances in Immunology, 81*, 254–330.

Urieli-Shoval, S., Linke, R. P., & Matzner, Y. (2000). Expression and function of serum amyloid A, a major acute-phase protein, in normal and disease states. *Current Opinion in Hematology, 7*(1), 64–69.

van der Vaart, M., Korbee, C. J., Lamers, G. E., Tengeler, A. C., Hosseini, R., Haks, M. C., … Meijer, A. H. (2014). The DNA damage-regulated autophagy modulator DRAM1 links mycobacterial recognition via TLP-MYD88 to autophagic defense. *Cell Host & Microbe, 15*(6), 753–767.

van der Vaart, M., van Soest, J. J., Spaink, H. P., & Meijer, A. H. (2013). Functional analysis of a zebrafish myd88 mutant identifies key transcriptional components of the innate immune system. *Disease Models & Mechanisms, 6*(3), 841–854.

van der Vaart, M., Spaink, H. P., & Meijer, A. H. (2012). Pathogen recognition and activation of the innate immune response in zebrafish. *Advances in Hematology, 2012*, 159807.

Vergne, I., Chua, J., Singh, S. B., & Deretic, V. (2004). Cell biology of *Mycobacterium tuberculosis* phagosome (Review) *Annual Review of Cell and Developmental Biology, 20*, 367–394.

Volkman, H. E., Clay, H., Beery, D., Chang, J. C., Sherman, D. R., & Ramakrishnan, L. (2004). Tuberculous granuloma formation is enhanced by a *Mycobacterium* virulence determinant. *PLoS Biology, 2*(11), e367.

Volkman, H. E., Pozos, T. C., Zheng, J., Davis, J. M., Rawls, J. F., & Ramakrishnan, L. (2010). Tuberculous granuloma induction via interaction of a bacterial secreted protein with host epithelium. *Science, 327*(5964), 466–469.

Walton, E. M., Cronan, M. R., Beerman, R. W., & Tobin, D. M. (2015). The macrophage-specific promoter mfap4 allows live, long-term analysis of macrophage behavior during mycobacterial infection in zebrafish. *PLoS One, 10*(10), e0138949.

Wang, Z., Gerstein, M., & Snyder, M. (2009). RNA-Seq: a revolutionary tool for transcriptomics (Review) *Nature Reviews. Genetics, 10*(1), 57–63.

Wang, Z., Zhang, S., Tong, Z., Li, L., & Wang, G. (2009). Maternal transfer and protective role of the alternative complement components in zebrafish *Danio rerio*. *PLoS One, 4*(2), e4498.

van der Wel, N., Hava, D., Houben, D., Fluitsma, D., van Zon, M., Pierson, J., ... Peters, P. J. (2007). *M. tuberculosis* and *M. leprae* translocate from the phagolysosome to the cytosol in myeloid cells. *Cell, 129*(7), 1287–1298.

Wellen, K. E., Fucho, R., Gregor, M. F., Furuhashi, M., Morgan, C., Lindstad, T., ... Hotamisligil, G. S. (2007). Coordinated regulation of nutrient and inflammatory responses by STAMP2 is essential for metabolic homeostasis. *Cell, 129*(3), 537–548.

Yang, C. T., Cambier, C. J., Davis, J. M., Hall, C. J., Crosier, P. S., & Ramakrishnan, L. (2012). Neutrophils exert protection in the early tuberculous granuloma by oxidative killing of mycobacteria phagocytosed from infected macrophages. *Cell Host & Microbe, 12*(3), 301–312.

Yang, K., He, Y. S., Wang, X. Q., Lu, L., Chen, Q. J., Liu, J., ... Shen, W. F. (2011). MiR-146a inhibits oxidized low-density lipoprotein-induced lipid accumulation and inflammatory response via targeting toll-like receptor 4. *FEBS Letters, 585*(6), 854–860.

Yang, S., Marin-Juez, R., Meijer, A. H., & Spaink, H. P. (2015). Common and specific downstream signaling targets controlled by Tlr2 and Tlr5 innate immune signaling in zebrafish. *BMC Genomics, 16*, 547.

Zhang, J., Yu, L. P., Li, M. F., & Sun, L. (2014). Turbot (*Scophthalmus maximus*) hepcidin-1 and hepcidin-2 possess antimicrobial activity and promote resistance against bacterial and viral infection. *Fish & Shellfish Immunology, 38*(1).

Zhang, S., & Cui, P. (2014). Complement system in zebrafish. *Developmental and Comparative Immunology, 46*(1).

# INDEX

'*Note:* Page numbers followed by "f" indicate figures and "t" indicate tables.'

## J



## K



## L



## M

## Q



## R

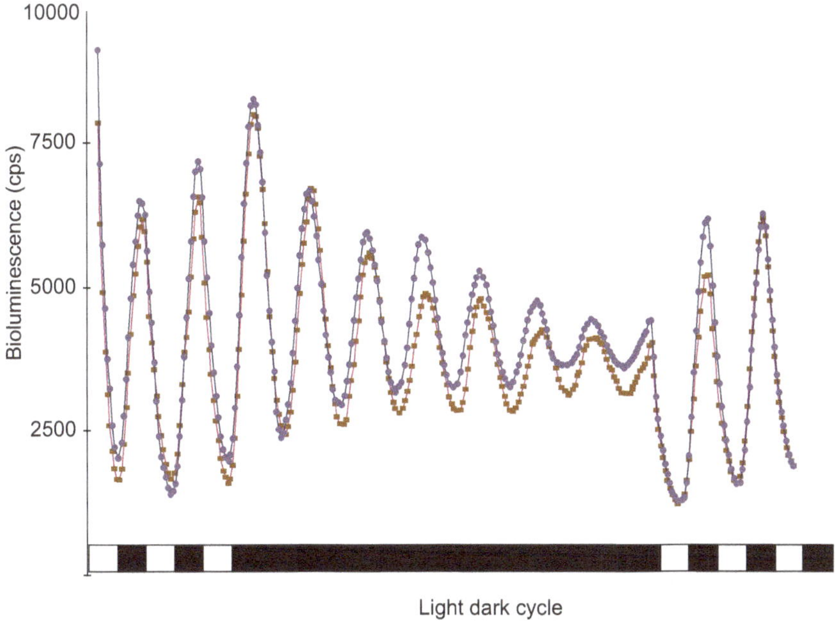


**N.S. Foulkes, *et al.*, Figure 1** Zebrafish cell lines models for studying clock and light-regulated gene expression. In vivo bioluminescence assay of Pac2 zebrafish cells transfected with the clock-regulated, luciferase reporter construct, *per1b:luc*. The bioluminescence emitted by two independently transfected sets of cells were monitored in real time after addition of luciferin to the culture medium and under various lighting conditions using an automated scintillation counter. White horizontal bars correspond to light and black bars, to dark periods.

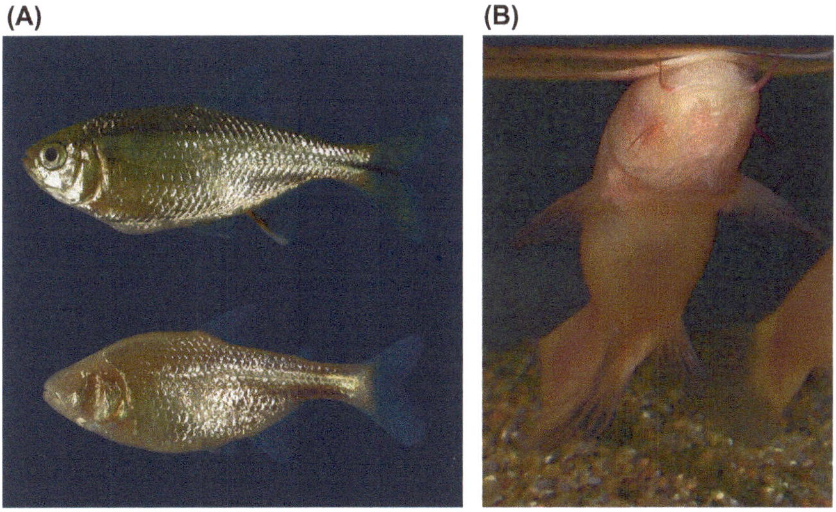


**N.S. Foulkes, *et al.*, Figure 2** Blind cavefish used in circadian clock studies. (A) Adult surface (upper) and blind cave (lower) forms of the Mexican tetra, *Astyanax mexicanus* (*Photograph by T. Katherine Tamai.*). (B) Adult *Phreatichthys andruzzii* (*Photograph by Andrea Margutti.*).

**N.S. Foulkes, *et al.*, Figure 3** Studying *Astyanax* in their natural habitat. (A) Dr. Victor Reynoso studying surface forms in the Micos River, NE Mexico. (B) Entrance to the Tinaja Cave. *(Photographs by David Whitmore)*

**J. Regneri, *et al.*, Figure 1** Transgenic medaka model for melanoma development. Adult *mitfa:xmrk* transgenic medaka showing malignant melanoma development in the tail and anal fin invading the adjacent body musculature. Scale bar represents 10 mm.

**J. Regneri, *et al.*, Figure 2** Wild-type (A) and transgenic (B) medaka at 3-4 weeks of age. The fish with the *mitfa:xmrk* transgene shows onset of melanoma development in the caudal fin, inside the belly and the trunk musculature. Scale bars represent 1 mm.

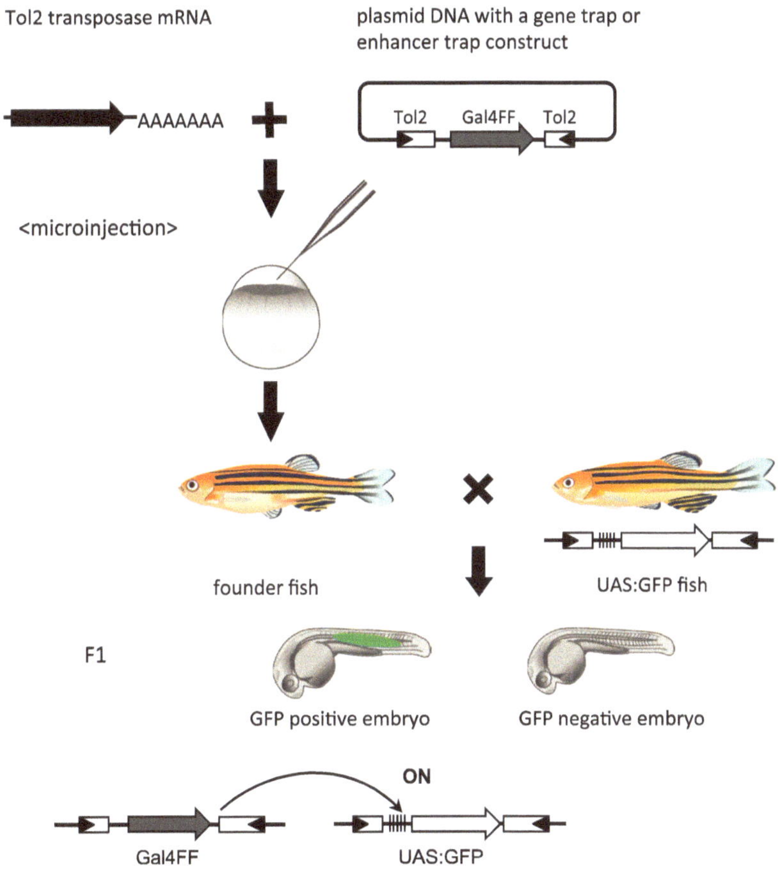


**K. Kawakami, *et al.*, Figure 2** ***Gene trapping and enhancer trapping with the Gal4-UAS system.*** A trap construct containing *gal4ff* is injected into fertilized eggs with the transposase mRNA. Injected fish are raised and mated with homozygous UAS:GFP reporter fish. Double transgenic F1 embryos expressing GFP in specific regions should be obtained.

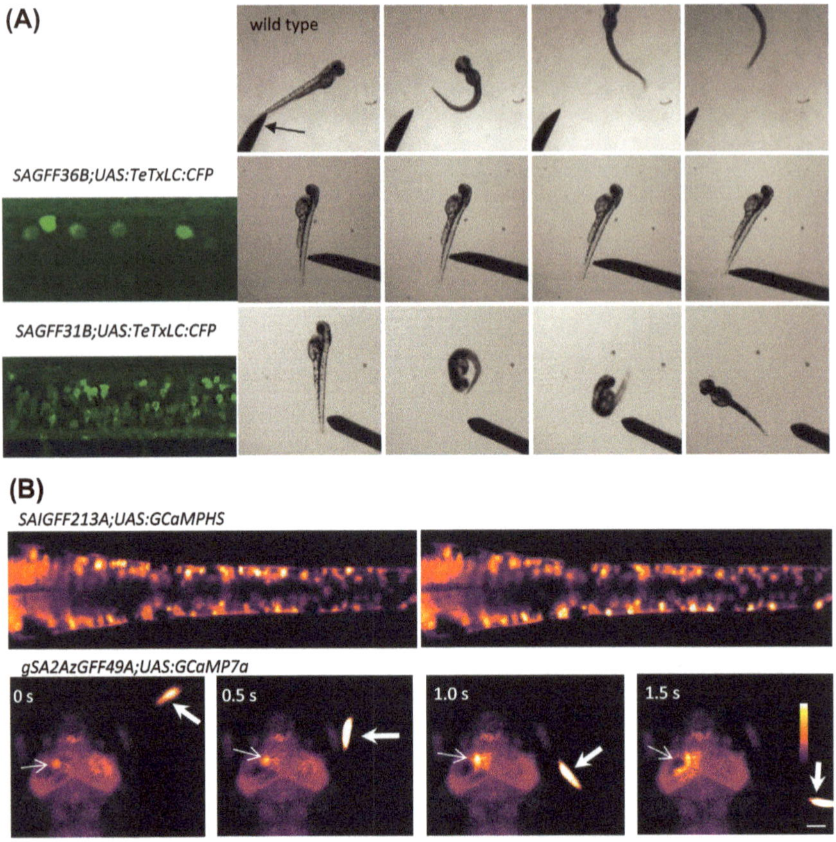


**K. Kawakami, *et al.*, Figure 3** ***Gal4FF driver transgenic fish and targeted expression via the Gal4FF-UAS system.*** (A) Inhibition of neuronal activity via targeted expression of a neurotoxin gene. A wild-type embryo at 2 dpf rapidly escapes from a gentle touch to the tail. The SAGFF36B;UAS:TeTxLC:CFP fish, which expresses TeTxLC:CFP in the sensory neurons, does not respond to touch. The SAGFF31B;UAS:TeTxLC:CFP fish, which expresses TeTxLC:CFP in subsets of interneurons and motor neurons, responds to the touch but shows abnormal escape swimming. (B) Imaging of the activity of CaP motor neurons in the SAIGFF213A;UAS:GCaMPHS double transgenic embryo. A top view of an immobilized embryo at 24 hpf. Anterior to the left. Simultaneous calcium signals on one side are detected alternately (pseudocolored). Imaging of the tectal neurons in the gSA2AzGFF49A;UAS:GCaMP7a double transgenic larva. Top view of an immobilized embryo at 6 dpf. A paramecium is moving from anterior to posterior on the right-hand side in a space around the larva, and calcium signals on the left tectum move from anterior to posterior (pseudocolored). (C) Genetic dissection of the motor system. HGj4A and mnGFF7;UAS:GFP at 48 hpf express GFP and Gal4FF in the ventral spinal cord, respectively. Single neurons are visualized in the mnGFF7 larva injected with the UAS:GFP plasmid. The hspGFF62A transgenic fish expresses Gal4FF in the Mauthner cells. (D) The architecture of cerebellar neural circuits. Top view of the cerebellum. *CCe*, corpus cerebelli; *EG*, eminentia granularis; *GC*, granule cells; *LCa*, lobus caudalis cerebelli; *PC*, Purkinje cells; *PF*, parallel fibers; *Va*, valvula cerebelli. The gSA2AzGFF152B, hspzGFFgDMC156A, and hspGFFDMC28C transgenic fish express Gal4FF in the granule cells, eurydendroid cells, and IO (inferior olive) neurons, respectively. An *arrow* indicates contralateral projections of the climbing fibers. (E) Transgenic fish that label the lateral line system. The HGn39D transgenic fish expresses Gal4FF in sensory axons (green) and

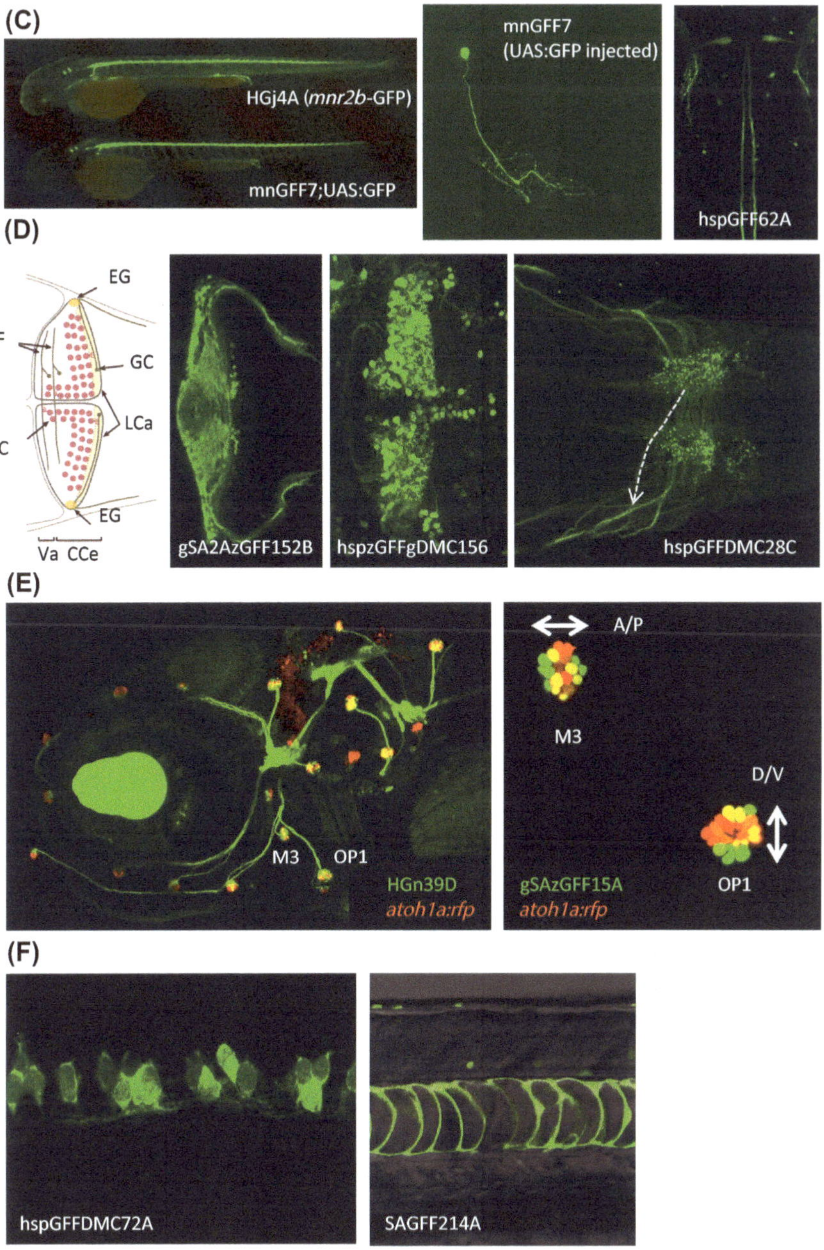


**K. Kawakami, *et al.*, Figure 3** (*continued*).

*atoh1a:rfp* labels hair cells (red) in the cranial region of an embryo at 5 dpf. Anterior to the left. gSAzGFF15A transgenic fish express Gal4FF in older hair cells (green) and *atoh1a:rfp* labels younger hair cells (red). *Arrows* indicate the direction of hair cell migration in each neuromast. (F) Transgenic fish used to study spatiotemporal functions of Notch signaling. The hspGFFDMC72A transgenic fish expresses Gal4FF in early differentiating primary neurons. The SAGFF214A transgenic fish expresses Gal4FF in vacuolated cells in the notochord.

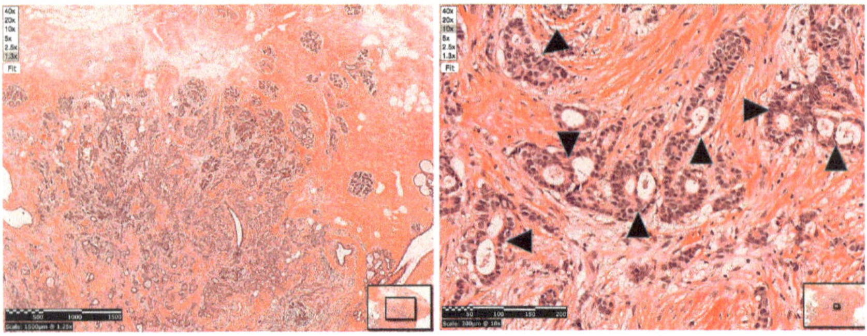

**K.C. Cheng, *et al.*, Figure 1** ***Infiltrating ductal carcinoma illustrating what is visible at different imaging resolutions.*** The nuclei of nests of invading tumor cells are evident at 1.3× magnification, at which pixels are ~7.7 × 7.7 μm, which makes cell types are difficult to discern. What is evident at this power, however, is that the cancer cells, generally represented by the purple cells, are surrounded by dense collections of collagen (stained pink) that cause the palpable hardness of tumors in the clinic and represents a reaction to the tumor. Virtually all cell types are distinguishable at 10× magnification, at which pixels are about 1 × 1 μm. *Arrowheads*: Invasive nests of cancer cells with multiple, "back-to-back" glandular spaces, typical of ductal adenocarcinoma. *Both panels are from the Penn State Histopathology Collection at bio-atlas.psu.edu, at http://bio-atlas.psu.edu/view.php?s=1817&atlas=107.*

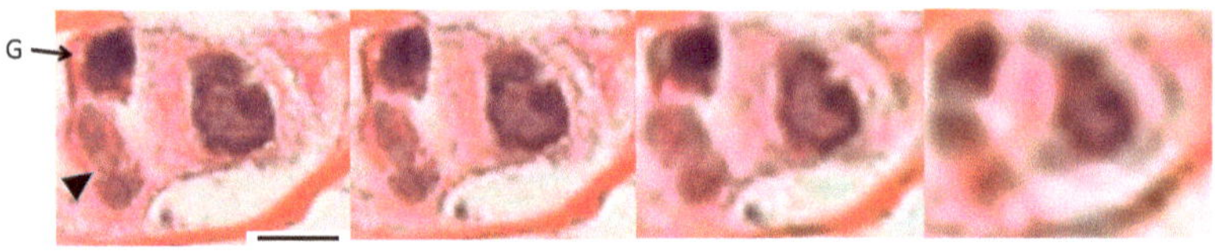


| | | | | |
|---|---|---|---|---|
| Objective | 40x | 20x | 10x | 5x |
| Pixel (μ x μ) | 0.25 | 0.5 | 1 | 2 |
| Chromosomes visible | Yes | Barely | No | No |
| Golgi visible | Yes | Yes | Barely | No |

**K.C. Cheng, *et al.*, Figure 2** ***Cellular features, including chromosomes and Golgi body at different imaging resolutions.*** Cells from the same infiltrating ductal carcinoma as in Fig. 1, pixels are about 1 × 1 μm. The cell at the *arrowhead* contains two clusters of chromosomes, whose threadlike appearance is visible at 40× (see http://bio-atlas.psu.edu/view.php?s=1817&atlas=107&z=0&c=59056,34025), particularly when focusing through optical sections on the microscope, and are barely discernable by individuals experienced in pathology. Individual chromosomes are not apparent when pixels reach about 1 μm, and very few cytological features are visible when pixels reach about 2 μm in dimension (about the size applicable to stereomicroscopy). *G*, Golgi apparatus, is the pale area next to the nucleus. Size bar, 20 μm.

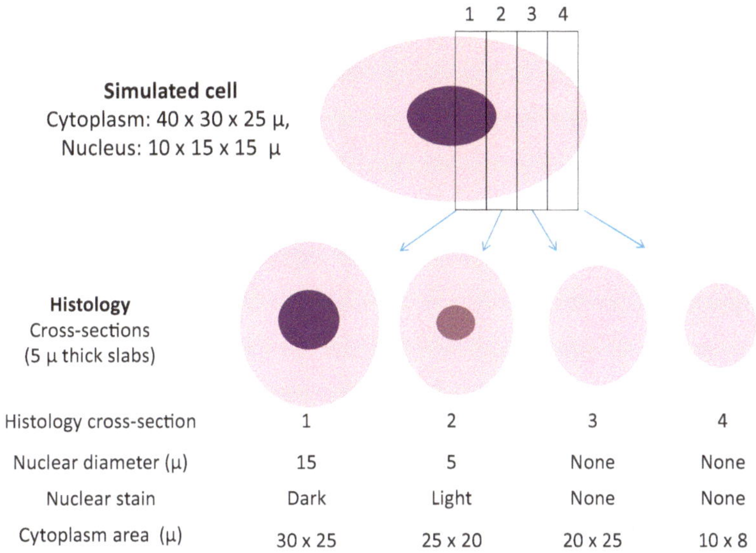


**K.C. Cheng, *et al.*, Figure 3 *Histology misrepresents the microanatomy of large cells.*** The simulated large cell is modeled after that shown in Fig. 2. 5-μm thick sections typical of histology yield strikingly different cell representations, including dimensions of cytoplasm and nucleus, and DNA staining, depending upon the plane of section.

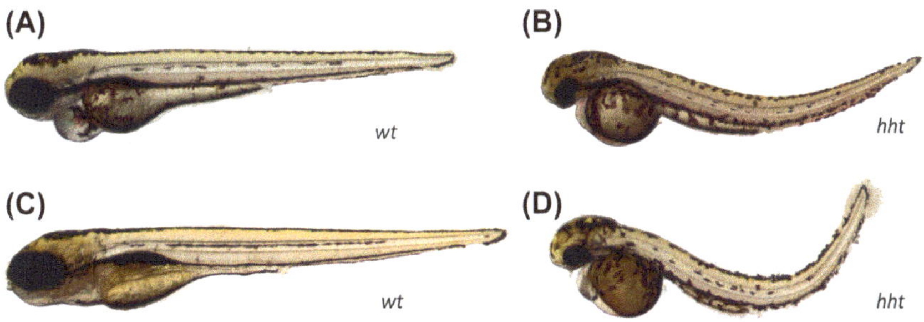


**K.C. Cheng, *et al.*, Figure 4 *Small-eye phenotype in 3 dpf and 5 dpf* hht *by stereomicroscope.*** 3 dpf (B) and 5 dpf (D) *hht* mutants both have "small eyes" compared to their 3 dpf (A) and 5 dpf (C) wild-type siblings. These views say nothing about the cellular mechanism underlying the small eye phenotype.

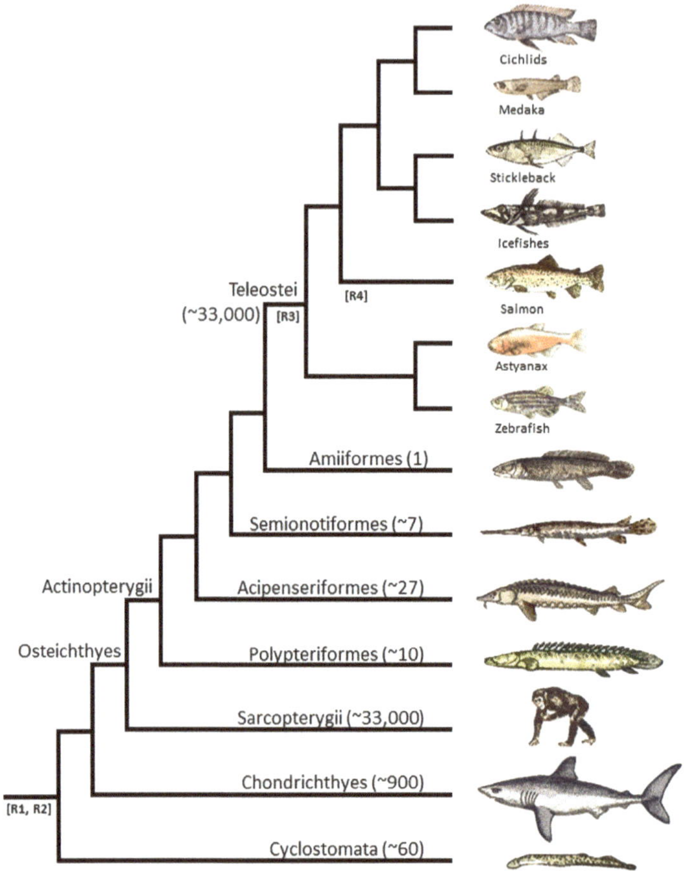


**D. Casane and S. Rétaux, Figure 1** ***Main teleost models for evolutionary genetics.*** Phylogenetic relationships are shown to the left. The number of species described in each clade is given between parentheses; R1, R2, R3, R4: whole genome duplications; Tree topology is based on *Near, T.J., Dornburg, A., Eytan, R.I., Keck, B.P., Smith, W.L., Kuhn, K.L., … Wainwright, P.C. (2013). Phylogeny and tempo of diversification in the superradiation of spiny-rayed fishes. Proceedings of the National Academy of Sciences of the United States of America, 110(31), 12738–12743; Near, T.J., Eytan, R.I., Dornburg, A., Kuhn, K.L., Moore, J.A., Davis, M.P., … Smith, W.L. (2012). Resolution of ray-finned fish phylogeny and timing of diversification.* Proceedings of the National Academy of Sciences of the United States of America, 109*(34), 13698–13703.*

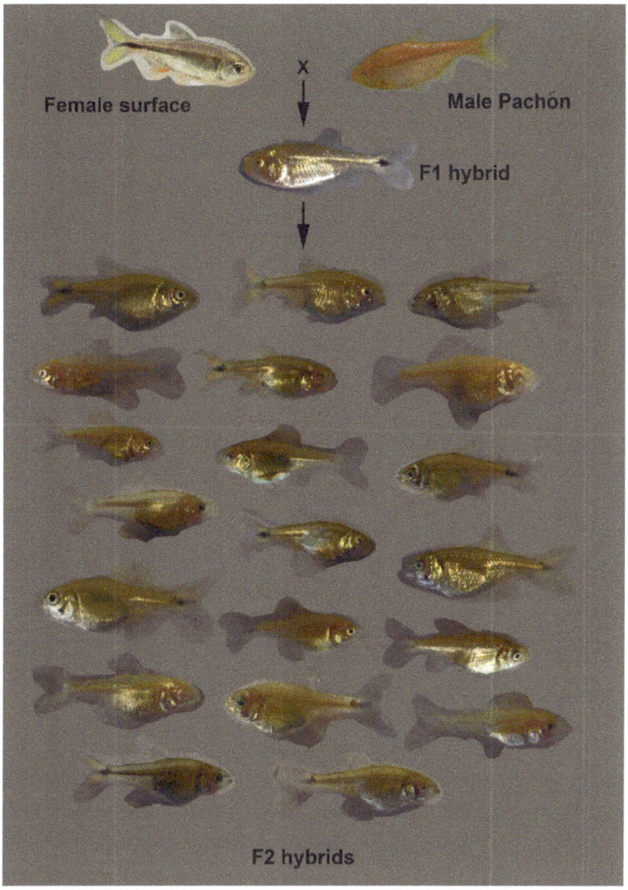


**D. Casane and S. Rétaux, Figure 2** ***Crosses for implementing quantitative trait loci (QTL) approach in* Astyanax.** Establishing crosses is the first step of the QTL approach. An example is shown here, with a Surface X Pachón cross giving rise to a first generation hybrid family (F1). These F1 fish are pigmented and have small eyes but are fully sighted. Subsequent crossing between two F1s (brother X sister cross) generates a second generation hybrid family (F2). Shown here are 20 randomly picked individuals from the F2 panel, photographed at 2 months of age. The segregation of traits is evident concerning pigmentation, eye and lens size. For example, a pigmented individual can have a small nearly degenerated eye, and a depigmented albino individual can have large eyes.

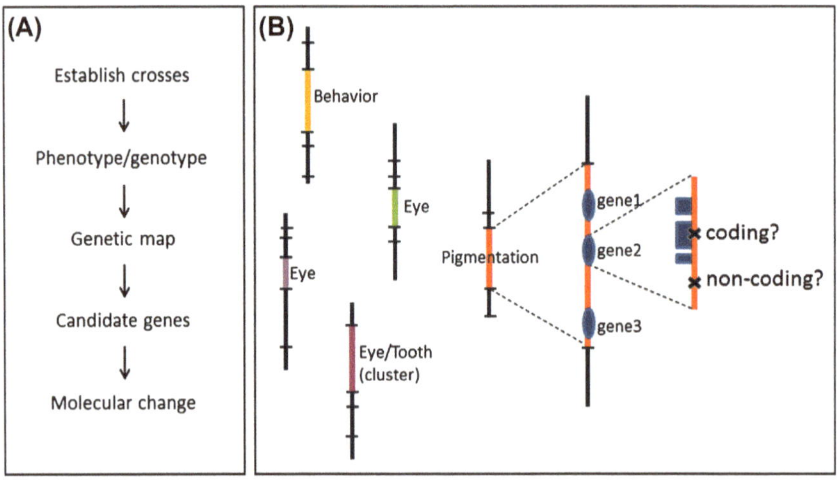


**D. Casane and S. Rétaux, Figure 3** ***Quantitative trait loci (QTL) approach in*** **Astyanax.** Scheme highlighting the steps of the quantitative genetics approach in this model system. *The figure is inspired from Peichel, C.L. (2005). Fishing for the secrets of vertebrate evolution in threespine sticklebacks.* Developmental Dynamics, *234(4), 815–823., reporting the same type of QTL pipeline analysis for sticklebacks.*

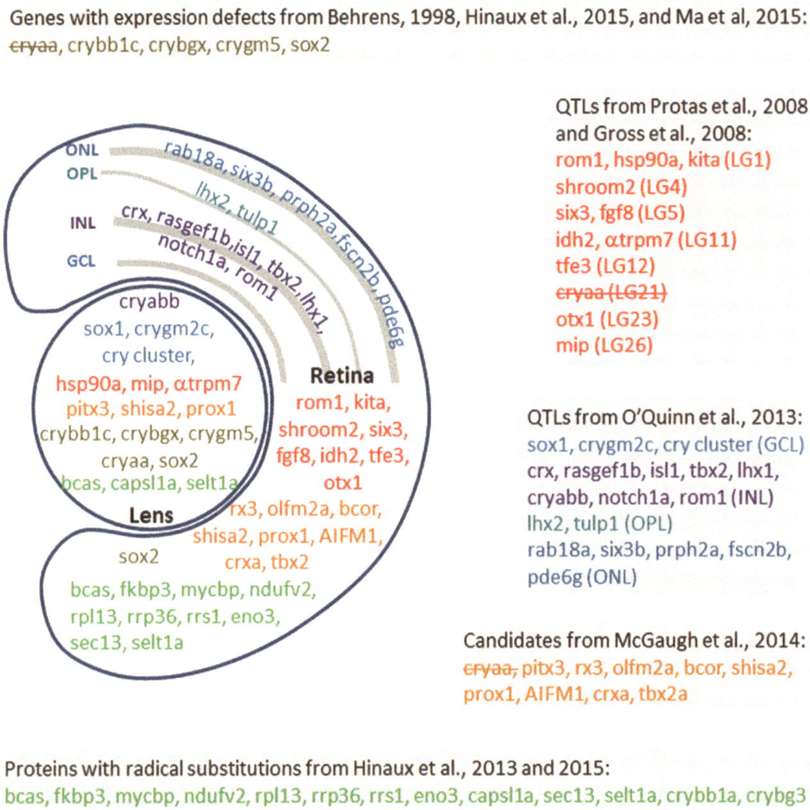


**D. Casane and S. Rétaux, Figure 4** ***Genetics of eye loss in cavefish.*** Candidate genes potentially involved in eye loss in cavefish are reported on a schematic eye, including a retina and a lens. The original papers from which data were taken are reported in *gray boxes*, and the candidate genes from these papers are color coded. Only genes that are expressed in the eye, according to the ZFIN zebrafish expression database, are reported. On the schematic eye, genes are indicated according to their expression patterns, ie, in the retina, in the lens, or both. Of note, cryaa (in red, orange, and brown) is crossed out in all cases because there is evidence that *cryaa* is not the causal gene in the identified quantitative trait loci (QTL) interval, although absence of *cryaa* expression in the cavefish lens is indeed deleterious to the lens (see text).

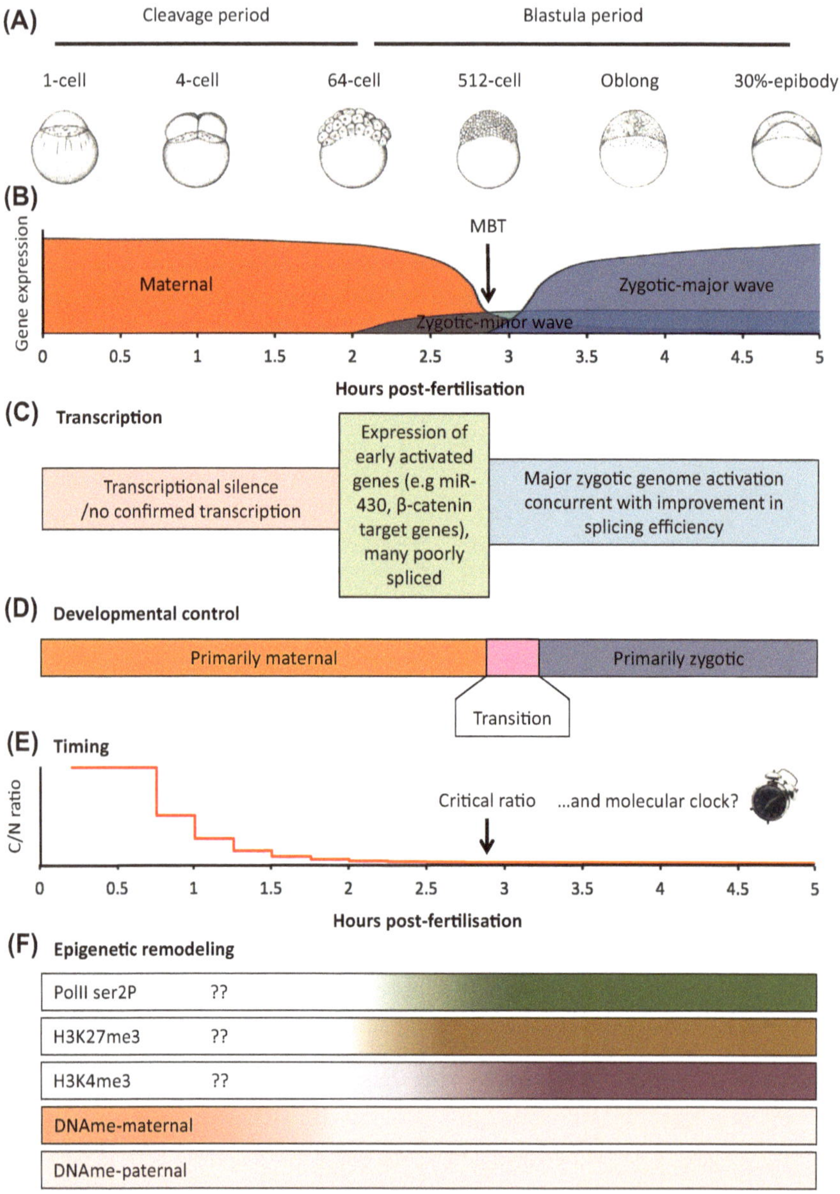
(A)
Cleavage period
Blastula period
1-cell
4-cell
64-cell
512-cell
Oblong
30%-epibody
(B)
Gene expression
MBT
Maternal
Zygotic-major wave
Zygotic-minor wave
0
0.5
1
1.5
2
2.5
3
3.5
4
4.5
5
Hours post-fertilisation
(C) Transcription
Transcriptional silence /no confirmed transcription
Expression of early activated genes (e.g miR-430, β-catenin target genes), many poorly spliced
Major zygotic genome activation concurrent with improvement in splicing efficiency
(D) Developmental control
Primarily maternal
Primarily zygotic
Transition
(E) Timing
C/N ratio
Critical ratio
...and molecular clock?
0
0.5
1
1.5
2
2.5
3
3.5
4
4.5
5
Hours post-fertilisation
(F) Epigenetic remodeling
PolII ser2P ??
H3K27me3 ??
H3K4me3 ??
DNAme-maternal
DNAme-paternal

**J. Wragg and F. Müller, Figure 1** Outline of the main events occurring during the process of zygotic genome activation. (A), Timeline of zygotic development, highlighting the major macroscopic changes occurring in the embryo during the cleavage and blastula periods. *(Adapted from (Kimmel et al., (1995), used with permission.)* (B), Gene expression level of maternal (*red*) and zygotic (*blue*) derived transcripts from 0 to 5 hours post fertilization. Mid-blastula transition (MBT) is marked. (C), Description of transitional events from 0 to 5 hours post fertilization. (D), Overview of developmental control from 0 to 5 hours post fertilization. (E), Plot of the cytoplasmic to nuclear (C/N) ratio within the embryo from 0 to 5 hours post fertilization. *(Plotted using data from (Kimmel et al., (1995).)* (F), Illustration of epigenetic remodeling events from 0 to 5 hours post fertilization (*darker color* indicates a greater level of DNA/chromatin modification, ??, status of modification unknown).

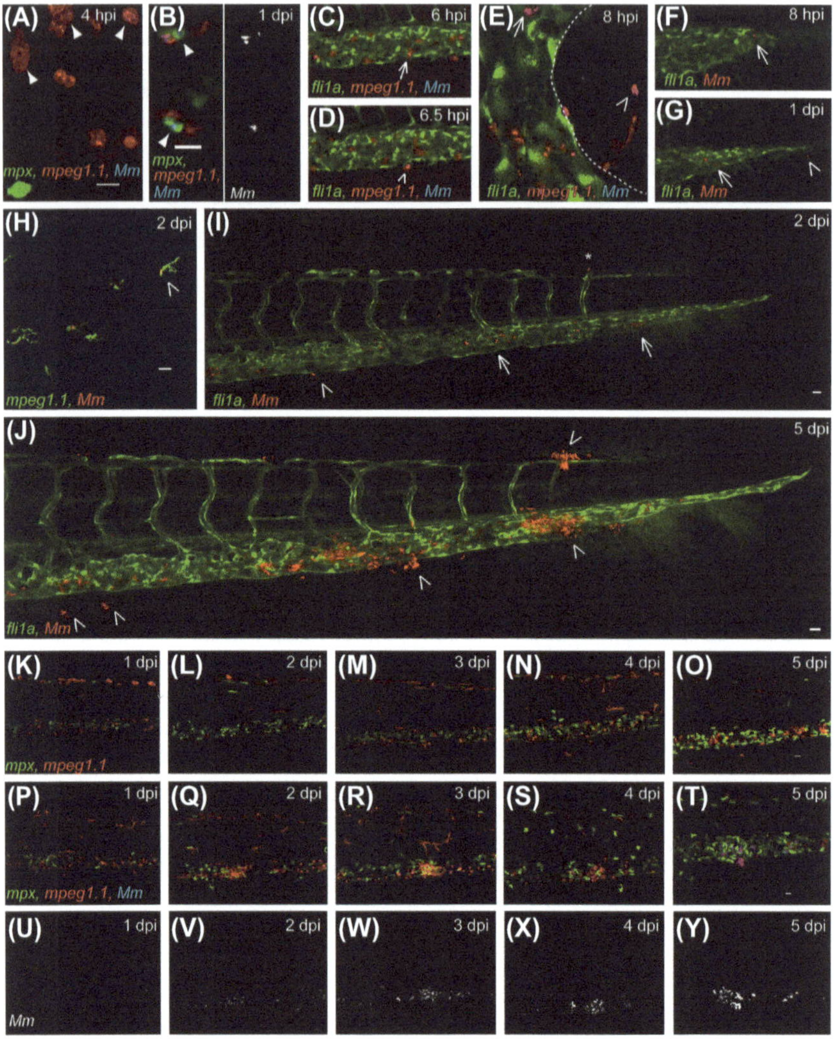


**E.L. Benard, *et al.*, Figure 1** ***Overview of the* M. marinum *infection process imaged in zebrafish embryos.*** Zebrafish embryos were injected at 28 hpf with 250 cfu of *M. marinum* in 2% PVP or with 2% PVP alone into the blood island. (A) E2-Crimson-expressing *M. marinum* (Swaim et al., 2006) bacterial infection in *Tg(mpeg1.1::mCherry-F)/Tg(mpx:GFP)* embryos at 4 hpi imaged at Duct of Cuvier site. *Solid arrowheads* indicate infected macrophages. (B) E2-Crimson-expressing *M. marinum* bacterial infection in *Tg(mpeg1.1::mCherry-F)/Tg(mpx:GFP)* embryos at caudal vein site. Left panel, overlay of all three fluorescence channels; right panel, E2-Crimson fluorescence channel with *M. marinum* fluorescence depicted as white pixels. *Solid arrowheads* indicate infected neutrophils. (C–E) E2-Crimson-expressing *M. marinum* bacterial infection in *Tg(fli1a:EGFP)/Tg(mpeg1.1::mCherry-F)* embryo imaged over time at (C and D) the caudal vein at 6 and 6.5 hpi and (E) the Duct of Cuvier site at 8 hpi. *M. marinum*-infected macrophages (purple color: double positive for mCherry and E2-Crimson) are observed both (*arrows*) inside and (*pointed arrowheads*) outside the vasculature. *Dotted line* in (E) indicates border of vasculature in Duct of Cuvier. (F and G) mCherry-expressing *M. marinum* (van der Sar et al., 2004) bacterial infection in a *Tg(fli1a:EGFP)* (Lawson & Weinstein, 2002) embryo at the transition of the caudal vein and artery at (F) 8 hpi and (G) 1 dpi. *M. marinum* can be observed (*arrow*) inside and (*pointed arrowhead*) outside the vasculature. (H) mCherry-expressing *M. marinum* bacterial infection in a *Tg(mpeg1.1:EGFP)* embryo imaged at the posterior tail area. Merged image of bright field, GFP, and mCherry channels show an infected macrophage (*pointed arrowhead*) located in the caudal fin. (I and J) mCherry-expressing *M. marinum* bacterial infection in a *Tg(fli1a:EGFP)* embryo imaged at the tail region at (I) 2 dpi and (J) 5 dpi. An *M. marinum* cluster can be seen (*asterisk*) in the dorsal longitudinal lymphatic vessel and various *M. marinum* clusters can be seen residing (*arrow*) within the caudal vein and (*pointed arrowhead*) outside the caudal vein at (I) 2 dpi. These clusters increase in size at (J) 5 dpi and clusters previously inside the caudal vein at 2 dpi are now observed (*pointed arrowhead*) outside the caudal vein. (K–O) Uninfected (mock injected) *Tg(mpeg1.1::mCherry-F)/Tg(mpx:GFP)* embryo tail region posterior to the urogenital opening imaged over a time course of (K) 1 dpi, (L) 2 dpi, (MI) 3 dpi, (N) 4 dpi, and (O) 5 dpi. Images are *z*-stack projections of the merged channels of GFP and mCherry. (P–Y) E2-Crimson-labeled *M. marinum* bacterial infection in a *Tg(mpeg1.1::mCherry-F)/Tg(mpx:GFP)* embryo imaged in identical area as (K–O) uninfected embryo over time course of (P) 1 dpi, (Q) 2 dpi, (R) 3 dpi, (S) 4 dpi, and (T) 5 dpi. Confocal images were taken on a Leica TCS SPE with an HC APO 20× 0.7 NA objective and an HCX APO 40× 0.8 NA objective or on a Zeiss LSM5 Exciter/AxioObserver with an HC APO 20× 0.7 NA objective. Images are *z*-stack projections of the merged channels of GFP, mCherry, and E2-Crimson. (U–Y) *z*-stack projections of the E2-Crimson channel isolated from P-T and depicted as white pixels. Scale bars: 20 μm.

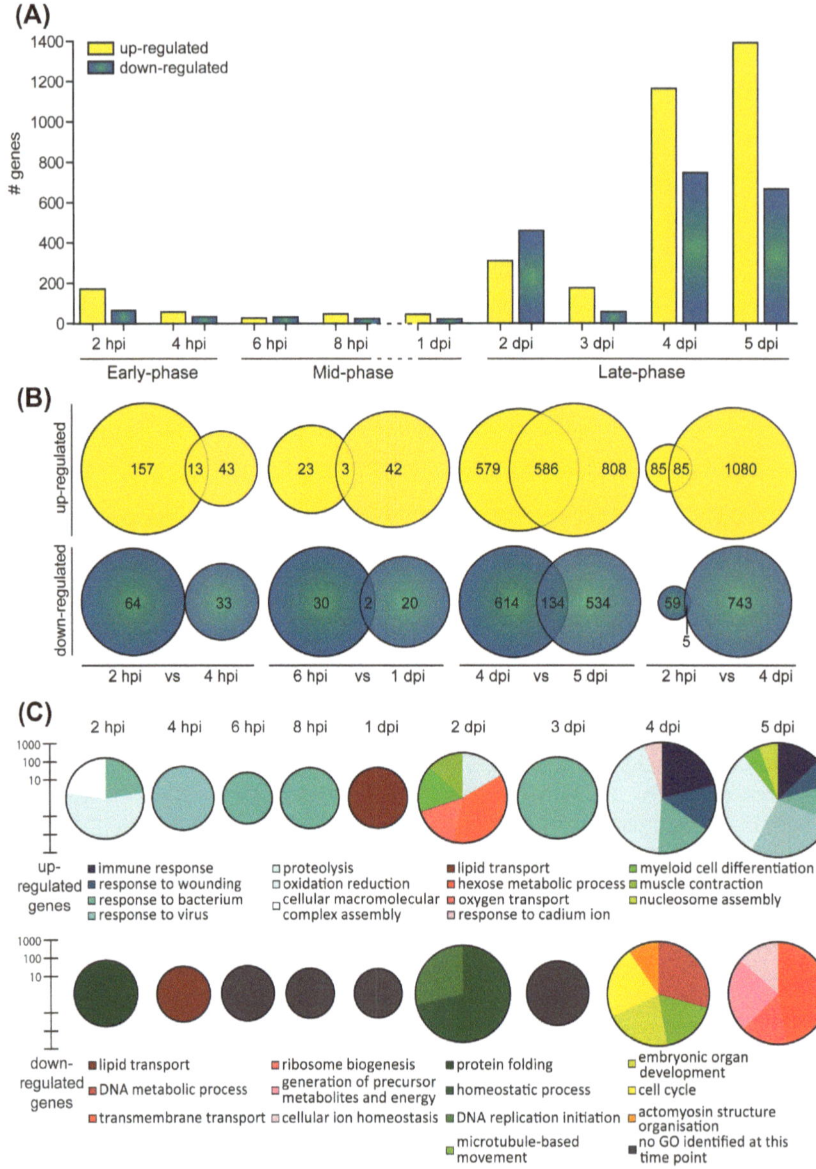


**E.L. Benard, *et al.*, Figure 2** ***Overview of total number of genes differentially regulated in* M. marinum*-infected embryos, their overlap between various time points, and gene ontology terms enriched per time point.*** AB/TL embryos were infected with *M. marinum* Mma20 strain at 1 day post fertilization (dpf) and RNA was isolated at 2, 4, 6, and 8 hpi and 1, 2, 3, 4, and 5 dpi. Four biological replicates each containing 30 embryos of each time point were used for RNA sequencing and DeSeq analysis of transcript count data was used for statistical comparison with mock-injected control embryos sampled at the same time points (* = fold change > 2, $p < 0.05$) (Anders & Huber, 2010). (A) Overview of the total number of genes either significantly upregulated (yellow) or downregulated (blue) per time point in the *M. marinum*-infected samples compared to the uninfected control embryos. (B) Area-proportionate Venn-diagrams representing the overlap of both upregulated and downregulated genes during *M. marinum* infection. The time points compared are 2 hpi versus 4 hpi, 6 hpi versus 1 dpi, 4 dpi versus 5 dpi, and 2 hpi versus 4 dpi (respectively left versus right circle for each comparison). (C) Gene ontology terms identified per time point represented as pie charts. The diameter of each chart is proportional to the total number of genes upregulated (top) or downregulated (bottom) per time point. Colors illustrate the relative proportion of each identified gene ontology term. Genes without associated gene ontology terms are not represented in the pie charts (except for time points with no gene ontologies identified). Proportional Venn diagrams were created with a Venn Diagram Generator (http://jura.wi.mit.edu/bioc/tools/venn.php). Gene ontology analysis was performed using DAVID v6.7 (http://david.abcc.ncifcrf.gov/home.jsp) (Huang da et al., 2009). Gene ontology terms are considered significant when the false-discovery rate was below 5%. Gene Expression Omnibus (GEO) database accession for RNA-Seq: GSE76499.

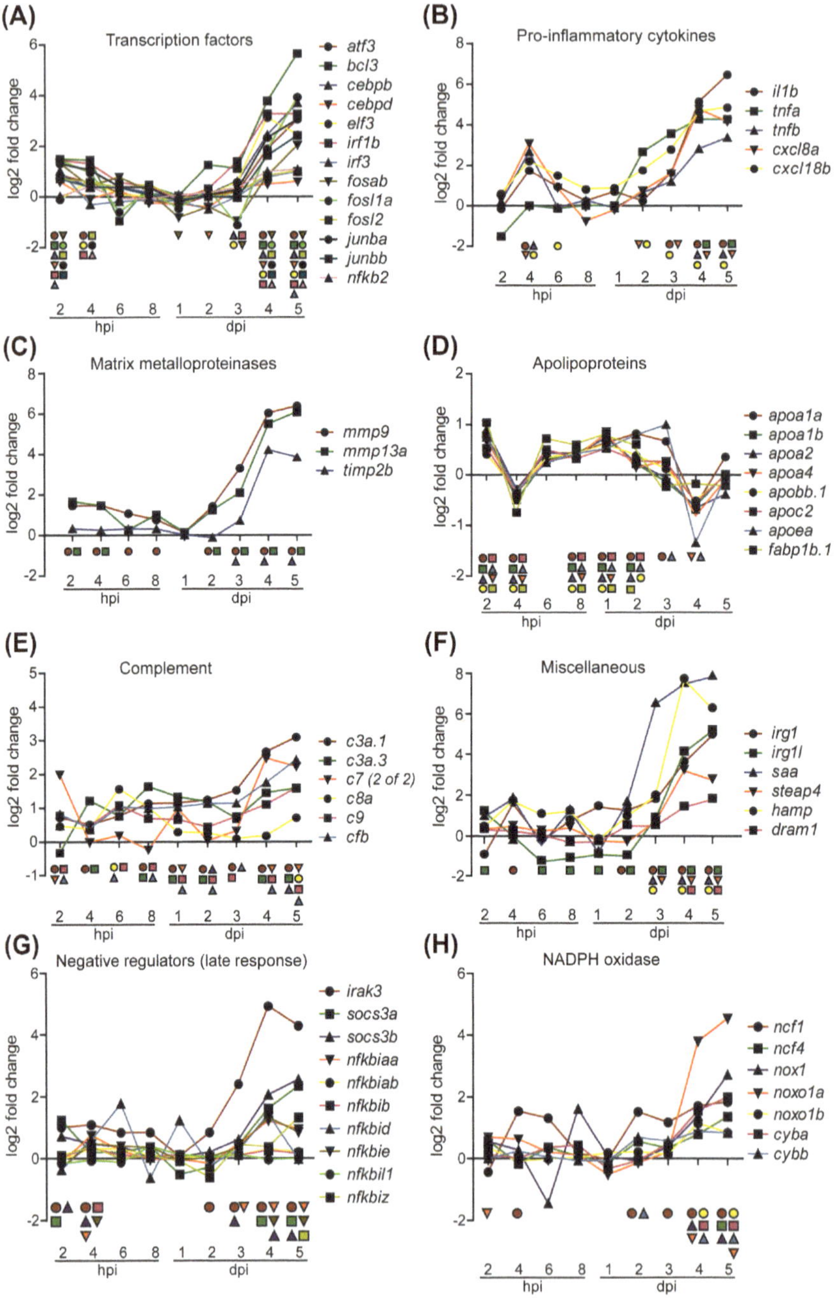
(A)
Transcription factors
log2 fold change
atf3
bcl3
cebpb
cebpd
elf3
irf1b
irf3
fosab
fosl1a
fosl2
junba
junbb
nfkb2
hpi
dpi
(B)
Pro-inflammatory cytokines
il1b
tnfa
tnfb
cxcl8a
cxcl18b
(C)
Matrix metalloproteinases
mmp9
mmp13a
timp2b
(D)
Apolipoproteins
apoa1a
apoa1b
apoa2
apoa4
apobb.1
apoc2
apoea
fabp1b.1
(E)
Complement
c3a.1
c3a.3
c7 (2 of 2)
c8a
c9
cfb
(F)
Miscellaneous
irg1
irg1l
saa
steap4
hamp
dram1
(G)
Negative regulators (late response)
irak3
socs3a
socs3b
nfkbiaa
nfkbiab
nfkbib
nfkbid
nfkbie
nfkbil1
nfkbiz
(H)
NADPH oxidase
ncf1
ncf4
nox1
noxo1a
noxo1b
cyba
cybb

**E.L. Benard, *et al.*, Figure 3** ***Overview of gene expression profiles during* M. marinum *infection.*** Gene expression levels of groups of selected genes during a time course of *M. marinum* infection are represented as log2 fold change values. Groups of genes represented encode: (A) transcription factors; (B) pro-inflammatory cytokines and chemokines; (C) matrix metalloproteinases and inhibitor; (D) apolipoproteins; (E) complement components; (F) miscellaneous late-phase responses; (G) negative regulators; (H) NADPH oxidase. Statistically significant data points (adjusted $P$-value $< 0.1$) are indicated below the $x$-axis as the shape filled with the color of the specific gene (see legend for gene shape and color). Gene symbols are according to ZFIN nomenclature, except for the following genes that are here identified by their ENSEMBL IDs: *c3a.1*, ENSDARG0000012694; *c3a.3*, ENSDARG00000052207; *c7 (2 of 2)*, ENSDARG00000057121. Only genes are shown for which differential expression was significant in at least one time point.

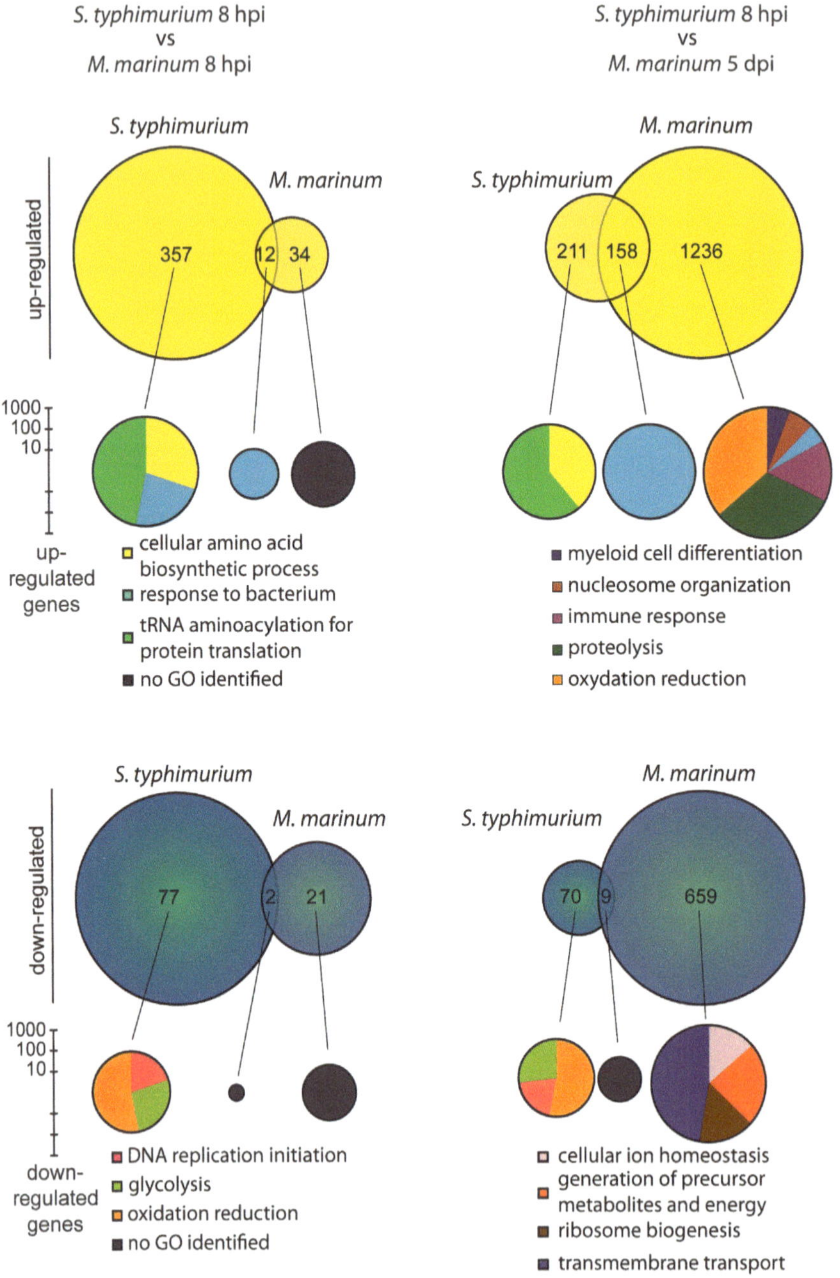


**E.L. Benard, *et al.*, Figure 4** ***Overview of total number of genes differentially regulated in* S. typhimurium- *and* M. marinum*-infected embryos and their overlap between various time points and gene ontology terms enriched per time point.*** RNA-Seq data from *M. marinum* infection (Fig. 2) was compared to RNA-Seq data from *S. typhimurium*-infected AB/TL embryos (Ordas et al., 2013) that has been re-analyzed with genome assembly ZV9, annotation version 71. Area proportionate Venn diagrams are shown representing significantly upregulated (yellow) and downregulated (blue) genes during *S. typhimurium* and *M. marinum* infection in zebrafish embryos. The *S. typhimurium* 8 hpi time point was compared to *M. marinum* 8 hpi (left) and 5 dpi (right). Gene ontology terms identified per time point per comparison are represented as pie charts below the Venn diagrams. Colors in the pie charts illustrate the relative proportion of each identified gene ontology term. Genes without associated gene ontology term are not represented in the pie charts (except for time points with no gene ontologies identified). Gene ontology analysis was performed using DAVID v6.7 (http://david.abcc.ncifcrf.gov/home.jsp) (Huang da et al., 2009). Gene ontology terms are considered significant when the false-discovery rate was below 5%.

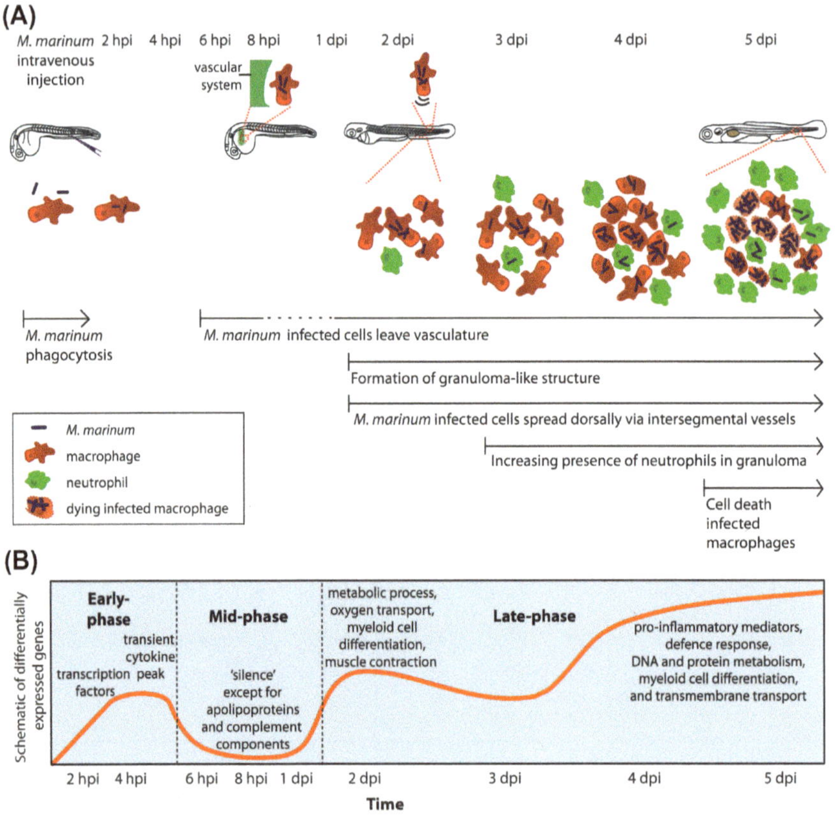


**E.L. Benard, *et al.*, Figure 5** ***Summary of the infection process of* M. marinum *in zebrafish embryos.*** A schematic overview representing (A) host–pathogen interactions and (B) the three distinct phases observed in gene expression profile responses (*orange line* represents total genes up- and downregulated during *M. marinum* infection).

FSC
www.fsc.org
MIX
Papier | Fördert
gute Waldnutzung
FSC® C083411

Kolibri 360 GmbH
Ferdinand-Jühlke-Straße 7
99095 Erfurt, Deutschland
produktsicherheit@kolibri360.de